Sir William Hamilton

Lectures on Metaphysics and Logic

Vol. 2

Sir William Hamilton

Lectures on Metaphysics and Logic
Vol. 2

ISBN/EAN: 9783742821188

Manufactured in Europe, USA, Canada, Australia, Japa

Cover: Foto ©Thomas Meinert / pixelio.de

Manufactured and distributed by brebook publishing software
(www.brebook.com)

Sir William Hamilton

Lectures on Metaphysics and Logic

LECTURES

ON

METAPHYSICS AND LOGIC

ON EARTH, THERE IS NOTHING GREAT BUT MAN;
IN MAN, THERE IS NOTHING GREAT BUT MIND.

LECTURES

ON

METAPHYSICS AND LOGIC

BY

SIR WILLIAM HAMILTON, BART.

PROFESSOR OF LOGIC AND METAPHYSICS IN THE
UNIVERSITY OF EDINBURGH

Advocate, A.M. (Oxon.) &c.; Corresponding Member of the Institute of France; Honorary Member of the
American Academy of Arts and Sciences; and of the Latin Society of Jena, &c.

EDITED BY THE

REV. H. L. MANSEL, B.D., LL.D.

WAYNFLETE PROFESSOR OF MORAL AND METAPHYSICAL PHILOSOPHY, OXFORD

AND

JOHN VEITCH, M.A., LL.D.

PROFESSOR OF LOGIC AND RHETORIC IN THE UNIVERSITY OF GLASGOW

IN FOUR VOLUMES

VOL. IV.

WILLIAM BLACKWOOD AND SONS
EDINBURGH AND LONDON
MDCCCLXXIV

LECTURES

ON

L O G I C

BY

SIR WILLIAM HAMILTON, BART.

EDITED BY THE

REV. H. L. MANSEL, B.D., LL.D.

WAYNFLETE PROFESSOR OF MORAL AND METAPHYSICAL PHILOSOPHY, OXFORD

AND

JOHN VEITCH, M.A., LL.D.

PROFESSOR OF LOGIC AND RHETORIC IN THE UNIVERSITY OF GLASGOW

VOL. II.

THIRD EDITION, REVISED

WILLIAM BLACKWOOD AND SONS
EDINBURGH AND LONDON
MDCCCLXXIV

CONTENTS OF VOL. II.

LECTURE XXXV.

APPENDIX.

LECTURES ON LOGIC.

LECTURE XXIV.

PURE LOGIC.

PART II.—METHODOLOGY.

SECTION I.—METHOD IN GENERAL.

SECTION II.—METHOD IN SPECIAL, OR LOGICAL METHODOLOGY.

I.—DOCTRINE OF DEFINITION.

GENTLEMEN,—We concluded, in our last Lecture, the consideration of Syllogisms, viewed as Incorrect or False ; in other words, the doctrine of Fallacies, in so far as the fallacy lies within a single syllogism. This, however, you will notice, does not exhaust the consideration of fallacy in general, for there are various species of false reasoning which may affect a whole train of syllogisms. These,—of which the *Petitio Principii*, the *Ignoratio Elenchi*, the *Circulus*, and the *Saltus in Concludendo*, are the principal,—will be appropriately considered in the sequel, when we come to treat of the Doctrine of Probation. With Fallacies terminated the one Great Division of Pure Logic,—the Doctrine of Elements, or Stoicheiology,—and I open the other

A

Great Division,—the Doctrine of Method, or Methodology,—with the following paragraph.

¶ LXXX. A Science is a complement of cognitions, having, in point of Form, the character
of Logical Perfection; in point of Matter, the
character of Real Truth.

The constituent attributes of Logical Perfec-
tion are the *Perspicuity*, the *Completeness*, the
Harmony, of Knowledge. But the Perspicuity,
Completeness, and Harmony of our cognitions
are, for the human mind, possible only through
Method.

Method in general denotes a procedure in the
treatment of an object, conducted according to
determinate rules. Method in reference to Science, denotes, therefore, the arrangement and elaboration of cognitions according to definite rules,
with the view of conferring on these a Logical
Perfection. The Methods by which we proceed
in the treatment of the objects of our knowledge
are two; or rather Method, considered in its integrity, consists of two processes,—*Analysis* and
Synthesis.

I. The Analytic or Regressive ;—in which, de-
parting from the individual and the determined,
we ascend always to the more and more general,
in order finally to attain to ultimate principles.

II. The Synthetic or Progressive ;—in which
we depart from principles or universals, and from
these descend to the determined and the individual.

Through the former we investigate and ascertain the reality of the several objects of science;

through the latter we connect the fragments of our knowledge into the unity of a system.

In its Stoicheiology or Doctrine of Elements, Logic considers the conditions of possible thought; for thought can only be exerted under the general laws of Identity, Contradiction, Excluded Middle, and Reason and Consequent, and through the general forms of Concepts, Judgments, and Reasonings. These, therefore, may be said to constitute the Elements of thought. But we may consider thought not merely as existing, but as existing well; that is, we may consider it not only in its possibility, but in its perfection: and this perfection, in so far as it is dependent on the form of thinking, is as much the object-matter of Logic as the mere possibility of thinking. Now that part of Logic which is conversant with the Perfection,—with the Well-being, of thought, is the Doctrine of Method,—Methodology.

Method in general is the regulated procedure towards a certain end; that is, a progress governed by rules which guide us by the shortest way straight towards a certain point, and guard us against devious aberrations.[a] Now the end of thought is truth,—knowledge,—science,—expressions which may here be considered as convertible. Science may, therefore,

(margin notes: LECT. XXIV. Explanation. Possibility and Perfection of Thought. Method in general,—what. Science,—what.)

a [On Method, see Alex. Aphrod., In Anal. Prior., f. 3 b, Ald. 1520; Ammonius, In Proem. Porphyrii, f. 21 b, Ald. 1546; Philoponus, In An. Prior., f. 4; In An. Post., f. 04; Eustratius, In An. Post., ff. 1 b, 53 b. See also Molinæus, Zabarella, Nunnesius, Timpler, Downam.] [Molinæus, Logica, L. ii., De Methodo, p. 245 et seq. Zabarella, Opera Logica, De Methodis, L. i. c. 2, p. 134.

Peter John Nunnesius, De Constitutione Artis Dialecticæ, p. 43 et seq., ed. 1554, with relative commentary. Timpler, Systema Logicæ, L. iv. c. viii. p. 716 et seq. G. Downam, Commentarii in P. Rami Dialecticam, L. ii. c. 17, p. 472 et seq. On the distinction between Method and Order, see Lectures on Metaphysics, vol. i. lect. vi. p. 96, and note.—ED.]

be regarded as the perfection of thought, and to the accomplishment of this perfection the Methodology of Logic must be accommodated and conducive. But Science, that is, a system of true or certain knowledge, supposes two conditions. Of these the first has a relation to the knowing subject, and supposes that what is known is known clearly and distinctly, completely, and in connection. The second has a relation to the objects known, and supposes that what is known has a true or real existence. The former of these constitutes the Formal Perfection of science, the latter is the Material.

Now, as Logic is a science exclusively conversant about the form of thought, it is evident that of these two conditions,—of these two elements, of science or perfect thinking, Logic can only take into account the formal perfection, which may, therefore, be distinctively denominated the *logical perfection* of thought. Logical Methodology will, therefore, be the exposition of the rules and ways by which we attain the formal or logical perfection of thought.

But Method, considered in general,—considered in its unrestricted universality,—consists of two processes, correlative and complementary of each other. For it proceeds either from the whole to the parts, or from the parts to the whole. As proceeding from the whole to the parts, that is, as resolving, as unloosing, a complex totality into its constituent elements, it is Analytic; as proceeding from the parts to the whole, that is, as recomposing constituent elements into their complex totality, it is Synthetic. These two processes are not, in strict propriety, two several methods, but together constitute only a single method. Each alone is imperfect ;—each is conditioned or consummated by the other : and, as I formerly observed,[a] Analysis and Synthesis are as necessary to themselves and to the life of science as expiration and inspiration in connection are necessary to each other and to the possibility of animal existence.

It is here proper to make you aware of the confusion which prevails in regard to the application of the terms *Analysis* and *Synthesis*.[b] It is manifest, in general, from the meaning of the words, that the term *analysis* can only be applied to the separation of a whole into its parts, and that the term *synthesis* can only be applied to the collection of parts into a whole. So far, no ambiguity is possible,—no room is left for abuse. But you are aware that there are different kinds of wholes and parts ; and that some of the wholes, like the whole of Comprehension (called also the *Metaphysical*), and the whole of Extension (called also the *Logical*), are in the inverse ratio of each other, so that what in the one is a part is necessarily in the

Marginal notes:

LECT.
XXIV.

Confusion in regard to the application of the terms Analysis and Synthesis.

These counter processes as applied to the counter wholes of Comprehension and Extension, correspond with each other.

a See *Lectures on Metaphysics*, vol. i. p. 99.—ED.

β [Zabarella, *Opera Logica, Liber de Regressu*, pp. 481, 489. See also, *In Anal. Post.*, L. ii. text 81, pp. 1212, 1213. Molinæus, *Logica*, L. ii. Appendix, p. 241 et sq., who notices that both the Analytic and Synthetic order may proceed from the general to the particular. See also to the same effect Hoffbauer, *Über die Analysis in der Philosophie*, p. 41 et sq., Halle, 1810 ; Gassendi, *Physica*, Sectio iii. Memb. Post., L. ix., *Opera*, t. ii. p. 460 ; Victorin, *Neue natürlichere Darstellung der Logik*, § 214 ; Trendelenburg, *Elementa Logices Aristotelicæ*, p. 89 ; Troxler, *Logik*, II. p. 100, a.** ; Krug, *Logik*, § 114, p. 408, n.**, and § 120, p. 431. Wyttenbach makes Synthetic method progress from particulars to universals ; other logicians generally the reverse.]—[See his *Præcepta Phil. Logicæ*, P. III. c. I. § 3, p. 84, ed. 1781 : " Mentem suapte natura Syntheticam Methodum sequi, atque ad ideas universales pervenire. . . . Contrarium est iter Analyticæ Methodi, quæ ab universalibus initium ducit et ad peculiaria progreditur, dividendo Genera in suas Formas." "Contra communem sensum et verborum naturam, Syntheticam vocant Methodum, quæ dividit, Analyticam contra, quæ componit." *Præl. sub fin.* In the edition of the *Præcepta* by Maass, Wyttenbach is made to say precisely the reverse of what he lays down in the original edition.—See *Præc. Phil. Log.*, ed. Maass, p. 64.—ED.]

other a whole. It is evident, then, that the counter processes of Analysis and Synthesis, as applied to these counter wholes and parts, should fall into one or correspond; inasmuch as each in the one quantity should be diametrically opposite to itself in the other. Thus Analysis as applied to Comprehension, is the reverse process of Analysis as applied to Extension, but a corresponding process with Synthesis; and *vice versa*. Now, should it happen that the existence and opposition of the two quantities are not considered,—that men, viewing the whole of Extension or the whole of Comprehension, each to the exclusion of the other, must define Analysis and Synthesis with reference to that single quantity which they exclusively take

Hence the
terms An-
alysis and
Synthesis
used in a
contrary
sense.
into account;—on this supposition, I say, it is manifest that, if different philosophers regard different wholes or quantities, we may have the terms *analysis* and *synthesis* absolutely used by different philosophers in a contrary or reverse sense. And this has actually happened. The ancients, in general, looking alone to the whole of Extension, use the terms *analysis* and *analytic* simply to denote a division of the genus into species,—of the species into individuals; the moderns, on the other hand, in general, looking only at the whole of Comprehension, employ these terms to express a resolution of the individual into its various attributes.[a] But though the contrast in this respect between the ancients and moderns holds in general, still it is exposed to sundry exceptions; for, in both periods, there are philosophers found at the same game of cross-purposes with their contemporaries as the ancients and moderns in general are with each other.

a [See Aristotle, *Physica*, L. iv. li. c. i. qu. 11, p. 248.]
c. 3. Timpler, *Logicæ Systema*, L.

This difference, which has never, so far as I know, been
fully observed and stated, is the cause of great confusion and mistake. It is proper, therefore, when we
use these terms, to use them not in exclusive relation
to one whole more than to another; and at the same
time to take care that we guard against the misapprehension that might arise from the vague and one-sided
view which is now universally prevalent. So much
for the meaning of the words *analytic* and *synthetic*,
which, by the way, I may notice, are, like most of our
logical terms, taken from Geometry.*

The Synthetic Method is likewise called the *Pro-*
gressive; the Analytic is called the *Regressive.* Now
it is plain that this application of the terms *progressive*
and *regressive* is altogether arbitrary. For the import
of these words expresses a relation to a certain point
of departure,—a *terminus a quo,* and to a certain point
of termination,—a *terminus ad quem;* and if these
have only an arbitrary existence, the correlative words
will, consequently, only be of an arbitrary application.
But it is manifest that the point of departure,—the
point from which the Progressive process starts,—may
be either the concrete realities of our experience,—the
principiata,—the *notiora nobis;* or the abstract gen-
eralities of intelligence,—the *principia,*—the *notiora*
natura. Each of these has an equal right to be regarded as the starting-point. The Analytic process is
chronologically first in the order of knowledge, and
we may, therefore, reasonably call it the *progressive,*
as starting from the primary data of our observation.
On the other hand the Synthetic process, as following

The Synthe-
tic Method
has been
called the
Progressive,
and the An-
alytic the
Regressive.
These desig-
nations
wholly arbi-
trary, and
of various
application.

---

a See above, vol. I. p. 279, n. β.
—ED. [On the Analysis of Geometry, see Plotinus, *Enneid.,* iv. L.
ix. c. 5; Philoponus, *In An. Post.,*
f. 36 a, Venet. 1534.]

the order of constitution, is first in the order of nature, and we may, therefore, likewise reasonably call it the *progressive*, as starting from the primary elements of existence. The application of these terms as synonyms of the analytic and synthetic processes, is, as wholly arbitrary, manifestly open to confusion and contradiction. And such has been the case. I find that the philosophers are as much at cross-purposes in their application of these terms to the Analytic and Synthetic processes, as in the application of analysis and synthesis to the different wholes.

In general, Synthesis has been designated the Progressive, and Analysis the Regressive Process.

In general, however, both in ancient and modern times, Synthesis has been called the *Progressive*, Analysis the *Regressive*, process; an application of terms which has probably taken its rise from a passage in Aristotle, who says, that there are two ways of scientific procedure,—the one from principles (ἀπὸ τῶν ἀρχῶν), the other to principles (ἐπὶ τὰς ἀρχάς.) From this and from another similar passage in Plato (?) the term *progressive* has been applied to the process of Comprehensive Synthesis, (*progrediendi a principiis ad principiata*), the term *regressive*, to the process of Comprehensive Analysis, (*progrediendi a principiatis ad principia.*)[a]

Method in special.

So much for the general relations of Method to thought, and the general constituents of Method itself. It now remains to consider what are the particular

a *Eth. Nic.*, i. 2 (4). The reference to Plato, whom Aristotle mentions as making a similar distinction, is probably to be found by comparing two separate passages in the *Republic*, B. iv. p. 435, vi. p. 504.— ED. [Plato is said to have taught Analysis to Leodamus the Thasian. See Laertius, L. iii. 24, and Proclus, quoted in Ia. Casanbon's note. On the views of Method of Aristotle and Plato, see Scheibler and Downam.] [Scheibler, *Opera Logica*, Pars iv., *Tract. Syllog.*, c. xviii., *De Methodo*, tit. 7, p. 603. Downam, *Comm. in P. Rami Dialecticam*, L. ii. c. 17, p. 482.—ED.]

applications of Method, by which Logic accomplishes the Formal Perfection of thought. In doing this, it is evident that, if the formal perfection of thought is made up of various virtues, Logic must accommodate its method to the acquisition of these in detail; and that the various processes by which these several virtues are acquired, will in their union constitute the system of Logical Methodology. On this I give you a paragraph.

¶ LXXXI. The Formal Perfection of thought is made up of the three virtues or characters :— 1°, Of *Clearness;* 2°, Of *Distinctness,* involving *Completeness;* and, 3°, Of *Harmony.* The character of Clearness depends principally on the determination of the Comprehension of our notions; the character of Distinctness depends principally on the development of the Extension of our notions; and the character of Harmony, on the mutual Concatenation of our notions. The rules by which these three conditions are fulfilled, constitute the Three Parts of Logical Methodology. Of these, the first constitutes the *Doctrine of Definition;* the second, the *Doctrine of Division;* and the third, the *Doctrine of Probation.*[a]

[a] Krug, *Logik,* § 121 a.—Ed. [Ramus was the first to introduce Method as a part of Logic under Syllogistic, (see his *Dialectica,* L. li. c. 17), and the Port Royalists (1662) made it a fourth part of logic. See *La Logique ou L'Art de Penser,* Prem. Dis., p. 26, pp. 47, 50; Quat. Part, p. 445 et seq. ed. 1775. Gassendi, in his *Institutio Logica,* has Pars iv., *De Methodo.* He died in 1655; his *Logic* appeared posthumously in 1658. John of Damascus speaks strongly of Method in his *Dialectic,* ch. 65, and makes four special logical methods, Division, Definition, Analysis, Demonstration. Eustachius treats of Method under Judgment, and Scheibler under Syllogistic.] [Eustachius, *Summa Philosophiæ, Logica,* P. ii. Tract. 2. *De Methodo,* p. 106, ed. Lugd. Batav., 1747. First edition, 1609. Scheibler, *Opera Logica,* Pars iv. c. xviii. p. 596 et seq.—Ed.]

LECT.
XXIV.
———
Explica-
tion.

" When we turn attention on our thoughts, and
deal with them to the end that they may be constituted into a scientific whole, we must perform a threefold operation. We must, first of all, consider what
we think, that is, what is comprehended in a thought.
In the second place, we must consider how many
things we think of, that is, to how many objects the
thought extends or reaches, that is, how many are
conceived under it. In the third place, we must consider why we think so and so, and not in any other
manner; in other words, how the thoughts are bound
together as reasons and consequents. The first consideration, therefore, regards the comprehension; the
second, the extension; the third, the concatenation of
our thoughts. But the comprehension is ascertained
by definitions; the extension by divisions; and the
concatenation by probations."[a] We proceed, therefore,
to consider these Three Parts of Logical Methodology
in detail; and first, of Declaration or Definition, in
regard to which I give the following paragraph.

Par. LXXXII.
1. The Doc-
trine of De-
claration or
Definition.

¶ LXXXII. How to make a notion Clear, is
shown by the logical doctrine of *Declaration*, or
Definition in its wider sense. A Declaration, (or
Definition in its wider sense), is a Categorical
Proposition, consisting of two clauses or members,
viz. of a Subject Defined (*membrum definitum*),
and of the Defining Attributes of the subject, that
is, those by which it is distinguished from other
things (*membrum definiens*). This latter member really contains the Definition, and is often
itself so denominated. Simple notions, as con-

taining no plurality of attributes, are incapable of definition. [a]

LECT.
XXIV.

The terms *declaration* and *definition*, which are here used as applicable to the same process, express it, however, in different aspects. The term *declaration* (*declaratio*) is a word somewhat vaguely employed in English; it is here used strictly in its proper sense of *throwing light upon,—clearing up.* The term *definition* (*definitio*) is employed in a more general, and in a more special, signification. Of the latter we are soon to speak. At present, it is used simply in the meaning of an *enclosing within limits,—*the *separating a thing from others.* Were the term *declaration* not of so vague and vacillating a sense, it would be better to employ it alone in the more general acceptation, and to reserve the term *definition* for the special signification.

Explica-
tion.
The terms
Declaration
and Defini-
tion express
the same
process in
different
aspects.

¶ LXXXIII. The process of Definition is founded on the logical relations of Subordination, Co-ordination, and Congruence. To this end we discriminate the constituent characters of a notion into the *Essential*, or those which belong to it in its unrestricted universality, and the *Unessential*, or those which belong to some only of its species. The Essential are again discriminated into *Original* and *Derivative*, a division which coincides with that into Internal or Proper, and External. In giving the sum of the original characters constituent of a notion, consists its *Definition* in the stricter sense. A Definition in the stricter sense must consequently

Par. LXXXIII.
Definition
in its stricter
sense,—
what.

a Krug, *Logik,* § 121 b.—ED.

afford at least two, and properly only two, original characters, viz. that of the *Genus* immediately superior (*genus proximum*), and that of the *Difference* by which it is itself marked out from its co-ordinates as a distinct species (*nota specialis, differentia specifica*.)[a]

Explication. Various names of Declaration. Declarations, (or definitions in the wider sense), obtain various denominations, according as the process is performed in different manners and degrees. A

Explicative. Declaration is called an *Explication* (*explicatio*), when the predicate or defining member indeterminately evolves only some of the characters belonging to the

Exposition. subject. It is called an *Exposition* (*expositio*), when the evolution of a notion is continued through several

Description. explications. It is called a *Description* (*descriptio*), when the subject is made known through a number of

Definition proper. concrete characteristics. Finally, it is called a *Definition Proper*, when, as I have said, two of the essential and original attributes of the defined subject are given, whereof the one is common to it with the various species of the same genus, and the other discriminates it from these.[β]

Definitions, —Nominal, Real, and Genetic. "Definitions are distinguished also into Verbal or Nominal, Real, and Genetic, (*definitiones nominales, reales, geneticæ*), according as they are conversant with the meaning of a term, with the nature of a thing, or with its rise or production.[γ] Nominal Definitions are, it is evident, merely explications. They are, therefore, in general only used as preliminary, in order to prepare the way for more perfect

a [Cf. Aristotle, *Topica*, L 8; Keckermann, *Systema Logicæ Minus*, L. i. c. 17,—*Opera*, t. L pp. 198, 656; Scheibler, *Topica*, c. 30; Richter, *Logik*, p. 94.]
β Cf. Krug, *Logik*, § 122.—Ed.
γ [Cf. Reusch, *Systema Logicum*, § 309 et seq.]

declarations. In Real Definitions the thing defined is considered as already there, as existing (ὄν), and the notion, therefore, as given, precedes the definition. They are thus merely analytic, that is, nothing is given explicitly in the predicate or defining member, which is not contained implicitly in the subject or member defined. In Genetic Definitions the defined subject is considered as in the progress to be, as becoming (γιγνόμενον); the notion, therefore, has to be made, and is the result of the definition, which is consequently synthetic, that is, places in the predicate or defining member more than is given in the subject or member defined. As examples of these three species, the following three definitions of a circle may suffice:—
1. The Nominal Definition,—The word *circle* signifies an uniformly curved line. 2. The Real Definition,—A circle is a line returning upon itself, of which all the parts are equidistant from a given point. 3. The Genetic Definition,—A circle is formed when we draw around, and always at the same distance from, a fixed point, a movable point which leaves its trace, until the termination of the movement coincides with the commencement.[a] It is to be observed that only those notions can be genetically defined, which relate to quantities represented in time and space. Mathematics are principally conversant with such notions, and it is to be noticed that the mathematician usually denominates such genetic definitions *real definitions*, while the others he calls without distinction *nominal definitions.*[β]

The laws of Definition are given in the following paragraph.

a This example is taken, with some alteration, from Wolf, *Philosophia Rationalis*, ¶ 191.—Ed.

β Krug, *Logik*, ¶ 122. Anm. 3, pp. 443, 442.—Ed.

¶ LXXXIV. I. A definition should be Adequate (*adequata*), that is, the subject defined, and the predicate defining, should be equivalent or of the same extension. If not, the sphere of the predicate is either less than that of the subject, and the definition Too Narrow (*angustior*), or greater, and the definition Too Wide (*latior*).

II. It should not define by Negative or Divisive attributes, (*Ne sit negans, ne fiat per disjuncta*).

III. It should not be Tautological,—what is contained in the defined, should not be repeated in the defining clause, (*Ne sit circulus vel diallelon in definiendo*).

IV. It should be Precise, that is, contain nothing unessential, nothing superfluous, (*Definitio ne sit abundans*).

V. It should be Perspicuous, that is, couched in terms intelligible, and not figurative, but proper and compendious.[a]

The First of these Rules:—That the definition should be adequate, that is, that the *definiens* and *definitum* should be of the same extension, is too manifest to require much commentary. Is the definition too wide?—then more is declared than ought to be declared; is it too narrow?—then less is declared than ought to be declared:—and, in either case, the definition does not fully accomplish the end which it proposes. To avoid this defect in definition, we must attend to two conditions. In the first place, that

a Cf. Krug, *Logik*, § 123.—Ed. [Victorin, *Logik*, § 223 et seq. Sigwart, *Handbuch zu Vorlesungen über die Logik*, § 371. Boethius, De Definitione, *Opera*, p. 648 et seq. Ruffier, *Veritez de Consequence*, § 45-51. Goclenius, *Lexicon Philosophicum*, v. *Definitio*, p. 500.]

attribute should be given which the thing defined has in common with others of the same class; and, in the second place, that attribute should be given which not only distinguishes it in general from all other things, but proximately from things which are included with it under a common class. This is expressed by Logicians in the rule—*Definitio constet genere proximo et differentia ultima,*—Let the definition consist of the nearest genus and of the lowest difference. But as the notion and its definition, if this rule be obeyed, are necessarily identical or convertible notions, they must necessarily have the same extent; consequently, everything to which the definition applies, and nothing to which it does not apply, is the thing defined. Thus;—if the definition, *Man is a rational animal,* be adequate, we shall be able to say —*Every rational animal is human—Nothing which is not a rational animal is human.* But we cannot say this, for though this may be true of this earth, we can conceive in other worlds rational animals which are not human. The definition is, therefore, in this case too wide; to make it adequate, it will be necessary to add *terrestrial* or some such term—as, *Man is a rational animal of this earth.* Again, were we to define Man,—*a rationally acting animal of this earth,*—the definition would be too narrow; for it would be false to say, *no animal of this earth not acting rationally is human,* for not only children, but many adult persons, would be excluded by this definition, which is, therefore, too narrow.[a]

The Second Rule is,—That the definition should not be made by negations, or disjunctions. In regard to the former,—negations,—that we should define a thing

---

[a] Cf. Krug, *Logik,* § 121. Anm. 1.—Ed.

by what it is, and not by what it is not,—the reason of the rule is manifest. The definition should be an affirmative proposition, for it ought to contain the positive, the actual, qualities of the notion defined, that is, the qualities which belong to it, and which must not, therefore, be excluded from or denied of it. If there are characters which, as referred to the subject, afford purely negative judgments ;—this is a proof that we have not a proper comprehension of the notion, and have only obtained a prelusory definition of it, enclosing it within only negative boundaries. For a definition which contains only negative attributions, affords merely an empty notion,—a notion which is to be called a *nothing*; for, as some think, it must at least possess one positive character, and its definition cannot, therefore, be made up exclusively of negative attributes. If, however, a notion stands opposed to another which has already been declared by positive characters, it may be defined by negative characters,—provided always that the genus is positively determined. Thus Cuvier and other naturalists define a certain order of animals by the negation of a spine or backbone,—the *invertebrata* as opposed to the *vertebrata ;* and many such definitions occur in Natural History.

For a similar reason, the definition must not consist of divisive or disjunctive attributions. The end of a definition is a clear and distinct knowledge. But to say that a thing is this or that or the other, affords us either no knowledge at all, or at best only a vague and obscure knowledge. If the disjunction be contradictory, its enunciation is, in fact, tantamount to zero ; for to say that a thing either is or is not so and so, is to tell us that of which we required no assertion

to assure us. But a definition by disparate alternatives is, though it may vaguely circumscribe a notion,
only to be considered as a prelusory definition, and
as the mark of an incipient and yet imperfect knowledge. We must not, however, confound definitions
by divisive attributes with propositions expressive of
a division.

The Third Rule is,—" The definition should not be *Third Rule.*
tautological; that is, what is defined should not be
defined by itself. This vice is called *defining in a* *Defining in a circle.*
circle. This rule may be violated either immediately
or mediately. The definition,—*Law is a lawful com-
mand,*—is an example of the immediate circle. A
mediate circle requires, at least, two correlative definitions, a principal and a subsidiary. For example,—
*Law is the expressed wish of a ruler, and a ruler is
one who establishes laws.* The circle, whether immediate or mediate, is manifest or occult according as
the thing defined is repeated in the same terms, or
with other synonymous words. In the previous
example it was manifest. In the following it is concealed :—*Gratitude is a virtue of acknowledgment,—
Right is the competence to do or not to do.* Such
declarations may, however, be allowed to stand as prelusory or nominal definitions. Concealed circular definitions are of very frequent occurrence, when they
are at the same time mediate or remote; for we are
very apt to allow ourselves to be deceived by the difference of expression, and fancy that we have declared
a notion when we have only changed the language.
We ought, therefore, to be strictly on our guard against
this besetting vice. The ancients called the circular
definition also by the name of *Diallelon,* as in this
case we declare the *definitum* and the *definiens* reci-

procally by each other (δὶ ἀλλήλων)[a]. In probation
there is a similar vice which bears the same names."[b]
We may, I think, call them by the homely English
appellation of the *Seesaw.*

The Fourth Rule is,—" That the definition should be
precise ; that is, should contain nothing unessential,
nothing superfluous.   Unessential or contingent attributes are not sufficiently characteristic, and as they are
now present, now absent, and may likewise be met with
in other things which are not comprehended under the
notion to be defined, they, consequently, if admitted
into a definition, render it sometimes too wide, sometimes too narrow.   The well-known Platonic definition,—' *Man is a two-legged animal without feathers,*'
—could, as containing only unessential characters, be
easily refuted, as was done by a plucked cock.[g]   And
when a definition is not wholly made up of such attributes, and when, in consequence of their intermixture
with essential characters, the definition does not absolutely fail, still there is a sin committed against logical
purity or precision, in assuming into the declaration
qualities such as do not determinately designate what
is defined.   On the same principle, all derivative characters ought to be excluded from the definition ; for
although they may necessarily belong to the thing
defined, still they overlay the declaration with superfluous accessories, inasmuch as such characters do not
designate the original essence of the thing, but are a
mere consequence thereof.   This fault is committed in
the following definition :—*The Circle is a curved line
returning upon itself, the parts of which are at an equal*

a Compare Sextus Empiricus,   —Ed.
Pyrrh. Hyp., i. 109, ii. 68.—Ed.   g Diog. Laert., vi. 40.—Ed.
β Krug, *Logik,* § 123. Anm. 3.

distance from the central point. Here precision is violated, though the definition be otherwise correct. For
that every line returning upon itself is curved, and
that the point from which all the parts of the line are
equidistant is the central point,—these are mere consequences of the returning on itself, and of the equi-
distance. Derivative characters are thus mixed up
with the original, and the definition, therefore, is not
precise." [a]

The Fifth Rule is,—" That the definition should be
perspicuous, that is, couched in terms intelligible, not
figurative, and compendious. That definitions ought
to be perspicuous, is self-evident. For why do we declare or define at all ? The perspicuity of the definition depends, in the first place, on the intelligible
character of the language, and this again depends on
the employment of words in their received or ordinary
signification. The meaning of words, both separate
and in conjunction, is already determined by conventional usage ; when, therefore, we hear or read these,
we naturally associate with them their ordinary meaning. Misconceptions of every kind must, therefore,
arise from a deviation from the accustomed usage ;
and though the definition, in the sense of the definer,
may be correct, still false conceptions are almost inevitable for others. If such a deviation becomes necessary, in consequence of the common meaning attached
to certain words not corresponding to certain notions,
there ought at least to be appended a comment or
nominal definition, by which we shall be warned that
such words are used in an acceptation wider or more
restricted than they obtain in ordinary usage. But, in
the second place, words ought not only to be used in

a Krug, *Logik,* § 123. Anm. 2. Etc

LECT.
XXIV.

their usual signification,—that signification, if the definition be perspicuous, must not be figurative but proper. Tropes and figures are logical hieroglyphics, and themselves require a declaration. They do not indicate the thing itself, but only something similar."[a] Such, for example, are the definitions we have of Logic as the *Pharus Intellectus*,—the *Lighthouse of the Understanding*,—the *Cynosura Veritatis*,—the *Cynosure of Truth*,—the *Medicina Mentis*,—the *Physic of the Mind*, &c.[b]

2. The meaning must be not figurative but proper.

"However, many expressions, originally metaphorical, (such as *conception, imagination, comprehension, representation*, &c. &c.), have by usage been long since reduced from figurative to proper terms, so that we may employ them in definitions without scruple,—nay frequently must, as there are no others to be found.

3. The definition must be brief.

"In the third place, the perspicuity of a definition depends upon its brevity. A long definition is not only burthensome to the memory, but likewise to the understanding, which ought to comprehend it at a single jet. Brevity ought not, however, to be purchased at the expense of perspicuity or completeness."[c]

The other kinds of Declaration.

"The rules hitherto considered, proximately relate to Definitions in the stricter sense. In reference to the other kinds of Declaration, there are certain modifications and exceptions admitted. These Dilucidations or Explications, as they make no pretence to logical perfection, and are only subsidiary to the discovery of more perfect definitions, are not to be very rigidly dealt with. They are useful, provided they contain even a single true character, by which we are con-

Dilucidations or Explications.

---

a Krug, *Logik*, § 121. Anm. 4.
—Ed.

β See above, vol. i. p. 35.—Ed.
γ Krug, *ibid*.—Ed.

ducted to the apprehension of others. They may, therefore, be sometimes too wide, sometimes too narrow. A contingent and derivative character may be also useful for the discovery of the essential and original. Even Circular Definitions are not here absolutely to be condemned, if thereby the language is rendered simpler and clearer. Figurative Expressions are likewise in them less faulty than in definitions proper, inasmuch as such expressions, by the analogics they suggest, contribute always something to the illustration of the notion.

"In regard to Descriptions, these must be adequate, and no circle is permitted in them. But they need not be so precise as to admit of no derivative or contingent characters. For descriptions ought to enumerate the characters of a thing as fully as possible; and, consequently, they cannot be so brief as definitions. They cannot, however, exceed a certain measure in point of length."[a]

a Krug. *Logik*, § 123. Anm. 5.—Ed.

LECTURE XXV.

METHODOLOGY.

SECTION II.—LOGICAL METHODOLOGY.

II.—DOCTRINE OF DIVISION.

LECT.
XXV.

Division.

I NOW proceed to the Second Chapter of Logical Methodology,—the Doctrine of Division,—the doctrine which affords us the rules of that branch of Method by which we render our knowledge more distinct and exhaustive. I shall preface the subject of Logical Division by some observations on Division in general.

Division in general.

" Under Division (*divisio,* διαίρεσις) we understand in general the sundering of a whole into its parts." The object which is divided is called the *divided whole* (*totum divisum*), and this whole must be a connected many,—a connected multiplicity, for otherwise no division would be possible. The divided whole must comprise at least one character, affording the condition of a certain possible splitting of the object, or through which a certain opposition of the object becomes recognised ; and this character must be an essential attribute of the object, if the division be not aimless and without utility. This point of view, from which alone the division is possible, is called the *principle of the division* (*principium sive fundamentum divisi-*

a [On Division and its various   *Voribus,* f. 6 a, AhL 1540.]
kinds, see Ammonius, *In Quinque*

onis); and the parts which, by the distraction of the whole, come into view, are called the *divisive members* (*membra dividentia*). When a whole is divided into its parts, these parts may, either all or some, be themselves still connected multiplicities; and if these are again divided, there results a *subdivision* (*subdivisio*), the several parts of which are called the *subdivisive members* (*membra subdividentia*). One and the same object may, likewise, be differently divided from different points of view, whereby *condivisions* (*condivisiones*) arise, which, taken together, are all reciprocally co-ordinated. If a division has only two members, it is called a *dichotomy* (*dichotomia*); if three, a *trichotomy* (*trichotomia*); if four, a *tetrachotomy*; if many, a *polytomy*, &c.

"Division, as a genus, is divided into two species, according to the different kind of whole which it sunders into parts." These parts are either contained in the divided whole, or they are contained under it. In the former case the division is called a *partition* (*partitio*, ἀπαρίθμησις),[β] in the latter it is named a *logical division*.[γ] Partition finds an application only when the object to be divided is a whole compounded of parts,—consequently, where the notion of the object is a complex one; Logical Division, on the other hand, finds its application only where the notion contains a plurality of characters under it, and where, consequently, the notion is an universal one. The simple

LECT.
XXV.

Division of
two species,
—Partition
and Logical
Division.

α [On various kinds of Wholes, see Caramuel], *Rationalis et Realis Philosophia*, L. iv. sect. iii. disp. iv. p. 277.) [and above, *Lectures on Metaphysics*, vol. ii. p. 340; *Lectures on Logic*, vol. i. p. 201. —Ed.]

β 'Απαρίθμησις is properly a rhetorical term, and signifies the division of a subject into successive heads, first, second, &c. See Hermogenes, Περὶ θεῶν,— *Rhetores Graeci*, i. p. 104, ed. Ald.—Ed.

γ [See Keckermann, *Systema Logicae*, L. L c. 3; *Opera*, t. i. p. 667; Drobisch, *Neue Darstellung der Logik*, § 112; Krug, *Logik*, § 124, Anm. 2.]

notion is thus the limit of Partition; and the individual or singular is thus the limit of Division.   Partition is divided into *physical* or *real*, (when the parts can actually be separated from each other), and *metaphysical* or *ideal*, (when the parts can only be sundered by abstraction).*   It may be applied in order to attain to a clear knowledge of the whole, or to a clear knowledge of the parts.   In the former case, the parts are given and the whole is sought; in the latter, the whole is given and the parts are sought. If the whole be given and the parts sought out, the object is first of all separated into its proximate, and, thereafter, into its remoter parts, until either any further partition is impossible, or the partition has attained its end.   To this there is, however, required an accurate knowledge of the object, of its parts proximate and remote, and of the connection of these parts together, as constituting the whole.   We must, likewise, take heed whether the partition be not determined from some particular point of view, in consequence of which the notions of more proximate and more remote may be very vague and undetermined.

a By Partition, *triangle* may be distinguished, 1°, Into a certain portion of space included within certain boundaries; 2°, Into sides and angles; 3°, Into two triangles, or into a trapezium and a triangle.   The first two partitions are ideal, they cannot be actually accomplished. The last is real, it may.

By Division, *triangle* is distinguished, 1°, Into the two species of rectilinear and curvilinear. 2°, Both of them are again subdivided (A) by reference to the sides, (B) by reference to the angles.   By reference to the sides, triangles are divided into the three species of equilateral, isosceles, and scalene. (The dichotomic division would, however, be here more proper.)   By reference to the angles, they are divided into the three species of rectangular, i.e. triangle which has one of its angles right; into amblygon, or triangle which has one of its angles obtuse; and into oxygon, i.e. triangle which has its three angles acute.

By Definition, *triangle* is distinguished into figure of three sides, equal to triangular figure; that is, into *figure*, the proximate genus, and *trilateral* or *three-sided*, the differential quality.

If the parts be given, and from them the whole sought out, this is accomplished when we have discovered the order,—the arrangement, of the parts; and this again is discovered when the principle of division is discovered; and of this we must obtain a knowledge, either from the general nature of the thing, or from the particular end we have in view. If, for example, a multitude of books of every various kind are arranged into the whole of a well-ordered library;—in this case the greater or lesser similarity of subject will afford, either exclusively or mainly, the principle of division. It happens, however, not unfrequently, that the parts are ordered or arranged according to different rules, and by them connected into a whole; and, in this case, as the different rules of the arrangement cannot together and at once accomplish this, it is proper that the less important arrangement should yield to the more important; as, for example, in the ordering of a library, when, besides the contents of the books, we take into account their language, size, antiquity, binding, &c."[a]

I now proceed to Logical Division, on which I give you the following paragraph :—

¶ LXXXV. The Distinctness and Completeness of our knowledge is obtained by that logical process which is termed *Division* (*divisio*, διαίρεσις). Division supposes the knowledge of the whole to be given through a foregone process of Definition or Declaration ; and proposes to discover the parts of this whole which are found and determined not by the development of the Comprehension, but by the development of the Extension.

a *Esser, Logik,* §§ 134, 135, p. 261-64.—Ed.

As Logical Definition, therefore, proposes to render the characters contained in an object, that is, the comprehension of a reality or notion, Clear; Logical Division proposes to render the characters contained under an object, that is, the extension of a notion, Distinct and Exhaustive. Division is, therefore, the evolution of the extension of a notion, and it is expressed in a disjunctive proposition, of which the notion divided constitutes the subject, and the notions contained under it, the predicate. It is, therefore, regulated by the law which governs Disjunctive Judgments, (the Principle of Excluded Middle), although it is usually expressed in the form of a Copulative Categorical Judgment. The rules by which this process is regulated are seven :—

1°. Every Division should be governed by some principle, (*Divisio ne careat fundamento*).

2°. Every Division should be governed by only a single principle.

3°. The principle of Division should be an actual and essential character of the divided notion, and the division, therefore, neither complex nor without a purpose.

4°. No dividing member of the predicate must by itself exhaust the subject.

5°. The dividing members, taken together, must exhaust, but only exhaust, the subject.

6°. The divisive members must be reciprocally exclusive.

7°. The divisions must proceed continuously from immediate to mediate differences, (*Divisio ne fiat per saltum*).

In this paragraph are contained, first, the general
Principles of Logical Division, and, secondly, the Laws
by which it is governed. I shall now illustrate these
in detail.

In the first place it is stated that "the distinctness and completeness of our knowledge is obtained
by that logical process which is termed *Division*
(*divisio, διαίρεσις*). Division supposes the knowledge of the whole to be given through a foregone
process of definition, and proposes to discover the
parts of this whole which are found and determined
not by the development of the comprehension, but
by the development of the extension. As logical
definition, therefore, proposes to render the characters
contained in a notion, that is, its comprehension, clear,
logical division proposes to render the characters contained under an object, that is, the extension of a
notion, distinct. Division is, therefore, the evolution
of the extension of a notion, and it is expressed in a
disjunctive proposition, of which the notion divided
constitutes the subject, and the notions contained
under it, the predicate. It is, therefore, regulated by
the law which governs disjunctive judgments (the
principle of excluded middle), although it be usually
expressed in the form of a copulative categorical
judgment."

The special virtue,—the particular element, of per
fect thinking which Division enables us to acquire,
is Distinctness, but, at the same time, it is evident
that it cannot accomplish this without rendering
our thinking more complete. This, however, is only
a secondary and collateral result; for the problem
which division proximately and principally proposes
to solve is,—to afford us a distinct consciousness of

the extension of a given notion through a complete
or exhaustive series of subordinate or co-ordinate
notions. This utility of Division, in rendering our
knowledge more complete, is, I find, stated by Aristotle,[a] though it has been overlooked by subsequent
logicians. He observes that it is only by a regular
division that we can be assured, that nothing has been
omitted in the definition of a thing.

As many
kinds of
Division
possible as
there are
characters
affording a
Principle of
Division.

"As it is by means of division that we discover
what are the characters contained under the notion of
an object, it follows that there must be as many kinds
of division possible as there are characters contained
under the notion of an object, which may afford the
principle of a different division. If the characters
which afford the principle of a division are only external and contingent, there is a division in the wider
sense; if, again, they are internal and constant, there
is a division in the stricter sense; if, finally, they are
not only internal but also essential and original, there

A universal
notion the
only object
of Logical
Division.

is a division in the strictest sense. From the very
conception of logical division it is manifest that it
can only be applied where the object to be divided is
a universal notion, and that it is wholly inapplicable
to an individual; for as the individual contains nothing under it, consequently it is not susceptible of

General
problem of
Division.

an ulterior division. The general problem of which
division affords the solution is,—To find the subordinate genera and species, the higher or generic notion
being given. The higher notion is always something
abstracted,—something generalised from the lower
notions, with which it agrees, inasmuch as it contains
all that is common to these inferior concepts, and from
which it differs, inasmuch as they contain a greater

a *Anal. Post.*, L. ii, c. 13.

complement of determining characters. There thus
subsists an internal connection between the higher and
the lower concepts, and there is thus afforded a transition from the superior notion to the subordinate,
and, consequently, an evolution of the lower notions
from the higher.   In order to discover the inferior
genera and species, we have only to discover those
characters which afford the proximate determinations,
by which the sphere or extension of the higher notion
is circumscribed.   But to find what characters are
wanted for the thorough-going determination of a
higher notion, we must previously know what characters the higher notion actually contains, and this
knowledge is only attainable by an analysis,—a sundering of the higher notion itself.   In doing this the
several characters must be separately drawn forth
and considered ; and in regard to each, we must
ascertain how far it must still be left undetermined,
and how far it is capable of opposite determinations.   But whether a character be still undetermined, and of what opposite determinations it is
capable,—on these points it is impossible to decide
a priori, but only *a posteriori*, through a knowledge
of this particular character and its relations to other
notions.   And the accomplishment of this is rendered easier by two circumstances ;—the one, that
the generic notion is never altogether abstract, but
always realised and held fast by some concrete form
of imagination ;—the other, that, in general, we are
more or less acquainted with a greater or a smaller
number of special notions in which the generic
notion is comprehended, and these are able to lead
us either mediately or immediately to other subordinate concepts.

"But the determinations or constituent characters
of a notion which we seek out, must not only be completely, but also precisely, opposed. Completely, inasmuch as all the species subordinate to the notions
ought to be discovered; and precisely, inasmuch as
whatever is not actually a subordinate species, ought
to be absolutely excluded from the notion of the
genus.

"In regard to the completeness of the opposition,
it is not, however, required that the notion should
be determined through every possible contradictory
opposition; for those at least ought to be omitted
concerning whose existence or non-existence the notion
itself decides. In regard to the opposition itself, it
is not required that the division should be carried
through by contradictory oppositions. The only opposition necessary is the reciprocal exclusion of the
inferior notions into which the higher notion is
divided."[a] In a mere logical relation, indeed, as we
know nothing of the nature of a thing more than that
a certain character either does or does not belong to
it, a strictly logical division can only consist of two
contradictory members, for example,—that angles are
either *right* or *not right*,—that men are either *white*
or *not white*. But looking to the real nature of the
thing known, either *a priori* or *a posteriori*, the division may be not only dichotomous but polytomous,
as for example,—*angles are right, or acute, or obtuse;
men are white, or black, or copper-coloured, or olive-
coloured*, &c.

We now come, in the second place, to the rules
dictated for Logical Division.

These Rules spring either, 1°, From the Principle of

a Esser, *Logik*, § 136.—Ed.

Division ; or, 2°, From the Relations of the Dividing
Members to the Divided Whole; or, 3°, From the
Relations of the several Dividing Members to each
other; or, 4°, From the Relations of the Divisions to
the Subdivisions.

The first of these heads,—the Principle of Division,
—comprehends the three first rules. Of these the
first is self-evident,—There must be some principle,
some reason, for every division; for otherwise there
would be no division determined, no division carried
into effect.

These
springing.
I. From the
Principle of
Division.
First Rule.

In regard to the second rule,—That every division
should have only a single principle,—the propriety of
this likewise is sufficiently apparent. In every division we should depart from a definite thought, which
has reference either to the notion as a unity or to some
single character. On the contrary, if we do not do
this, but carry on the process by different principles,
the series of notions in which the division is realised,
is not orderly and homogeneous, but heterogeneous
and perplexed.

Second.

The third rule,—That the principle of division should
be an actual and essential character of the divided
notion,—is not less manifest. "As the ground of division is that which principally regulates the correctness
of the whole process, that is, the completeness and
opposition of the division,—it follows that this ground
must be of notoriety and importance, and accommodated to the end for the sake of which the division is
instituted. Those characters of an object are best
adapted for a division, whose own determinations
exert the greatest influence on the determinations of
other characters, and, consequently, on those of the
notion itself; but such are manifestly not the external

Third.

and contingent, but the internal and essential, characters, and, of these, those have the pre-eminence through whose determination the greater number of others are determined, or, what is the same thing, from which, as fundamental and original attributes, the greater number of the others are derived. The choice of character is, however, for the most part, regulated by some particular end ; so that, under certain circumstances, external and contingent characters may obtain a preponderant importance. Such ends cannot, however, be enumerated. The character affording the principle of division must likewise be capable of being clearly and definitely brought out ; for unless this be possible, we can have no distinct consciousness of the completeness and contrast of the determination of which it is susceptible. We ought, therefore, always to select those characters for principles of division, which are capable of a clear and distinct recognition." [a]

The second part of the rule,—That the division be not, therefore, too complex, and without a purpose,—is a corollary of the first. " In dividing, we may go on to infinity. For while, as was formerly shown, there is, in the series of higher and lower notions, no one which can be conceived as absolutely the lowest ; so in subdividing, there is no necessary limit to the process. In like manner, the co-ordinations may be extended *ad infinitum*. For it is impossible to exhaust all the possible relations of notions, and each of these may be employed as the principle of a new division. Thus we can divide men by relation to their age, to their sex, to their colour, to their stature, to their knowledge, to their riches, to their rank, to their man-

a Esser, *Logik*, § 137.—Ed.

ner of life, to their education, to their costume, &c. &c.
It would, however, be ridiculous, and render the divisions wholly useless, if we multiplied them in this fashion without end. We, therefore, intentionally restrict them, that is, we make them comparatively limited, inasmuch as we only give them that completeness which is conducive to a certain end. In this manner divisions become relatively useful, or acquire the virtue of adaptation. In the selection of a principle of division, we must take heed whether it be fertile and pertinent. A ground of division is fertile, when it affords a division out of which again other important consequences may be drawn ; it is pertinent, when these consequences have a proximate relation to the end, on account of which we were originally induced to develop the extension of a concept. A principle of division may, therefore, be useful with one intent, and useless with another. *Soldiers*, for example, may be conveniently divided into *cavalry* and *infantry*, as this distinction has an important influence on their determination as soldiers. But in considering man in general and his relations, it would be ludicrous to divide men into *foot* and *horsemen* ; while, on the contrary, their division would be here appropriate according to principles which in the former case would have been absurd. Seneca [a] says well,—' Quicquid in majus crevit facilius agnoscitur, si discessit in partes ; quas innumerabiles esse et parvas non oportet. Idem enim vitii habet nimia, quod nulla divisio. Simile confuso est, quicquid usque in pulverem sectum est.' " [β]

Under the second head, that is, as springing from the relations of the Dividing Members to the Divided Wholes, there are included the fourth and fifth laws.

a *Epist.*, 00.   β Krug, *Logik*, § 126, Anm. 4.—ED.

"As the notion and the notions into which it is divided, stand to each other in the relation of whole and parts, and as the whole is greater than the part, the fourth rule is manifestly necessary, viz. That no dividing member of the predicate must by itself exhaust the subject. When this occurs, the division is vicious, or, more properly, there is no division. Thus the division of *man* into *rational animals* and *uncultivated nations*, would be a violation of this law.

"On the other hand, as the notions into which a notion is divided, stand to each other in the relation of constituting parts to a constituted whole, and as the whole is only the sum of all the parts, the necessity of the fifth rule is manifest,—That the dividing members of the predicate, taken together, must exhaust the subject. For if this does not take place, then the division of the principal notion has been only partial and imperfect. We transgress this law, in the first place, when we leave out one or more members of division ; as for example,—*The actions of men are either good or bad*,—for to these we should have added *or indifferent*. And in the second place, we transgress it when we co-ordinate a subdivision with a division ; as for example,—*Philosophy is either theoretical philosophy or moral philosophy :* here the proper opposition would have been *theoretical philosophy* and *practical philosophy*." [a] On the other hand, the dividing members, taken together, must not do more than exhaust the subject. The definition of the whole must apply to every one of its parts, but this condition is not fulfilled if there be a dividing member too much, that is, if there be a notion brought as a dividing member, which, however, does not stand in subordination to

a Esser, *Logik*, § 137.— Ed.

the divided whole. For example,—*Mathematical fig-*
ures are either solids or surfaces [or lines or points].
Here the two last members (*lines and points*) are redundant and erroneous, for lines and points, though
the elements of mathematical figures, are not themselves figures.

Under the third head, as springing from the relations of the several Dividing Members to Each Other,
there is a single law, the sixth, which enjoins,—That
the dividing members be reciprocally exclusive.

" As a division does not present the same but the different determinations of a single notion, (for otherwise
one and the same determination would be presented
twice), the dividing members must be so constituted
that they are not mutually coincident, so that they
either in whole or in part contain each other. This
law is violated when, in the first place, a subdivision
is placed above a division, as,—*Philosophy is either
theoretical philosophy, or moral philosophy, or practical philosophy ;* here *moral philosophy* falls into
practical philosophy as a subordinate part ; or when,
in the second place, the same thing is divided in different points of view, as,—*Human actions are either
necessary, or free, or useful, or detrimental.*" [a]

Under the fourth and last head, as arising from the
relations of the Divisions to the Subdivisions, there is
contained one law, the seventh, which prescribes,—
That the divisions proceed continuously from immediate to mediate differences, (*Divisio ne fiat per saltum
vel hiatum*).

" As divisions originate in the character of a notion,
capable of an opposite determination, receiving this
determination, and as the subdivisions originate in

these opposite determinations being themselves again capable of opposite determinations, in which gradual descent we may proceed indefinitely onwards,—from this it is evident, that the divisions should, as far as possible, be continuous, that is, the notion must first be divided into its proximate, and then into its remoter parts, and this without overleaping any one part; or in other words, each part must be immediately subordinated to its whole." [a] Thus, when some of the ancients divided *philosophy* into *rational, and natural, and moral*, the first and second members are merely subdivisions of *theoretical philosophy*, to which *moral* as *practical philosophy* is opposed. Sometimes, however, such a spring,—such a *saltus*,—is, for the sake of brevity, allowed; but this only under the express condition, that the omitted members are interpolated in thought. Thus, many mathematicians say, *angles are either right, or acute, or obtuse*, although, if the division were continuous,—without hiatus, it would run, *angles are either right or oblique; and the oblique, again, either acute or obtuse.*

a Esser, *Logik*, § 137.—ED.

LECTURE XXVI.

METHODOLOGY.

SECTION II.—LOGICAL METHODOLOGY.

III.—DOCTRINE OF PROBATION.

We now proceed to the Third Part of Pure Methodology, that which guides us to the third character or virtue of Perfect Thinking,—the Concatenation of Thought;—I mean Probation, or the Leading of Proof. I commence with the following paragraph :—

¶ LXXXVI. When there are propositions or judgments which are not intuitively manifest, and the truth of which is not admitted, then their validity can only be established when we evolve it, as an inference, from one or more judgments or propositions. This is called *Probation*, *Proving*, or the *Leading of Proof* (*probatio, argumentatio*, or *demonstratio* in its wider sense). A Probation is thus a series of thoughts, in which a plurality of different judgments stand to each other, in respect of their validity, in the dependence of determining and determined, or of antecedents and consequents. In every Probation there are three things to be distinguished,— 1°, The Judgment to be Proved, (*thesis*); 2,° The

LECT. XXVI.

Probation.

Par. LXXXVI. Probation, —its Nature and Elements.

Ground or Principle of Proof, (*argumentum*) ; and, 3°, The Cogency of this principle to necessitate the connection of antecedents and consequents, (*vis demonstrationis* or *nervus probandi*). From the nature of Probation, it is evident that Probation without inference is impossible ; and that the Thesis to be proved and the Principles of Proof stand to each other as conclusion and premises, with this difference, that, in Probation, there is a judgment (thesis) expressly supposed, which in the Syllogism is not, at least necessarily, the case.[a]

In regard to the terms here employed, it is to be noticed that the term *argumentation* (*argumentatio*) is applied not only to a reasoning of many syllogisms, but likewise to a reasoning of one. The term *argument* (*argumentum*), in like manner, is employed not only for the ground of a consecutive reasoning, but for the middle term of a single syllogism. But it is, moreover, vulgarly employed for the whole process of argumentation. [b]

The term *demonstration* (*demonstratio*) is used in a looser, and in a stricter, signification. In the former sense, it is equivalent to *probation*, or *argumentation in general*; in the latter, to *necessary probation*, or *argumentation from intuitive principles*.

The expression *leading of proof* might, perhaps, be translated by the term *deduction*, but then this term must be of such a latitude as to include induction, to which it is commonly opposed ; for Probation may be

fang der Logik, § 32 et seq.]
β See above, vol. 1. p. 278.—
ED.

either a process of Deduction, that is, the leading of proof out of one higher or more general proposition, or a process of Induction, that is, the leading of proof out of a plurality of lower or less general judgments.

LECT.
XXVI.

To prove, is to evince the truth of a proposition not admitted to be true, from other propositions the truth of which is already established. In every probation there are three things to be distinguished:—1°, The Proposition to be Proved,—the Thesis; 2°, The Grounds or Principle of Proof,—the Argument; and, 3°, The Degree of Cogency with which the thesis is inferred by the argumentum or argumenta,—the *vis* or *nervus probandi*. All probation is thus syllogistic; but all syllogism is not probative. The peculiarity of probation consists in this,—that it expressly supposes a certain given proposition, a certain thesis, to be true; to the establishment of this proposition the proof is relative; this proposition constitutes the conclusion of the syllogism or series of syllogisms of which the probation is made up: whereas, in the mere syllogistic process, this supposition is not necessarily involved. It is also evident that the logical value of a probation depends,—1°, On the truth of its principles or arguments, 2°, On their connection with each other and with the thesis or proposition to be proved, and, 3°, On the logical formality of the inference of the thesis from its arguments. No proposition can be for another the principle of proof, which is not itself either immediately or mediately certain. A proposition is immediately certain, or evident at first hand, when, by the very nature of thought, we cannot but think it to be true, and when it, therefore, neither requires nor admits of proof. A proposition is mediately certain, or evident at second hand, when it is not at

Probation
in general.

How distinguished from Syllogism.

Whereon depends the logical value of a probation.

once and in itself thought as necessarily true, but when we are able to deduce it, with a consciousness of certainty, from a proposition which is evident at first hand. The former of these certainties is called *self-evident, intuitive, original, primary, ultimate, &c.,* and the latter, *demonstrative, derivative, secondary, &c.*

Ground of Proof either Absolute or Relative.

According to this distinction, the Ground or Principle of Proof is either absolute or relative. Absolute, when it is an intuitive; relative, when it is a demonstrative, proposition. That every proposition must ultimately rest on some intuitive truth,—on some judgment at first hand, is manifest, if the fact of probation itself be admitted ; for otherwise the regress would extend to infinity, and all probation, consequently, be impossible. When, for example, in the series of grounds H, G, F, E, D, C, B, there is no ultimate or primary A, and when, consequently, every A is only relatively, in respect of the consequent series, but not absolutely and in itself, first ;—in this case, no sufficient and satisfactory probation is possible, for there always remains the question concerning a still higher principle. But positively to show that such primary judgments are actually given, is an exposition which, as purely metaphysical, lies beyond the sphere of Logic.[a]

Distinction of Propositions in respect of the general form of a system of Proof. Theoretical and Practical.

To the general form of a system of Proof belong the following distinctions of propositions, to which I formerly alluded,[b] and which I may again recall to your remembrance. Propositions are either *Theoretical* or *Practical.* Practical, when they enounce the way in which it is possible to effectuate or produce

---

a Compare Esser, *Logik.* § 138.— Ed.   b See above, vol. I. p. 265.— Ed.

something; Theoretical, when they simply enunciate LECT.
XXVI.
a truth, without respect to the way in which this may
be realised or produced.[a]

A Theoretical proposition, if a primary or intuitive Axiom.
principle, is styled an *Axiom.* Examples of this are
given in the four Fundamental Laws of Logic, and in
the mathematical common notions—*The whole is greater
than its part,—If equals be added to equals, the wholes
are equal,* &c. A Practical proposition, if a primary or Postulate.
intuitive principle, is styled a *Postulate.* Thus Geometry postulates the possibility of drawing lines,—of
producing them *ad infinitum,* of describing circles, &c.

A Theoretical proposition, if mediate and demon- Theorem.
strable, is called a *Theorem.* This is laid down as a
Thesis,—as a judgment to be proved,—and is proved
from intuitive principles, theoretical and practical.
A Practical proposition, if mediate and demonstrable, Problem.
is called a *Problem.* In the probation, the Problem
itself is first enounced; it is then shown in the solution how that which is required is to be done,—is
to be effected; and, finally, in the proof, it is demonstrated that through this procedure the solution of
the problem is obtained. For example, in the geometrical problem,—to describe an equilateral triangle
on a given straight line;—there this problem is first
stated; the solution then shows that, with this given
line as a semi-diameter, we are to describe from each
of its points of termination a circle; the two circles
will intersect each other, and we are then, from the
point of intersection, to draw straight lines to each
point of termination; this being done, the proof
finally demonstrates that these circles must intersect
each other, that the drawn straight lines necessarily

a [Fries, *System der Logik*, § 73.]

LECT.
XXVI.

constitute a triangle, and that this triangle is necessarily equilateral.

Corollaries.

Corollaries or *Consectaries* are propositions which, as flowing immediately as collateral result of others, require no separate proof. *Empeiremata* or *Empirical Judgments* are propositions, the validity of which reposes upon observation and experience. *Scholia* or *Comments* are propositions which serve only for illustration. *Lemmata* or *Sumptions* are propositions borrowed either from a different part of the system we treat of, or from sciences other than that in which we now employ them. Finally, *Hypotheses* are propositions of two different significations. For, in the first place, the name is sometimes given to the arbitrary assumption or choice of one out of various means of accomplishing an end ; when, for example, in the division of the periphery of the circle, we select the division into 360 degrees, or when, in Arithmetic, we select the decadic scheme of numeration. But, in the second place, the name of *hypothesis* is more emphatically given to provisory suppositions, which serve to explain the phenomena in so far as observed, but which are only asserted to be true, if ultimately confirmed by a complete induction. For example, the supposition of the Copernican solar system in Astronomy.[a]

Empeiremata.

Scholia.

Lemmata.

Hypotheses.

Now these various kinds of propositions are mutually concatenated into system by the Leading of Proof, —by Probation.

So much for the character of this process in general. The paragraph, already dictated, contains a summary of the various particular characters by which Probations are distinguished. Before consid-

a [Fries, *System der Logik*, § 73. Krug, *Logik*, §§ 67, 68.]

ering these in detail, I shall offer some preparatory observations.

"The differences of Probations are dependent partly on their Matter, and partly on the Form in which they are expressed.

"In respect of the former ground of difference,— the Matter,—Probations are distinguished into Pure or *a priori* and Empirical or *a posteriori*, according as they are founded on principles which we must recognise as true, as constituting the necessary conditions of all experience, or which we do recognise as true, as particular results given by certain applications of experience. In respect of the latter ground of difference,—the Form,—Probations fall into various classes according to the difference of the form itself, which is either External or Internal.

"In relation to the Internal Form, probations are divided into Direct or Ostensive and Indirect or Apagogical, according as they are drawn from the thing itself or from its opposite, in other words, according as the principles of probation are positive or are negative."[a]—Under the same relation of Internal Form, they are also distinguished by reference to their order of procedure,—this order being either Essential or Accidental. The essential order of procedure regards the nature of the inference itself, as either from the whole to the part, or from the parts to the whole. The former constitutes Deductive Probation, the latter Inductive. The accidental order of procedure regards only our point of departure in considering a probation. If, commencing with the highest principle, we descend step by step to the conclusion, the process is Synthetic or Progressive; here the conclusion is evolved out

The differences of Probations depend partly on their Matter and partly on their Form.

1. In respect of their Matter, Probations are Pure and Empirical.

2. In respect of their Form.

a. In relation to the Internal Form, Probations are Direct or Ostensive and Indirect or Apagogical.

Synthetic or Progressive, and Analytic or Regressive.

a Esser, *Logik*, § 141.—Ed.

LECT.
XXVI. of the principle. If again, starting from the conclusion, we ascend step by step to the highest principle, the process is Analytic or Regressive; here the principle is evolved out of the conclusion.

b. External
Form.
Probations
are Simple
and Com-
posite.
Regular and
Irregular.
Perfect and
Imperfect. In respect to the External Form, Probations are Simple or Monosyllogistic if they consist of a single reasoning, Composite or Polysyllogistic if they consist of a plurality of reasonings. Under the same relation of external form, they are also divided into Regular and Irregular, into Perfect and Imperfect.

8. Accord-
ing to their
Degree of
Cogency,
Probations
are Apodeic-
tic and
Probable. Another division of Probations is by reference to their Cogency, or the Degree of Certainty with which their inference is drawn. But their cogency is of various degrees, and this either objectively considered, that is, as determined by the conditions of the proof itself, or subjectively considered, that is, by reference to those on whom the proof is calculated to operate conviction. In the former or objective relation, probations are partly Apodeictic, or Demonstrative in the stricter sense of that term,—when the certainty they necessitate is absolute and complete, that is, when the opposite alternative involves a contradiction; partly Probable,—when they do not produce an invincible assurance, but when the evidence in favour of the conclusion preponderates over that which is opposed Universally
and Parti-
cularly
Valid. to it. In the latter or subjective relation, probations are either Universally Valid, when they are calculated to operate conviction on all reasonable minds, or Particularly Valid, when they are fitted to convince only certain individual minds.

Par. LXXXVII.
Probations,
—their
Divisions. ¶ LXXXVII. Probations are divided by reference to their Matter, to their Form, and to their Degree of Cogency.

In relation to their Matter, they are partly *Pure* or *a priori*, partly *Empirical* or *a posteriori*.

As to their Form,—this is either Internal or External. In respect to their Internal Form, they are, 1°, By reference to the Manner of Inference, *Direct* or *Ostensive* (δεικτικαί, *ostensivæ*), and *Indirect* or *Apagogical* (*probationes apagogicæ, reductiones ad absurdum*); 2°, By reference to their Essential or Internal Order of Procedure, they are either *Deductive* or *Inductive*; 3°, By reference to their Accidental or External Order of Procedure, they are partly *Synthetic* or *Progressive*, partly *Analytic* or *Regressive*. In respect to their External Form, they are, 1°, *Simple* or *Monosyllogistic*, and *Composite* or *Polysyllogistic*; 2°, *Perfect* and *Imperfect*; 3°, *Regular* and *Irregular*.

In respect to their Degree of Cogency, they are, 1°, As objectively considered, either *Apodeictic* or *Demonstrative* in the stricter signification of the term, (ἀποδείξεις, *demonstrationes stricte dictæ*), or *Probable*, (*probationes sensu latiori*); 2°, As subjectively considered, they are either *Universally Valid*, (κατ' ἀληθείαν, *secundum veritatem*), or *Particularly Valid* (κατ' ἄνθρωπον, *ad hominem*).[a]

To speak now of these distinctions in detail. In the first place, "Probations," we have said, " in relation to their matter, are divided into Pure or *a priori* and Empirical or *a posteriori*. Pure or *a priori*

Explication. Probations, 1. In respect of their Matter, are Pure and Empirical.

a Cf. Krug, *Logik*, §§ 128, 129,   —En. [Cf. Degerando, *Des Signes*, 130, 131, 132; Esser, *Logik*, § 130. t. iv. ch. 7, p. 211.]

proofs are those that rest on principles which, although
rising into consciousness only on occasion of some
external or internal observation,—of some act of experience, are still native, are still original, contributions
of the mind itself, and a contribution without which no
act of experience becomes possible. Proofs again are
called *Empirical* or *a posteriori*, if they rest on principles which are exclusively formed from experience
or observation, and whose validity is cognisable in
no other way than that of experience or observation.
When the principles of Probation are such as are not
contingently given by experience, but spontaneously
engendered by the mind itself, these principles are
always characterised by the qualities of necessity and
universality ; consequently, a proof supported by them
is elevated altogether above the possibility of doubt.
When, on the other hand, the principles of Probation
are such as have only the guarantee of observation
and experience for their truth,—(supposing even that
the observation be correct and the experience stable
and constant),—these principles, and, consequently,
the probation founded on them, can pretend neither
to necessity nor to universality ; seeing that what produces the observation or experience, has only a relation to individual objects, and is only competent to
inform us of what now is, but not of what always is,
of what necessarily must be. Although, however,
these empirical principles are impressed with the character neither of necessity nor of universality, they
play a very important part in the theatre of human
thought."[a] This distinction of Proofs, by reference
to the matter of our knowledge, is one, indeed, which
Logic does not take into account. Logic, in fact, con-

This distinc-
tion of Pro-
bations not
taken into
account by
Logic.

a Esser, *Logik*, § 140.—Ed.

siders every inference of a consequent from an antecedent as an inference *a priori*, supposing even that the
antecedents themselves are only of an empirical character. Thus we may say, that, from the general relations of distance found to hold between the planets,
Kant and Olbers proved *a priori* that between Mars
and Jupiter a planetary body must exist, before Ceres,
Pallas, Juno, and Vesta were actually discovered.[a]
Here, however, the *a priori* principle is in reality only
an empirical rule,—only a generalisation from experience. But with the manner in which these empirical rules (Bacon would call them *axioms*) are
themselves discovered or evolved,—with this Pure
Logic has no concern. This will fall to be considered
in Modified Logic, when we treat of the concrete
Doctrine of Induction and Analogy.

In the second place, " in respect of their Form, and
that the Internal, Probations are, as we said, first of
all, divided into Direct or Ostensive and Indirect or
Apagogical. A proof is Direct or Ostensive, when
it evinces the truth of a thesis through positive principles, that is, immediately ; it is Indirect or Apagogical,
when it evinces the truth of a thesis through the falsehood of its opposite, that is, mediately. The indirect
is specially called the *apagogical*, (*argumentatio apa-
gogica* sive *deductio ad impossibile*), because it shows
that something cannot be admitted, since, if admitted,
consequences would necessarily follow impossible or
absurd. The Indirect or Apagogical mode of proof is
established on the principle, that that must be con-
ceded to be true whose contradictory opposite contains within itself a contradiction. This principle

2. In respect
of their
Form,—
a. Direct
and In-
direct.

Principle
of Indirect
Proof.

a See Kant's *Vorlesungen über* vi. p. 449. --ED.
physische Geographie, 1802; *Werke*,

manifestly rests on the Law of Contradiction and on the Law of Excluded Middle; for what involves a contradiction it is impossible for us to think, and if a character must be denied of an object,—and that it must be so denied the probation has to show,—then the contradictory opposite of that character is of necessity to be affirmed of that object. The Direct mode of probation has undoubtedly this advantage over the Indirect,—that it not only furnishes the sought-for truth, but also clearly develops its necessary connection with its ultimate principles; whereas the Indirect demonstrates only the repugnance of some proposition with certain truths, without, however, positively evincing the truth of its opposite, and thereby obtaining for it a full and satisfactory recognition. It is, therefore, usually employed only to constrain a troublesome opponent to silence, by a display of the absurdities which are implied in, and which would flow out of, his assertions. Nevertheless the indirect probation establishes the proposition to be proved not less certainly than the direct; nay, it still more precisely excludes the supposition of the opposite alternative, and, consequently, affords an intenser consciousness of necessity. We ought, however, to be on our guard against the paralogisms to which it is peculiarly exposed, by taking care—1°, That the opposites are contradictory and not contrary; and, 2°, That an absurdity really is, and not merely appears to be. The differences of Apagogical Probations correspond to the different kinds of propositions which may be indirectly demonstrated; and these are, in their widest generality, either Categorical, or Hypothetical, or Disjunctive. Is the thesis a categorical proposition? Its contradictory opposite is supposed, and from this counter proposition conclusions are deduced, until we obtain one of so absurd a character, that we are able to argue back to the falsehood of the original proposition itself.   Again, is the thesis an hypothetical judgment?   The contradictory opposite of the consequent is assumed, and the same process to the same end is performed as in the case of a categorical proposition.   Finally, is the thesis a disjunctive proposition?   In that case, if its *membra disjuncta* are contradictorily opposed, we cannot, either directly or indirectly, prove it false as a whole; all that we can do being to show that one of these disjunct members cannot be affirmed of the subject, from which it necessarily follows that the other must."[a]

Under the Internal Form, Probations are, in the second place, in respect of their Essential or Internal Order of procedure, either Deductive or Inductive, according as the thesis is proved by a process of reasoning descending from generals to particulars and individuals, or by a process of reasoning ascending from individuals and particulars to generals.   On this subject it is not necessary to say anything, as the rules which govern the formal inference in these processes have been already stated in the Doctrine of Syllogisms ; and the consideration of Induction, as modified by the general conditions of the matter to which it is applied, can only be treated of when, in the sequel, we come to Modified or Concrete Methodology.

"Under the Internal Form, Probations are, however, in the third place, in respect of their External or Accidental Order of procedure, Synthetic or Progressive, and Analytic or Regressive.   A probation

a. Deductive and Inductive.

a. Synthetic and Analytic.

a Esser, *Logik*, § 142.—Ed.

is called *synthetic* or *progressive*, when the conclusion is evolved out of the principles,—*analytic* or *regressive*, when the principles are evolved out of the conclusion. In the former case, the probation goes from the subject to the predicate; in the latter case, from the predicate to the subject. Where the probation is complex,—if synthetic, the conclusion of the preceding syllogism is the subsumption of that following; if analytic, the conclusion of the preceding syllogism is the sumption of that following. In respect of certainty, both procedures are equal, and each has its peculiar advantages; in consequence of which the combination of these two modes of proof is highly expedient. But the Analytic Procedure is often competent where the Synthetic is not; whereas the Synthetic is never possible where the Analytic is not, and this is never possible where we have not a requisite stock of propositions already verified. When the Probation is partly analytic, partly synthetic, it is called *Mixed.*"[a]

Par. LXXXVIII.
Formal
Legitimacy
of a Proba-
tion,—its
Rules.
¶ LXXXVIII. The Formal Legitimacy of a Probation is determined by the following rules.

1°, Nothing is to be begged, borrowed, or stolen; that is, nothing is to be presupposed as proved, which itself requires a demonstration. The violation of this rule affords the vice called the *Petitio principii*, or *Fallacia quæsiti medii* (τὸ ἐν ἀρχῇ αἰτεῖσθαι).[b]

2°, No proposition is to be employed as a principle of proof, the truth of which is only to be

---

a Esser, *Logik*, § 142.—Ed.

β [On error of this term, see Pacius, *Commentarius in Org.*] [*In Anal. Prior.*, ii. 16: "Non est petitio τῆς ἀρχῆς, id est, principii, vel ἐν τῇ ἀρχῇ, id est, in principio; sed τοῦ ἐν ἀρχῇ προσειμένου, id est, ejus problematis, quod initio fuit propositum et in disquisitionem vocatum." *Ibid.* ii. 24.—Ed.

evinced as a consequence of the proposition
which it is employed to prove. The violation of
this rule is the vice called *ὕστερον πρότερον*.

3°, No circular probation is to be made ; that
is, the proposition which we propose to prove
must not be used as a principle for its own probation. The violation of this rule is called the
Orbis vel circulus in demonstrando,—diallelus,—
ὁ δὲ ἀλλήλων τρόπος.[a]

4°, No leap, no hiatus, must be made ; that is,
the syllogisms of which the probation is made
up, must stand in immediate or continuous connection. From the transgression of this rule
results the vice called the *Saltus* vel *Hiatus in
demonstrando*.

5°, The scope of the probation is not to be
changed ; that is, nothing is to be proved other
than what it was proposed to prove. The violation of this rule gives the *Heterozetesis, Ignoratio*
vel *Mutatio elenchi*, and the *Transitus in aliud
genus* vel *a genere ad genus,—μετάβασις εἰς ἄλλο
γένος*.[b]

In this paragraph, I have given, as different rules,
those canons which are opposed to vices not absolutely identical, and which have obtained different
denominations. But you must observe, that the first
three rules are all manifestly only various modifications
—only special cases—of one general law. To this law,
likewise, the fourth rule may with perfect propriety

a See Sextus Empiricus, *Pyrrh.
Hyp.*, i. 109, ii. 68 ; Laertius, L. ix.
§§ 88, 89. [Cf. Facciolati, *Acroases*,
p. 69 et seq.]
β [See Reinhold, *Die Logik oder
die allgemeine Denkformenlehre*, §
150, p. 407, Jena, 1827.] [Cf. Krug,
Logik, § 133 ; Esser, *Logik*, § 144.—
En.]

These rules
reduced to
two.

be reduced, for the *saltus* or *hiatus in probando* is, in fact, no less the assumption of a proposition as a principle of probation which itself requires proof, than either the *petitio principii*, the *hysteron proteron*, or the *circulus in probando*. These five laws, therefore, and the correspondent vices, may all be reduced to two ; the one of which regards the means, the principles of proof ; the other the end, the proposition to be proved. The former of these laws prescribes,—That no proposition be employed as a principle of probation, which stands itself in want of proof ; the latter,—That nothing else be proved than the proposition for whose proof the probation was instituted. You may, therefore, add to the last paragraph the following supplement :—

¶ LXXXIX. These rules of the logicians may, however, all be reduced to two.

1°, That no proposition be employed as a Principle of Probation, which stands itself in need of proof.

2°, That nothing else be proved than the Proposition for whose proof the Probation was instituted.

Of these two, the former comprehends the first four rules of the logicians, the latter the fifth. I shall now, therefore, proceed to illustrate the five rules in detail.

The First Rule,—Nothing is to be begged, borrowed, or stolen, that is, nothing is to be presupposed as proved, which itself requires a demonstration,—is, in fact, an enunciation of the first general rule I gave you, and to this, therefore, as we shall see, the second,

third, and fourth are to be reduced as special applications. But, in considering this law in its universality, it is not to be understood as if every probation Limitation
under which
this Rule is
to be under-
stood.
were at once to be rejected as worthless, in which
anything is presupposed and not proved. Were this
its sense, it would be necessary in every probation to
ascend to the highest principles of human knowledge,
and these themselves as immediate and, consequently,
incapable of proof, might be rejected as unproved
assumptions. Were this the meaning of the law,
there could be no probation whatever. But it is not
to be understood in this extreme rigour. That probation alone is a violation of this law, and, consequently, alone is vicious, in which a proposition is
assumed as a principle of proof, which may be doubted
on the ground on which the thesis itself is doubted,
and where, therefore, we prove the uncertain by the
equally uncertain. The probation must, therefore,
depart from such principles as are either immediately
given as ultimate, or mediately admit of a proof from
other sources than the proposition itself in question.
When, for example, it was argued that the Newtonian
theory is false, which holds colours to be the result
of a diversity of parts in light, on the ground, admitted by the ancients, that the celestial bodies, and,
consequently, their emanations, consist of homogeneous elements ;—this reasoning was inept, for the
principle of proof was not admitted by modern
philosophers. Thus, when Aristotle defends the in-
stitution of slavery as a natural law, on the ground
that the barbarians, as of inferior intellects, are the
born bondsmen of the Greeks, and the Greeks, as of
superior intellect, the born masters of the barbarians,[a]

a *Polit.*, l. 2.—ED.

—(an argument which has, likewise, been employed in modern times in the British Parliament, with the substitution of negroes for barbarians, and whites for Greeks),—this argument is invalid, as assuming what is not admitted by the opponents of slavery. It would be a *petitio principii* to prove to the Mohammedan the divinity of Christ from texts in the New Testament, for he does not admit the authority of the Bible; but it would be a valid *argumentum ad hominem* to prove to him from the Koran the prophetic mission of Jesus, for the authority of the Koran he acknowledges.

Second
Rule. The Second Rule,—That no proposition is to be employed as a principle of proof, the truth of which is only to be evinced as a consequence of the proposition which it is employed to prove,—is only a special case of the preceding. For example, if we were to argue that man is a free agent, on the ground that he is morally responsible for his actions, or that his actions can be imputed to him, or on the ground that vice and virtue are absolutely different,—in these cases, the *hysteron proteron* is committed; for only on the ground that the human will is free, can man be viewed as a morally responsible agent, and his actions be imputed to him, or can the discrimination of vice and virtue, as more than a merely accidental relation, be maintained. But we must pause before we reject a reasoning on the ground of *hysteron proteron;* for the reasoning may still be valid, though this logical fault be committed. Nay, it is frequently necessary for us to reason by such a regress. In the very example given, if we be unable to prove directly that the will of man is free, but are able to prove that he is a moral agent, responsible for his actions, as sub-

jected to the voluntary but unconditioned Law of
Duty, and if the fact of this law of duty and its unqualified obligation involve, as a postulate, an eman-
cipation from necessity,—in that case, no competent
objection can be taken to this process of reasoning.
This, in fact, is Kant's argument. From what he calls
the *categorical imperative*, that is, from the fact of
the unconditioned law of duty as obligatory on man,
he postulates, as conditions, the liberty of the human
will, and the existence of a God, as the moral governor of a moral universe.[a]

The Third Law,—That no circular probation is to
be made, that is, the proposition which we propose to
prove must not be used as a principle for its own probation,—this, in like manner, is only a particular case
of the first. " To the Circle there are required properly
two probations, which are so reciprocally related that
the antecedent in the one is proved by its own consequent in the other. The proposition A is true because the proposition B is true; and the proposition
B is true because the proposition A is true. A circle
so palpable as this would indeed be committed by no
one. The vice is usually concealed by the interpolation of intermediate propositions, or by a change in
the expression."[b] Thus Plato, in his *Phædo*,[y] demonstrates the immortality of the soul from its simplicity;
and, in the *Republic*,[δ] he demonstrates its simplicity
from its immortality.

In relation to the Hysteron Proteron and the Circle,
I must observe that these present some peculiar difficulties for the systematic arrangement of our know-

Third Rule.

Regressive and Progressive Proofs not to be em-

a *Kritik der reinen Vernunft*, Me-
thodenlehre, Hauptst. ii., Abschn. 2.
Kritik der praktischen Vernunft, p.
274, ed. Rosenkranz.—Ed.

β *Krug, Logik*, § 123, Anm. 3.—
Ed.

y P. 78.—Ed.

δ B. x. p. 611.—Ed.

LECT.
XXVI.

founded
with the
tautological
Circle.

ledge. Through the Circle, (the result of which is only the proof of an assertion),—through the Circle by itself, nothing whatever is gained for the logical development of our knowledge. But we must take care not to confound the connection of Regressive and Progressive Proofs with the tautological Circle. When, in the treatment of a science out of the observed facts, we wish to generalise universal laws, we lead, in the first place, an inductive probation, that (ὅτι) certain laws there are. Having assured ourselves of the existence of these laws by this regressive process, we then place them in theory at the head of a progressive or synthetic probation, in which the facts again recur, reversed and illustrated from the laws, which, in the antecedent process, they had been employed to establish; that is, it is now shown why (διότι) these facts exist.

The Fourth Rule,—No leap, no gap, must be made, that is, the syllogisms of which the probation is made up must stand in immediate or continuous connection, —may be, likewise, reduced to the first. For here the only vice is, that by an ellipsis of an intermediate link in the syllogistic chain we use a proposition which is actually without its proof, and it is only because this proposition is as yet unproved, that its employment is illegitimate. The *Saltus* is, therefore, only a special case of the *Petitio.*

The *Saltus* is committed when the middle term of one of the syllogisms in a probation is not stated. If the middle term be too manifest to require statement, then is the *saltus* not to be blamed, for it is committed only in the expression and not in the thought. If the middle term be not easy of discovery, then the *saltus* is a fault; but if there be no middle term to be found, then the *saltus* is a vice LECT. XXVL which invalidates the whole remainder of the proba- tion. The proper *saltus,*—the real violation of this law, is, therefore, when we make a transition from one proposition to another, the two not being connected together as reason and consequent." The (vulgar) Enthymeme and the Sorites do not, therefore, it is evident, involve violations of this law.

The Fifth Rule,—The scope of the probation is not Fifth Rule. to be changed, that is, nothing is to be proved other than what was proposed to be proved,—corresponds to the second of the two rules which I gave, and of which it is only a less explicit statement. It evidently Admits of three de- admits of three kinds or degrees. In the first case, grees. the proposition to be proved is changed by the change of its subject or predicate into different no- tions. Again, the proposition may substantially re- main the same, but may be changed into one either of a wider or of a narrower extension,—the second and third cases.

The first of these cases is the *Mutatio Elenchi,* or First *Transitus ad aliud genus,* properly so called. "When Degree,— Mutatio Elenchi. a probation does not demonstrate what it ought to demonstrate, it may, if considered absolutely or in itself, be valid; but if considered relatively to the pro- position which it behoves us to prove, it is of no value. We commute by this procedure the whole scope or pur- port of the probation; we desert the proper object of inquiry,—the point in question. If a person would prove the existence of ghosts, and to this end prove by witnesses the fact of unusual noises and appear- ances during the night, he would prove something very different from what he proposed to establish;

a Cf. Krug, *Logik,* § 133, Anm. 4.—Ed.

for this would be admitted without difficulty by those
who still denied the apparition of ghosts : it, therefore,
behoved him to show that the unusual phænomena
were those of a spirit good or bad." [a]

Second De-
gree,—in
which too
little is
proved.

The two other cases,—when the proposition actually
proved is either of a smaller or of a greater extension
than the proposition which ought to have been proved,
—are not necessarily, like the preceding, altogether
irrelevant. They are, however, compared together, of
various degrees of relevancy. In the former case,
where too little is proved,—here the end proposed is,
to a certain extent at least, changed, and the probation results in something different from what it was
intended to accomplish. For example, if we propose
to prove that Sempronius is a virtuous character,
and only prove the legality of his actions, we here
prove something less than, something different from,
what we professed to do ; for we proposed to prove
the internal morality, and not merely the external law-
fulness, of his conduct. Such a proof is not absolutely
invalid ; it is not even relatively null, for the external legality is always a concomitant of internal morality. But the existence of the latter is not evinced
by that of the former, for Sempronius may conform
his actions to the law from expediency and not from
duty. [b]

Third De-
gree,—in
which too
much is
proved.

In the other case, in which there is proved too much,
the probation is lawful, and only not adequate and
precise. For example, if we propose to prove that the
soul does not perish with the body, and actually prove
that its dissolution is absolutely impossible,—here
the proof is only superabundant. The logical rule,—

[a] Krug, *Logik*, § 133, Anm. 2.—
ED.
[b] Cf. Krug, *Logik*, § 133, Anm. 5.
—ED.

Qui nimium probat, nihil probat, is, therefore, in its universal or unqualified expression, incorrect. The proving too much is, however, often the sign of a *saltus* having been committed. For example,—when a religious enthusiast argues from the strength of his persuasion, that he is, therefore, actuated by the Holy Spirit, and his views of religion consequently true,— there is here too much proved, for there is implied the antecedent, omitted by a *saltus*, that whoever is strongly persuaded of his inspiration is really inspired, —a proposition too manifestly absurd to bear an explicit enouncement. In this case, the apparent too much is in reality a too much which, when closely examined, resolves itself into a nothing.[a]

We have thus terminated the consideration of Pure or Abstract Logic in both its Parts, and now enter on the Doctrine of Modified or Concrete Logic.

a [Cf. Sigwart, *Handbuch zu Vorlesungen über die Logik,* § 407, p. 252.]

LECTURE XXVII.

MODIFIED LOGIC.

PART I.—MODIFIED STOICHEIOLOGY.

SECTION I.—DOCTRINE OF TRUTH AND ERROR.

TRUTH.—ITS CHARACTER AND KINDS.

LECT.
XXVII.

Modified
Logic,—
its object.

HAVING now terminated the Doctrine of Pure or Abstract Logic, we proceed to that of Modified or Concrete Logic. In entering on this subject, I have to recall to your memory what has formerly been stated in regard to the object which Modified Logic proposes for consideration. Pure Logic takes into account only the necessary conditions of thought, as founded on the nature of the thinking process itself. Modified Logic, on the contrary, considers the conditions to which thought is subject, arising from the empirical circumstances, external and internal, under which exclusively it is the will of our Creator that man should manifest his faculty of thinking. Pure Logic is thus exclusively conversant with the form, Modified Logic is, likewise, occupied with the matter, of thought. And as their objects are different, so, likewise, must be their ends. The end of Pure Logic is formal truth, — the harmony of thought with thought; the end of Modified Logic is the harmony of thought with existence. Of these ends, that which

Pure Logic proposes is less ambitious, but it is fully
and certainly accomplished; the end which Modified
Logic proposes is higher, but it is far less perfectly
attained.   The problems which Modified Logic has to
solve may be reduced to three: 1°, What is Truth
and its contradictory opposite,—Error? 2°, What
are the Causes of Error and the Impediments to Truth,
by which man is beset in the employment of his faculties, and what are the Means of their Removal? And,
3°, What are the Subsidiaries by which Human
Thought may be strengthened and guided in the
exercise of its functions?

From this statement it is evident that Concrete And distributed be-
tween its
Stoicheio-
logy and its
Method-
ology.
Logic might, like Pure Logic, have been divided into
a Stoicheiology and a Methodology,—the former com-
prising the first two heads,—the latter the third.   For
if to Modified Stoicheiology we refer the consideration of the nature of concrete truth and error, and of
the conditions of a merely not erroneous employment
of thought,—this will be exhausted in the First and
Second Chapters; whereas if we refer to Methodology
a consideration of the means of employing thought
not merely without error but with a certain positive
perfection,—this is what the Third Chapter professes
to expound.

I commence the First Chapter, which proposes to
answer the question,—What is Truth? with its correlatives,—by the dictation of the following paragraph.

¶ XC. The end which all our scientific efforts
are exerted to accomplish, is *Truth* and *Cer-*
tainty.   Truth is the correspondence or agreement of a cognition with its object; its Criterion
is the necessity determined by the laws which

govern our faculties of knowledge ; and Certainty is the consciousness of this necessity.* Certainty, or the conscious necessity of knowledge, absolutely excludes the admission of any opposite supposition. Where such appears admissible, doubt and uncertainty arise. If we consider truth by relation to the degree and kind of Certainty, we have to distinguish *Knowledge, Belief,* and *Opinion.* Knowledge and Belief differ not only in degree but in kind. Knowledge is a certainty founded upon insight ; Belief is a certainty founded upon feeling. The one is perspicuous and objective ; the other is obscure and subjective. Each, however, supposes the other ; and an assurance is said to be a knowledge or a belief, according as the one element or the other preponderates. Opinion is the admission of something as true, where, however, neither insight nor feeling is so intense as to necessitate a perfect certainty. What prevents the admission of a proposition as certain is called *Doubt.* The approximation of the imperfect certainty of opinion to the perfect certainty of knowledge or belief is called *Probability.*

If we consider Truth with reference to Knowledge, and to the way in which this knowledge arises, we must distinguish *Empirical* or *a posteriori* from *Pure* or *a priori Truth.* The former has reference to cognitions which have their source in the presentations of Perception, External and Internal, and which obtain their form by the elaboration of the Understanding or Faculty of Relations (διάνοια.) The latter is con-

* Cf. Twesten, *Die Logik, insbesondere die Analytik,* § 300.—Ed.

tained in the necessary and universal cognitions afforded by the Regulative Faculty,—Intellect Proper, or Common Sense, (νοῦς.) *LECT. XXVII.*

This paragraph, after stating that Truth and Certainty constitute the end of all our endeavours after knowledge, because only in the attainment of truth and certainty can we possibly attain to knowledge or science ;—I say, after the statement of this manifest proposition, it proceeds to define what is meant by the two terms *Truth* and *Certainty*; and,—to commence with the former,—Truth is defined, the correspondence or agreement of a cognition or cognitive act of thought with its object. *Explication.*

The question—What is Truth? is an old and celebrated problem. It was proposed by the Roman Governor,—by Pontius Pilate,—to our Saviour ; and it is a question which still recurs, and is still keenly agitated, in the most recent schools of Philosophy. In one respect, all are nearly agreed in regard to the definition of the term, for all admit that by truth is understood a harmony,—an agreement,—a correspondence between our thought and that which we think about. This definition of truth we owe to the schoolmen. "Veritas intellectus," says Aquinas, "est adæquatio intellectus et rei, secundum quod intellectus dicit esse, quod est, vel non esse, quod non est."[a] From the schoolmen this definition has been handed down to modern philosophers, by whom it is currently employed, without, in general, a suspicion of its origin. It is not, therefore, in regard to the meaning of the term *truth*, that there is any difference of opinion *Truth,—what. Definition of the term.*

a [*Contra Gentiles,* lib. I. c. 59. See Biunde, *Über Wahrheit in Erkennen,* p. 11. On Truth in general, see Reix, *Commentl. de Scientia, de Ideis, de Veritate, &c.,* Disp. lxxxv., p. 871 *et seq.*]

LECT.
XXVII.
among philosophers. The questions which have pro-
voked discussion, and which remain, as heretofore,

Questions
in debate
regarding
Truth.
without a definitive solution, are not whether truth be
the harmony of thought and reality, but whether this
harmony, or truth, be attainable, and whether we possess any criterion by which we can be assured of its
attainment. Considering, however, at present only the
meaning of the term, philosophers have divided Truth,
(or the harmony of thought and its object), into different species, to which they have given diverse names ;
but they are at one neither in the division nor in the
nomenclature.

For man
only two
kinds of
Truth,—
Formal and
Real.
It is plain that for man there can be conceived only
two kinds of Truth, because there are for human thought
only two species of object.   For that about which we
think, must either be a thought, or something which a
thought contains.   On this is founded the distinction
of Formal Knowledge and Real Knowledge,—of Formal Truth and Real Truth.   Of these in their order.

I. Formal
Truth.
I. In regard to the former, a thought abstracted
from what it contains, that is, from its matter or what
it is conversant about, is the mere form of thought.
The knowledge of the form of thought is a formal
knowledge, and the harmony of thought with the form

Formal
Truth of
two kinds,
—Logical
and Mathe-
matical.
of thought is, consequently, Formal Truth. Now Formal
Knowledge is of two kinds ; for it regards either the
conditions of the Elaborative Faculty,—the Faculty
of Thought Proper,—or the conditions of our Presentations or Representations of external things, that is,
the intuitions of Space and Time. The former of these
sciences is Pure Logic,—the science which considers
the laws to which the Understanding is astricted in its
elaborative operations, without inquiring what is the
object,—what is the matter, to which these operations

are applied. The latter of these sciences is Mathematics, or the science of Quantity, which considers the relations of Time and Space, without inquiring whether there be any actual reality in space or time. Formal truth will, therefore, be of two kinds,—Logical and Mathematical. Logical truth is the harmony or agreement of our thoughts with themselves as thoughts, in other words, the correspondence of thought with the universal laws of thinking. These laws are the object of Pure or General Logic, and in these it places the criterion of truth. This criterion is, however, only the negative condition,—only the *conditio sine qua non*,—of truth. Logical truth is supposed in supposing the possibility of thought; for all thought presents a combination, the elements of which are repugnant or congruent, but which cannot be repugnant and congruent at the same time. Logic might be true, although we possessed no truth beyond its fundamental laws, although we knew nothing of any real existence beyond the formal hypothesis of its possibility.

But were the Laws of Logic purely subjective, that is, were they true only for our thought alone, and without any objective validity, all human sciences, (and Mathematics among the rest), would be purely subjective likewise; for we are cognisant of objects only under the forms and rules of which Logic is the scientific development. If the true character of objective validity be universality, the laws of Logic are really of that character, for these laws constrain us, by their own authority, to regard them as the universal laws not only of human thought, but of universal reason.

The case is the same with the other formal science, the science of Quantity, or Mathematics. Without inquiring into the reality of existences, and without borrowing from or attributing to them anything. Arithmetic, the science of Discrete Quantity, creates its numbers, and Geometry, the science of Continuous Quantity, creates its figures; and both operate upon these their objects in absolute independence of all external actuality. The two mathematical sciences are dependent for their several objects only on the notion of time and the notion of space,—notions under which alone matter can be conceived as possible, for all matter supposes space, and all matter is moved in space and in time. But to the notions of space and time the existence or non-existence of matter is indifferent,— indifferent, consequently, to Geometry and Arithmetic, so long at least as they remain in the lofty regions of pure speculation, and do not descend to the practical application of their principles. If matter had no existence, nay, if space and time existed only in our minds, mathematics would still be true; but their truth would be of a purely formal and ideal character,—would furnish us with no knowledge of objective realities.[a]

So much for Formal Truth, under its two species of Logical and Mathematical.

II. Real
Truth.
The other genus of truth,—(the end which the Real Sciences propose),—is the harmony between a thought and its matter. The Real Sciences are those which
Real and
Formal
Sciences.
have a determinate reality for their object, and which are conversant about existences other than the forms of thought. The Formal Sciences have a superior certainty to the real; for they are simply ideal combinations, and they construct their objects without inquiring about their objective reality. The real sciences are sciences of fact, for the point from which

a Cf. Esser, Logik, § 172.—Ed. [Fries, Logik, § 124.]

they depart is always a fact,—always a presentation.
Some of these rest on the presentations of Self-consciousness, or the facts of mind; others on the presentations of Sensitive Perception, or the facts of
nature. The former are the Mental Sciences, the
latter the Material. The facts of mind are given
partly as contingent, partly as necessary ; the latter,—
the necessary facts,—are universal virtually and in
themselves, the former,—the contingent facts,—only
obtain a factitious universality by a process of generalisation. The facts of nature, however necessary in
themselves, are given to us only as contingent and
isolated phœnomena ; they have, therefore, only that
conditional, that empirical, generality, which we bestow
on them by classification.

Real truth is, therefore, the correspondence of our
thoughts with the existences which constitute their
objects. But here a difficulty arises ;—How can we
know that there is, that there can be, such a correspondence ? All that we know of the objects is through
the presentations of our faculties; but whether these
present the objects as they are in themselves we can
never ascertain, for to do this it would be requisite to
go out of ourselves,—out of our faculties,—to obtain a
knowledge of the objects by other faculties, and thus
to compare our old presentations with our new. But
all this, even were the supposition possible, would be
incompetent to afford us the certainty required. For
were it possible to leave our old, and to obtain a new,
set of faculties, by which to test the old, still the
veracity of these new faculties would be equally obnoxious to doubt as the veracity of the old. For
what guarantee could we obtain for the credibility in
the one case, which we do not already possess in the

other ? The new faculties could only assert their own truth ;—but this is done by the old, and it is impossible to imagine any presentations of the non-ego by any finite intelligence, in regard to which a doubt might not be raised, whether these presentations were not merely subjective modifications of the conscious ego itself. All that could be said in answer to such a doubt is, that if such be true, our whole nature is a lie,—a supposition which is not, without the strongest evidence, to be admitted ; and the argument is as competent against the sceptic in our present condition, as it would be were we endowed with any other conceivable form of acquisitive and cognitive faculties. But I am here trenching on what ought to be reserved for an explanation of the Criterion of Truth.

Real
Truth,—
its subdivi-
sions.

Such, as it appears to me, is the only rational division of Truth, according to the different character of the objects to which thought is relative,—into Formal and Real Truth. Formal Truth, as we have seen, is subdivided into Logical and Mathematical. Real Truth might likewise be subdivided, were this requisite, into various species. For example, Metaphysical Truth might denote the harmony of thought with the necessary facts of mind ; Psychological Truth, the harmony of thought with the contingent facts of mind ; and Physical Truth, the harmony of thought with the phænomena of external experience.

Metaphysical.

Psychological.

Physical.

Various applications of the term Truth.

It now remains to say a word in regard to the confusion which has been introduced into this subject, by the groundless distinctions and contradictions of philosophers. Some have absurdly given the name of *truth* to the mere reality of existence, altogether abstracted from any conception or judgment relative to it, in any intelligence human or divine. In this sense *physical truth* has been used to denote the actual existence of a thing. Some have given the name of *metaphysical truth* to the congruence of the thing with its idea in the mind of the Creator. Others again have bestowed the name of *metaphysical truth* on the mere logical possibility of being thought; while they have denominated by *logical truth* the metaphysical or physical correspondence of thought with its objects. Finally, the term *moral* or *ethical truth* has been given to veracity, or the correspondence of thought with its expression. In this last case, truth is not, as in the others, employed in relation to thought and its object, but to thought and its enouncement. So much for the notion, and the principal distinctions, of Truth.

But returning to the paragraph, I take the next clause, which is,—'The Criterion of truth is the necessity determined by the laws which govern our faculties of knowledge; and the consciousness of this necessity is Certainty.' That the necessity of a cognition, that is, the impossibility of thinking it other than as it is presented,—that this necessity, as founded on the laws of thought, is the criterion of truth, is shown by the circumstance, that where such necessity is found, all doubt in regard to the correspondence of the cognitive thought and its object must vanish; for to doubt whether what we necessarily think in a certain manner, actually exists as we conceive it, is nothing less than an endeavour to think the necessary as the not necessary or the impossible, which is contradictory.

What has just been said also illustrates the truth of the next sentence of the paragraph,—viz., 'Certainty, or the conscious necessity of a cognition, absolutely excludes the admission of any opposite supposition.

When such is found to be admissible, doubt and uncertainty arise.' This sentence requiring no explanation, I proceed to the next—viz., 'If we consider truth by relation to the degree and kind of Certainty, we have to distinguish Knowledge, Belief, and Opinion. Knowledge and Belief differ not only in degree but in kind. Knowledge is a certainty founded on intuition. Belief is a certainty founded upon feeling. The one is perspicuous and objective, the other is obscure and subjective. Each, however, supposes the other, and an assurance is said to be a knowledge or a belief, according as the one element or the other preponderates.'

In reference to this passage, it is necessary to say something in regard to the difference of Knowledge and Belief. In common language the word *Belief* is often used to denote an inferior degree of certainty.

That the
certainty of
all know-
ledge is
ultimately
resolvable
into a cer-
tainty of
Belief,
maintained
by Luther.
We may, however, be equally certain of what we believe as of what we know, and it has, not without ground, been maintained by many philosophers, both in ancient and in modern times, that the certainty of all knowledge is, in its ultimate analysis, resolved into a certainty of belief. "All things," says Luther, "stand in a belief, in a faith, which we can neither see nor comprehend. The man who would make these visible, manifest, and comprehensible, has vexation and heartgrief for his reward. May the Lord increase Belief in you and in others."[a] But you may perhaps think that the saying of Luther is to be taken theologically, and that, philosophically considered, all belief ought to be founded on knowledge, not all knowledge on belief. But the same doctrine is held even by those philo-

---

[a] *Weishit*, Th. iii. Abth. 2. *Works*, p. 778.—ED.
Quoted by Sir W. Hamilton, *Reid's*

sophers who are the least disposed to mysticism or blind faith. Among these Aristotle stands distinguished. He defines science, strictly so called, or the knowledge of indubitable truths, merely by the intensity of our conviction or subjective assurance;[a] and on a primary and incomprehensible belief he hangs the whole chain of our comprehensible or mediate knowledge. The doctrine which has been called *The Philosophy of Common Sense*, is the doctrine which founds all our knowledge on belief; and, though this has not been signalised, the doctrine of Common Sense is perhaps better stated by the Stagirite than by any succeeding thinker. "What," he says, "appears to all men, that we affirm to be, and he who rejects this belief (πίστις) will assuredly advance nothing better worthy of credit." This passage is from his *Nicomachean Ethics.*[β] But, in his Physical Treatises, he founds in belief the knowledge we have of the reality of motion, and by this, as a source of knowledge paramount to the Understanding, he supersedes the contradictions which are involved in our conception of motion, and which had so acutely been evolved by the Eleatic Zeno, in order to show that motion was impossible.[γ] In like manner, in his Logical Treatises, Aristotle shows that the primary or ultimate principles of knowledge must be incomprehensible; for if comprehensible, they must be comprehended in some higher notion, and this again, if not itself incomprehensible, must be comprehended in a still higher, and so on in a progress *ad infinitum*, which is absurd.[δ] But what is given as an ultimate and incomprehen-

a Various passages from Aristotle to this effect are cited by the Author, *Reid's Works*, p. 771.—Ed.

β B. x. c. 2.—Ed.

γ B. viii. c. 3. See *Reid's Works*, p. 773.—Ed.

δ *Metaphys.*, iii. (iv.) c. 4. Cf. *Anal. Post.*, i. 2, 3.—Ed.

sible principle of knowledge, is given as a fact, the existence of which we must admit, but the reasons of whose existence we cannot know,—we cannot understand. But such an admission, as it is not a knowledge, must be a belief: and thus it is that, according to Aristotle, all our knowledge is in its root a blind, a passive, faith, in other words, a feeling. The same doctrine was subsequently held by many of the acutest thinkers of ancient times, more especially among the Platonists; and of these Proclus is perhaps the philosopher in whose works the doctrine is turned to the best account.[a] In modern times we may trace it in silent operation, though not explicitly proclaimed, or placed as the foundation of a system. It is found spontaneously recognised even by those who might be supposed the least likely to acknowledge it without compulsion. Hume, for example, against whose philosophy the doctrine of Common Sense was systematically arrayed, himself pointed out the weapons by which his adversaries subsequently assailed his scepticism; for he himself was possessed of too much philosophical acuteness not to perceive that the root of knowledge is belief. Thus, in his *Inquiry*, he says,—" It seems evident that men are carried by a natural instinct or prepossession to repose faith in their senses : and that, without any reasoning, or even almost before the use of reason, we always suppose an external universe which depends not on our perception, but would exist though we and every sensible creature were absent or annihilated. Even the animal creation are governed by a like opinion, and preserve this belief,—the belief of external objects,—in all their thoughts, designs, and actions. . . . This very table, which we see

The Platonists.
Proclus.

Hume.

a *In Platonis Theologiam,* i. c. 25. Quoted in *Reid's Works,* p. 776.—ED.

white, and which we feel hard, is believed to exist independent of our perception, and to be something external to our mind which perceives it."[a]

But, on the other hand, the manifestation of this belief necessarily involves knowledge; for we cannot believe without some consciousness or knowledge of the belief, and, consequently, without some consciousness or knowledge of the object of the belief. Now, the immediate consciousness of an object is called an *intuition,*—an *insight.* It is thus impossible to separate belief and knowledge,—feeling and intuition. They each suppose the other.

The consideration, however, of the relation of Belief and Knowledge does not properly belong to Logic, except in so far as it is necessary to explain the nature of Truth and Error. It is altogether a metaphysical discussion; and one of the most difficult problems of which Metaphysics attempts the solution.

The remainder of the paragraph contains the statement of certain distinctions and the definition of certain terms, which it was necessary to signalise, but which do not require any commentary for their illustration. The only part that might have required an explanation is the distinction of Truth into Pure, or *a priori,* and Empirical, or *a posteriori.* The explanation of this division has been already given more than once in the course of the Lectures,[b] but the following may now be added.

Experience presents to us only individual objects, and as these individual objects might or might not have come within our sphere of observation, our whole

a *Inquiry concerning the Human Understanding,* sect. 12. *Philosophical Works,* iv. p. 177.—E.D.

β See above, *Lectures on Meta-* physics, vol. ii. p. 194 et seq. Cf. Esser, *Logik,* §§ 4, 171.—Ed. [Fries, *Logik,* § 124.]

knowledge of, and from, these objects might or might not exist;—it is merely accidental or contingent. But as our knowledge of individual objects affords the possibility, as supplying the whole contents, of our generalised or abstracted notions, our generalised or abstracted notions are, consequently, not more necessary to thought, than the particular observations out of which they are constructed. For example, every horse I have seen I might not have seen, and I feel no more necessity to think the reality of a horse than the reality of a hippogriff; I can, therefore, easily annihilate in thought the existence of the whole species. I can suppose it not to be,—not to have been. The case is the same with every other notion which is mediately or immediately the datum of observation. We can think away each and every part of the knowledge we have derived from experience; our whole empirical knowledge is, therefore, a merely accidental possession of the mind.

But there are in the mind notions of a very different character,—notions which we cannot but think, if we think at all. These, therefore, are notions necessary to the mind; and, as necessary, they cannot be the product of experience. For example, I perceive something to begin to be. I feel no necessity to think that this thing must be at all, but, thinking it existent, I cannot but think that it has a cause. The notion, or rather the judgment, of Cause and Effect is, therefore, necessary to the mind. If so, it cannot be derived from experience.

LECTURE XXVIII.

MODIFIED STOICHEIOLOGY.

SECTION I.—DOCTRINE OF TRUTH AND ERROR.

SECTION II.—ERROR,—ITS CAUSES AND REMEDIES.

A.—GENERAL CIRCUMSTANCES—SOCIETY.

I NOW proceed to the consideration of the opposite of Truth,—Error, and, on this subject, give you the following paragraph.

LECT.
XXVIII.

¶ XCI. Error is opposed to Truth; and Error arises, 1°, From the commutation of what is subjective with what is objective in thought; 2°, From the repugnance of a supposed knowledge with its laws; or, 3°, From a want of adequate activity in our cognitive faculties.

Error is to be discriminated from Ignorance and from Illusion: these, however, along with Arbitrary Assumption, afford the most frequent occasions of error.[a]

Par. XCI.
Error,—its character
and sources.

This paragraph consists of two parts, and these I shall successively consider. The first is—'Error is

Explica-
tion.

a Twesten, *Die Logik, insbesondere die Analytik*, §§ 308, 309.—En. [Cf. Rulz, *Commentarius de Scientia*, &c. Disp. xcii. p. 925.]

LECT.
XXVIII.
opposed to Truth ; and Error arises, 1°, From the commutation of what is subjective with what is objective in thought; 2°, From the repugnance of a supposed knowledge with its laws; or, 3°, From a want of adequate activity in our cognitive faculties.'

Error,—
what.
"In the first place, we have seen that Truth is the agreement of a thought with its object. Now, as Error is the opposite of truth, Error must necessarily consist in a want of this agreement. In the second place, it has been shown, that the criterion or standard of truth is the necessity founded on the laws of our cognitive faculties ; and from this it follows that the essential character of error must be, either that it is not founded on these laws, or that it is repugnant to them. But these two alternatives may be viewed as only one ; for, inasmuch as, in the former case, the judgment remains undecided and can make no pretence to certainty, it may be thrown out of account no less than in the latter, where, as positively contradictory of the laws of knowledge, it is neces-

As Material.
sarily false. Of these statements the first, that is, the non-agreement of a notion with its object, is error viewed on its material side; and as a notion is the common product, the joint result, afforded by the reciprocal action of object and subject, it is evident that whatever the notion contains not correspondent to the object, must be a contribution by the thinking subject alone, and we are thus warranted in saying that Material Error consists in the commuting of what is subjective with what is objective in thought, —in other words, in mistaking an ideal illusion for a

As Formal.
real representation. The second of these statements, that is, the incongruence of the supposed cognition with the laws of knowledge, is error viewed on its

formal side. Now here the question at once presents
itself,—How can an act of cognition contradict its
own laws? The answer is that it cannot; and error, Arises from
the want of
adequate
activity of
the Cogni-
tive Facul-
ties.
when more closely scrutinised, is found to consist not
so much in the contradictory activity of our cognitive faculties as in their want of activity. And this
may be in consequence of one or other of two causes.
For it may arise from some other mental power,—the
will, for example,—superseding,—taking the place of,
the defective cognition, or, by its intenser force, turning it aside and leading it to a false result; or it may
arise from some want of relative perfection in the object, so that the cognitive faculty is not determined by
it to the requisite degree of action.

" What is actually thought, cannot but be correctly
thought. Error first commences when thinking is re-
mitted, and can in fact only gain admission in virtue
of the truth which it contains ;—every error is a perverted truth. Hence Descartes* is justified in the
establishment of the principle,—that we would never
admit the false for the true, if we would only give
assent to what we clearly and distinctly apprehend.—
'Nihil nos unquam falsum pro vero admissuros, si
tantum iis assensum præbeamus, quæ clare et dis-
tincte percipimus.'"ß   In this view the saying of the
Roman poet,—

" Nam neque decipitur ratio, nec decipit unquam,"ɣ

—is no longer a paradox : for the condition of error
is not the activity of intelligence, but its inactivity.

So much for the first part of the paragraph. The
second is—' Error is to be discriminated from Ignor-

---

a *Principia Philosophiæ,* L. 43.   ß Twesten, *Logik,* ¶ 308.—Ed.
Cf. *Med.* iv. *De Vero et Falso.*   ɣ Manilius, ii. 131.—Ed.

LECT.
XXVIII.

ance and

Illusion.

Ignorance.

ance and from Illusion, which, however, along with Arbitrary Assumption, afford the usual occasions of error."

"Ignorance is a mere negation,—a mere not-knowledge; whereas in error there lies a positive pretence to knowledge. Hence a representation, be it imperfect, be it even without any correspondent objective reality, is not in itself an error. The imagination of a hippogriff is not in itself false; the Orlando Furioso is not a tissue of errors. Error only arises when we attribute to the creations of our minds some real object, by an assertory judgment; we do not err and deceive either ourselves or others, when we hold and enounce a subjective or problematic supposition only for what it is. Ignorance,—not-knowledge,—however, leads to error, when we either regard the unknown as non-existent, or when we falsely fill it up. The latter is, however, as much the result of will, of arbitrary assumption, as of ignorance; and, frequently, it is the result of both together. In general, the will has no inconsiderable share in the activity by which knowledge is realised. The will has not immediately an influence on our judgment, but mediately it has. Attention is an act of volition, and attention furnishes to the Understanding the elements of its decision. The will determines whether we shall carry on our investigations, or break them off, content with the first apparent probability; and whether we shall apply our observations to all, or, only partially, to certain, moments of determination.

Illusion.

"The occasions of Error which lie in those qualities of Presentation, Representation, and Thought arising from the conditions and influences of the thinking subject itself, are called *Illusions*. But the existence of illusion does not necessarily imply the existence of error. Illusion becomes error only when we attribute to it objective truth; whereas illusion is no error when we regard the fallacious appearance as a mere subjective affection. In the jaundice, we see everything tinged with yellow, in consequence of the suffusion of the eye with bile. In this case, the yellow vision is illusion; and it would become error, were we to suppose that the objects we perceive were really so coloured. All the powers which co-operate to the formation of our judgments, may become the sources of illusion, and, consequently, the occasions of error. The Senses,[a] the Presentative Faculties, External and Internal, the Representative, the Retentive, the Reproductive, and the Elaborative, Faculties, are immediate, the Feelings and the Desires are mediate, sources of illusion. To these must be added the Faculty of Signs, in all its actual manifestations in language. Hence we speak of sensible, psychological, moral, and symbolical, illusion."[β] In all these relations the causes of illusion are partly general, partly particular; and though they proximately manifest themselves in some one or other of these forms, they may ultimately be found contained in the circumstances by which the mental character of the individual is conformed. Taking, therefore, a general view of all the possible Sources of Error, I think they may be reduced to the

---

a La Fontaine. See Mamre, *Cours de Philosophie*, ii. 241. ["Toutes les sciences naturelles ne sont autre chose qu'une guerre ouverte de la raison contre les déceptions de la sensibilité; ... c'est-à-dire, qu'elles ont pour objet de réformer les erreurs de nos sens, et de substituer les réalités de la science aux apparences fallacieuses que nos sens nous suggèrent. C'est ce que La Fontaine a très bien exprimé dans les vers suivant:

'Quand l'eau courbe un bâton, ma raison le redresse,' " &c.—ED.]

β [Twesten, *Logik*, § 300, p. 288, 289. Cf. Sigwart, *Logik*, §§ 484, 485.]

LECT.
XXVIII.
following classes, which, as they constitute the heads and determine the order of the ensuing discussion, I shall comprise in the following paragraph, with which commences the consideration of the Second Chapter of Modified Logic. Before, however, proceeding to consider these several classes in their order, I may observe that Bacon is the first philosopher who attempted a systematic enumeration of the various sources of error;[a] and his quaint classification of these, under the significant name of *idols*, into the four genera of Idols of the Tribe (*idola tribus*), Idols of the Den (*idola specus*), Idols of the Forum (*idola fori*), which may mean either the market-place, the bar, or the place of public assembly, and Idols of the Theatre (*idola theatri*), he thus briefly characterises.

Bacon's
classifica-
tion of the
sources of
error.

Par. XCII.
Error,—its
sources.
¶ XCII. The Causes and Occasions of Error are comprehended in one or other of the four following classes. For they are found either, 1°, In the General Circumstances which modify the intellectual character of the individual; or, 2°, In the Constitution, Habits, and Reciprocal Relations of his powers of Cognition, Feeling, and Desire; or, 3°, In the Language which he employs, as an Instrument of Thought and a Medium of Communication; or, 4°, In the nature of the Objects themselves, about which his knowledge is conversant.

Par. XCIII.
1. General
circumstan-
ces which
modify the
character
of the indi-
vidual.
¶ XCIII. Under the General Circumstances which modify the character of the individual, are comprehended 1°, The particular degree of Cultivation to which his nation has attained; for its

a *Novum Organum*, i. Aph. xxxix.—ED.

rudeness, the partiality of its civilisation, and its over-refinement are all manifold occasions of error; and this cultivation is expressed not merely in the state of the arts and sciences, but in the degree of its religious, political, and social advancement; 2°, The Stricter Associations, in so far as these tend to limit the freedom of thought, and to give it a one-sided direction: such are Schools, Sects, Orders, Exclusive Societies, Corporations, Castes, &c.—[a]

In the commencement of the Course, I had occasion to allude to the tendency there is in man to assimilate in opinions and habits of thought to those with whom he lives.[β] Man is by nature, not merely by accidental necessity, a social being. For only in society does he find the conditions which his different faculties require for their due development and application. But society, in all its forms and degrees, from a family to a State, is only possible under the condition of a certain harmony of sentiment among its members; and as man is by nature destined to a social existence, he is by nature determined to that analogy of thought and feeling which society supposes, and out of which society springs. There is thus in every association, great and small, a certain gravitation of opinions towards a common centre. As in our natural body every part has a necessary sympathy with every other, and all together form, by their harmonious conspiration, a healthy whole; so, in the social body, there is always a strong predisposition in each of its members to act and think in unison with the rest.

Explication. Man by nature social, and influenced by the opinions of his fellows.

---

[a] Bachmann, *Logik*, §§ 402, 403.
—Ed.

[β] See *Lectures on Metaphysics*, vol. i. p. 48.—Ed.

This universal sympathy or fellow-feeling is the principle of the different spirit dominant in different ages, countries, ranks, sexes, and periods of life. It is the cause why fashions, why political and religious enthusiasm, why moral example either for good or evil, spread so rapidly and exert so powerful an influence. As men are naturally prone to imitate others, they, consequently, regard as important or insignificant, as honourable or disgraceful, as true or false, as good or bad, what those around them consider in the same light.[a]

Pascal quoted on the power of custom.

Of the various testimonies I formerly quoted, of the strong assimilating influence of man on man, and of the power of custom to make that appear true, natural, and necessary, which in reality is false, unnatural, and only accidentally suitable, I shall adduce only that of Pascal. "In the just and the unjust," says he, "we find hardly anything which does not change its character in changing its climate. Three degrees of an elevation of the pole reverse the whole of jurisprudence. A meridian is decisive of truth, and a few years, of possession. Fundamental laws change. Right has its epochs. A pleasant justice which a river or a mountain limits! Truth on this side the Pyrenees, error on the other!"[b] It is the remark of an ingenious philosopher, "that if we take a survey of the universe, all nations will be found admiring only the reflection of their own qualities, and contemning in others whatever is contrary to what they are accustomed to meet with among themselves. Here is the Englishman accusing the French of frivolity; and

a [Meiners, *Untersuchungen über die Denkkräfte und Willenskräfte des Menschen*, ii. 372.]
β *Pensées*, partie I. art. vi. § 8,

(vol. ii. p. 128, ed. Faugere.) Compare *Lectures on Metaphysics*, vol. i. p. 80.—ED.

here the Frenchman reproaching the Englishman with selfishness and brutality.   Here is the Arab persuaded of the infallibility of his Caliph, and deriding the Tartar who believes in the immortality of the Grand Lama.   In every nation we find the same congratulation of their own wisdom, and the same contempt of that of their neighbours.

" Were there a sage sent down to earth from heaven, who regulated his conduct by the dictates of pure reason alone, this sage would be universally regarded as a fool.   He would be, as Socrates says, like a physician accused by the pastry-cooks, before a tribunal of children, of prohibiting the eating of tarts and cheesecakes; a crime undoubtedly of the highest magnitude in the eyes of his judges.   In vain would this sage support his opinions by the clearest arguments,—the most irrefragable demonstrations; the whole world would be for him like the nation of hunchbacks, among whom, as the Indian fabulists relate, there once upon a time appeared a god, young, beautiful, and of consummate symmetry.   This god, they add, entered the capital; he was there forthwith surrounded by a crowd of natives; his figure appeared to them extraordinary; laughter, hooting, and taunts manifested their astonishment; and they were about to carry their outrages still further, had not one of the inhabitants (who had undoubtedly seen other men), in order to snatch him from the danger, suddenly cried out—' My friends! my friends!   What are we going to do?   Let us not insult this miserable monstrosity.   If heaven has bestowed on us the general gift of beauty,—if it has adorned our backs with a mount of flesh, let us with pious gratitude repair to the temple and render our acknowledgment to the immortal gods.'" This fable is the history of human vanity. Every nation admires its own defects, and contemns the opposite qualities in its neighbours. To succeed in a country one must be a bearer of the national hump of the people among whom he sojourns.

There are few philosophers who undertake to make their countrymen aware of the ridiculous figure they cut in the eye of reason ; and still fewer the nations who are able to profit by the advice. All are so punctiliously attached to the interests of their vanity, that none obtain in any country the name of wise, except those who are fools of the common folly. There is no opinion too absurd not to find nations ready to believe it, and individuals prompt to be its executioners or its martyrs. Hence it is that the philosopher declared, that if he held all truths shut up within his hand, he would take especial care not to show them to his fellow-men. In fact, if the discovery of a single truth dragged Galileo to the prison, to what punishment would he not be doomed who should discover all ? Among those who now ridicule the folly of the human intellect, and are indignant at the persecution of Galileo, there are few who would not, in the age of that philosopher, have clamoured for his death. They would then have been imbued with different opinions, and opinions not more passively adopted than those which they at present vaunt as liberal and enlightened. To learn to doubt of our opinions, it is sufficient to examine the powers of the human intellect, to survey the circumstances by which it is affected, and to study the history of human follies. Yet in modern Europe six centuries elapsed from the foundation of Universities until the appearance of that

extraordinary man,—I mean Descartes,—whom his LECT. XXVIII.
age first persecuted, and then almost worshipped as a
demi-god, for initiating men in the art of doubting,—
of doubting well,—a lesson at which, however, both
their scepticism and credulity show that, after two
centuries, they are still but awkward scholars. Socrates
was wont to say—"All that I know is that I know
nothing."[a] In our age it would seem that men know
everything except what Socrates knew. Our errors
would not be so frequent were we less ignorant; and
our ignorance more curable, did we not believe ourselves to be all-wise.

Thus it is that the influence of Society, both in
its general form of a State or Nation, and in its particular forms of Schools, Sects, &c., determines a
multitude of opinions in its members, which, as they
are passively received, are often altogether erroneous.

Among the more general and influential of these
there are two, which, though apparently contrary, are, Two general forms of the influence of Example.
however, both, in reality, founded on the same in-
capacity of independent thought,—on the same influ- 1. Prejudice in favour of the Old.
ence of example,—I mean the excessive admiration of
the Old, and the excessive admiration of the New.
The former of these prejudices,[β]—under which may be
reduced the prejudice in favour of Authority,—was at
one time prevalent to an extent of which it is difficult
for us to form a conception. This prejudice is prepared by the very education not only which we do,

a Plato, *Apol.*, p. 23.—ED.

β [On Prejudice in general see the following works:—Damaraais, *Essai sur les Préjugés*, new ed., Paris, 1822—*Examen de l'Essai sur les Préjugés*, Berl. 1777—*Essai sur les Préjugés*, Neuchâtel, 1796; J. B. Salgues, *Des Erreurs et des Préjugés répandus dans la Société*, Paris, 1810-1813, 3 vols. 8vo; J. L. Castillon, *Essai sur les Erreurs et les Superstitions, Anciennes et Modernes*, Amsterdam, 1765; Paris, 1767; Sir Thomas Browne, *Vulgar Errors*; Glanvil, *Essays*.]

but which we all must, receive. The child necessarily
learns everything at first on credit,—he believes upon
authority. But when the rule of authority is once
established, the habit of passive acquiescence and
belief is formed, and, once formed, it is not again
always easily thrown off. When the child has grown
up to an age in which he might employ his own reason,
he has acquired a large stock of ideas; but who can
calculate the number of errors which this stock contains? and by what means is he able to discriminate
the true from the false! His mind has been formed
to obedience and uninquiry ; he possesses no criterion
by which to judge; it is painful to suspect what has
been long venerated, and it is felt even as a kind of
personal mutilation to tear up what has become irra-
dicated in his intellectual and moral being. *Ponere
difficile est quæ placuere diu.* The adult does not,
therefore, often judge for himself more than the child ;
and the tyranny of authority and foregone opinion
continues to exert a sway during the whole course of
his life. In our infancy and childhood the credit
accorded to our parents and instructors is implicit ;
and if what we have learned from them be confirmed
by what we hear from others, the opinions thus recommended become at length stamped in almost in-
delible characters upon the mind. This is the cause
why men so rarely abandon the opinions which vulgarly pass current; and why what comes as new is
by so many, for its very novelty, rejected as false.
And hence it is, as already noticed, that truth is as it
were geographically and politically distributed, what
is truth on one side of a boundary being error and
absurdity on the other. What has now been said of
the influence of society at large, is true also of the

lesser societies which it contains, all of which impose with a stronger or feebler, a wider or more contracted, authority, certain received opinions upon the faith of the members. Hence it is that whatever has once obtained a recognition in any society, large or small, is not rejected when the reasons on which it was originally admitted, have been proved erroneous. It continues, even for the reason that it is old and has been accepted, to be accepted still ; and the title which was originally defective, becomes valid by continuance and prescription.

But opposed to this cause of error, from the prejudice in favour of the Old, there is the other, directly the reverse,—the prejudice in favour of the New. *2. Prejudice in favour of the New.* This prejudice may be, in part at least, the result of sympathy and fellow-feeling. This is the cause why new opinions, however erroneous, if they once obtain a certain number of converts, often spread with a rapidity and to an extent which, after their futility has been ultimately shown, can only be explained on the principle of a kind of intellectual contagion. But the principal cause of the prejudice in favour of novelty lies in the Passions, and the consideration of these does not belong to the class of causes with which we are at present occupied.

Connected with and composed of both these prejudices,—that in favour of the old and that in favour of the new,—there is the prejudice of Learned Authority ; *Prejudice of Learned Authority.* for this is usually associated with the prejudices of Schools and Sects. "As often as men have appeared, who, by the force of their genius, have opened up new views of science, and thus contributed to the progress of the human intellect, so often have they, likewise, afforded the occasion of checking its advancement,

and of turning it from the straight path of improvement. Not that this result is to be imputed as a re-
proach to them, but simply because it is of the nature
of man to be so affected. The views which influenced
these men of genius, and which, consequently, lie at
the foundation of their works, are rarely comprehended
in their totality by those who have the names of these
authors most frequently in their mouths. The many
do not concern themselves to seize the ideal which
a philosopher contemplated, and of which his actual
works are only the imperfect representations; they
appropriate to themselves only some of his detached
apophthegms and propositions, and of these compound,
as they best can, a sort of system suited to their understanding, and which they employ as a talisman in
their controversies with others. As their reason is
thus a captive to authority, and, therefore, unable to
exert its native freedom, they, consequently, catch up
the true and the false without discrimination, and
remain always at the point of progress where they
had been placed by their leaders. In their hands a
system of living truths becomes a mere petrified or-
ganism; and they require that the whole science shall
become as dead and as cold as their own idol. Such
was Plato's doctrine in the hands of the Platonists;
such was Aristotle's philosophy in the hands of the
Schoolmen; and the history of modern systems affords
equally the same result."[a]

So much for the first genus into which the Sources
of Error are divided.

a Bachmann, *Logik*, § 404, p. 550.—ED.

LECTURE XXIX.

MODIFIED STOICHEIOLOGY.

SECTION II.—ERROR—ITS CAUSES AND REMEDIES.

A.—GENERAL CIRCUMSTANCES—SOCIETY.

B.—AS IN POWERS OF COGNITION, FEELING, AND DESIRE.

1.—AFFECTIONS.—PRECIPITANCY—SLOTH—HOPE AND FEAR—SELF-LOVE.

In our last Lecture, we entered on the consideration of the various sources of Error. These, I stated, may be conveniently reduced to four heads, and consist, 1°, In the General Circumstances which modify the intellectual character of the individual; 2°, In the Constitution, Habits, and Reciprocal Relations of his powers of Cognition, Feeling, and Desire; 3°, In the Language which he employs as an Instrument of Thought and a Medium of Communication; and, 4°, In the nature of the Objects themselves about which his knowledge is conversant.

I then gave you a general view of the nature of those occasions of Error, which originate in the circumstances under the influence of which the character and opinions of man are determined for him as a member of society. Under this head I stated, that, as man is destined by his Creator to fulfil the end of his existence in society, he is wisely furnished with a disposition to imitate those among whom his lot is cast, and thus to conform himself to whatever section of human society he may by birth belong, or of which he may afterwards become a member. The education we receive, nay the very possibility of receiving education at all, supposes to a certain extent the passive infusion of foreign and traditionary opinions. For as man is compelled to think much earlier than he is able to think for himself, all education necessarily imposes on him many opinions which, whether in themselves true or false, are, in reference to the recipient, only prejudices; and it is even only a small number of mankind who, at a later period, are able to bring these obtruded opinions to the test of reason, and by a free exercise of their own intelligence to reject them if found false, or to acknowledge them if proved true.

But while the mass of mankind thus remain, during their whole lives, only the creatures of the accidental circumstances which have concurred to form for them their habits and beliefs, the few who are at last able to form opinions for themselves, are still dependent, in a great measure, on the unreasoning judgment of the many. Public opinion, hereditary custom, despotically impose on us the capricious laws of propriety and manners. The individual may possibly, in matters of science, emancipate himself from their servitude; in the affairs of life he must quietly submit himself to the yoke. The only freedom he can here prudently manifest, is to resign himself with a consciousness that he is a slave not to reason but to conventional accident. And while he conforms himself to the usages of his own society, he will be tolerant

of those of others. In this respect his maxim will be
that of the Scythian prince :—" With you such may
be the custom, with us it is different."

So much for the general nature of the influence to
which we are exposed from the circumstances of Society; it now remains to say what are the means by
which this influence, as a source of error, may be
counteracted.

It has been seen that, in consequence of the manner in which our opinions are formed for us by the
accidents of society, our imposed and supposed knowledge is a confused medley of truths and errors.
Here it is evidently necessary to institute a critical
examination of the contents of this knowledge. Descartes proposes that, in order to discriminate, among
our prejudiced opinions, the truths from the errors, we
ought to commence by doubting all.[a] This has exposed him to much obloquy and clamour, but most
unjustly. The doctrine of Descartes has nothing
sceptical or offensive; for he only maintains that it
behoves us to examine all that has been inculcated on
us from infancy, and under the masters to whose
authority we have been subjected, with the same attention and circumspection which we accord to dubi-
ous questions. In fact there is nothing in the precept
of Descartes, which had not been previously enjoined
by other philosophers. Of these I formerly quoted to
you several, and among others the remarkable testimonies of Aristotle, St Augustin, and Lord Bacon.[b]

But although there be nothing reprehensible in the
precept of Descartes, as enounced by him, it is of
less practical utility in consequence of no account

---

[a] *Discours de la Méthode,* Partie
ii.—Ed.

[b] See *Lectures on Metaphysics,* vol.
i. p. 90 et seq.—Ed.

being taken of the circumstances which condition and modify its application. For, in the first place, the judgments to be examined ought not to be taken at random, but selected on a principle, and arranged in due order and dependence. But this requires no ordinary ability, and the distribution of things into their proper classes is one of the last and most difficult fruits of philosophy. In the second place, there are among our prejudices, or pretended cognitions, a great many hasty conclusions, the investigation of which requires much profound thought, skill, and acquired knowledge. Now, from both of these considerations, it is evident that to commence philosophy by such a review, it is necessary for a man to be a philosopher before he can attempt to become one. The precept of Descartes is, therefore, either unreasonable, or it is too unconditionally expressed. And this latter alternative is true.

A gradual and progressive abrogation of prejudices all that can be required of the student of philosophy. What can be rationally required of the student of philosophy, is not a preliminary and absolute, but a gradual and progressive, abrogation of prejudices. It can only be required of him, that, when, in the course of his study of philosophy, he meets with a proposition which has not been already sufficiently sifted, (whether it has been elaborated as a principle or admitted as a conclusion), he should pause, discuss it without prepossession, and lay aside for future consideration all that has not been subjected to a searching scrutiny. The precept of Descartes, when rightly explained, corresponds to that of St Paul:[a] "If any man among you seemeth to be wise in this world, let him become a fool, that he may be wise;" that is, let him not rely more on the opinions in which he has

a 1 Cor. iii. 18.

been brought up, and in favour of which he and those
around him are prejudiced, than on so many visions
of imagination, and let him examine them with the
same circumspection as if he were assured that they
contain some truth among much falsehood and many
extravagancies.[a]

Proceeding now to the second class of the Sources
of Error, which are found in the Mind itself, I shall
commence with the following paragraph :—

¶ XCIV. The Sources of Error which arise *Par. XCIV. II. Source of Error arising from the powers of Cognition, Feeling, and Desire,- of two kinds.*
from the Constitution, Habits, and Reciprocal
Relations of the powers of Cognition, Feeling,
and Desire, may be subdivided into two kinds.
The first of these consists in the undue prepon-
derance of the Affective Elements of mind, (the
Desires and Feelings), over the Cognitive : the
second, in the weakness or inordinate strength
of some one or other of the Cognitive Faculties
themselves.

Affection is that state of mind in which the Feel- *Explica-tion. 1. Prepon-derance of Affection over Cog-nition.*
ings and Desires exert an influence not under the con-
trol of reason ; in other words, a tendency by which
the intellect is impeded in its endeavour to think an
object as that object really is, and compelled to think
it in conformity with some view prescribed by the
passion or private interest of the subject thinking.

The human mind, when unruffled by passion, may
be compared to a calm sea. A calm sea is a clear
mirror, in which the sun and clouds, in which the

---

a This criticism of the precept   gique, t. III. part ii., ch. 6, p. 263
of Descartes is, with some slight   et seq. — ED.
changes, taken from Crousaz, Lo-

forms of heaven and earth, are reflected back precisely as they are presented. But let a wind arise,
and the smooth clear surface of the water is lifted
into billows and agitated into foam. It no more reflects the sun and clouds, the forms of heaven and
earth, or it reflects them only as distorted and broken

Influence of Passion on the Mind.
images. In like manner, the tranquil mind receives
and reflects the world without as it truly is ; but let
the wind of passion blow, and every object is represented, not as it exists, but in the colours and aspects
and partial phases in which it pleases the subject to
regard it. The state of passion and its influence

Boethius quoted.
on the Cognitive Faculties are truly pictured by
Boethius.[a]

" Nubibus atris	. . . . . .
Condita nullum	Tu quoque si vis
Fundere possunt	Lumine claro
Sidera lumen.	Cernere verum,
Si mare volvens	Tramite recto
Turbidus auster	Carpere callem ;
Misceat æstum,	Gaudia pelle,
Vitrea dudum,	Pelle timorem,
Parque serenis	Spemque fugato,
Unda diebus,	Nec dolor adsit.
Mox resoluto	Nubila mens est,
Sordida cœno,	Vinctaque frenis,
Visibus obstat.	Hæc ubi regnant."

Error limited to Probable Reasoning.
Every error consists in this,—that we take something for non-existent, because we have not become
aware of its existence, and that, in place of this existent something, we fill up the premises of a probable
reasoning with something else.

I have here limited the possibility of error to Probable Reasoning, for in Intuition and Demonstration
there is but little possibility of important error.

a De Consol. Phil., L. i. Metr. 7.—Ed.

Hobbes indeed asserts that had it been contrary to the interest of those in authority, that the three angles of a triangle should be equal to two right angles, this truth would have been long ago proscribed as heresy, or as high treason.[a] This may be an ingenious illustration of the blind tendency of the passions to subjugate intelligence; but we should take it for more than was intended by its author, were we to take it as more than an ingenious exaggeration. Limiting, therefore, error to probable inference, (and this constitutes, with the exception of a comparatively small department, the whole domain of human reasoning), we have to inquire, How do the Passions influence us to the assumption of false premises? To estimate the amount of probability for or against a given proposition, requires a tranquil, an unbiassed, a comprehensive, consideration, in order to take all the relative elements of judgment into due account. But this requisite state of mind is disturbed when any interest, any wish, is allowed to interfere.

¶ XCV. The disturbing Passions may be reduced to four :—Precipitancy, Sloth, Hope and Fear, Self-love.

1°, A restless anxiety for a decision begets impatience, which decides before the preliminary inquiry is concluded. This is Precipitancy.

2°, The same result is the effect of Sloth, which dreams on in conformity to custom, without subjecting its beliefs to the test of active observation.

3°, The restlessness of Hope or Fear impedes observation, distracts attention, or forces it only

a *Leviathan*, Part I. ch. 11.—ED.

on what interests the passion;—the sanguine looking only on what harmonises with his hopes, the diffident only on what accords with his fears.

4°, Self-love perverts our estimate of probability by causing us to rate the grounds of judgment, not according to their real influence on the truth of the decision, but according to their bearing on our personal interests therein.

In regard to Impatience or Precipitation,—"all is the cause of this which determines our choice on one side rather than another. An imagination excites pleasure, and because it excites pleasure we yield ourselves up to it. We suppose, for example, that we are all that we ought to be, and why? Because this supposition gives us pleasure. This, in some dispositions, is one of the greatest obstacles to improvement; for he who entertains it, thinks there is no necessity to labour in order to become what he is already. 'I believe,' says Seneca,[a] 'that many had it in their power to have attained to wisdom, had they not been impeded by the belief that wisdom they had already attained.' 'Multos puto ad sapientiam potuisse pervenire, nisi putassent se pervenisse.'"[β] Erasmus gives the following as the principal advice to a young votary of learning in the conduct of his studies: "To read the most learned books; to converse with the most learned men; but, above all, never to conceit that he himself was learned."[γ]

---

a *De Tranquillitate Animi*, c. 1.—
—Ed.

β Crousaz, *Logique*, t. iii. part ii. ch. 7, p. 297.—Ed.

γ "Joannes Alexander Brassica-

nus rogavit Erasmum, qua ratione doctus posset fieri. Respondit ex tempore: si doctis assiduus conviveret, si doctos audiret non minus submisse quam honorifice, si doctos strenue

" From the same cause, men flatter themselves with the hope of dying old, although few attain to longevity. The less probable the event the more certain are they of its occurrence; and why? Because the imagination of it is agreeable. 'Decrepiti senes paucorum annorum accessionem votis mendicant; minores natu seipsos esse fingunt: mendacio sibi blandiuntur: et tam libenter fallunt, quam si fata una decipiant.'" [a]

" Preachers," says Montaigne, " are aware that the emotion which arises during their sermons animates themselves to belief; and we are conscious that when roused to anger we apply ourselves more intently to the defence of our thesis, and embrace it with greater vehemence and approbation, than we did when our mind was cool and unruffled. You simply state your case to an advocate, he replies with hesitation and doubt; you are aware that it is indifferent to him whether he undertakes the defence of the one side or of the other. But have you once fee'd him well to take your case in hand, he begins to feel an interest in it, his will is animated. His reason and his science become also animated in proportion. Your case presents itself to his understanding as a manifest and indubitable truth; he now sees it in a wholly different light, and really believes that you have law and justice on your side." [β] It is proper to observe that Montaigne was himself a lawyer,—he had been a counsellor of the Parliament of Bordeaux.

It might seem that precipitate dogmatism and an

legeret, si doctos diligenter edisceret, denique si se doctum unquam putaret." Motto to G. J. Vossius, Opuscula de Studiorum Ratione. See Crenius, Consilia et Methodus, &c., p. 886, 1692.—Ed.

a Seneca, De Brevitate Vitæ, c. 11. Crousaz, Logique, t. iii. p. ll. ch. 7, p. 297, ed. 1725.—Ed.

β Essais, L. ii. ch. 12. Quoted by Crousaz. l. c.—Ed.

LECT.
XXIX.

Precipitate
Dogmatism
and Scepti-
cism, phases
of the same
disposition.

inclination to scepticism were opposite characters of mind. They are, however, closely allied, if not merely phases of the same disposition. This is indeed confessed by the sceptic Montaigne.[a] "The most uneasy condition for me is to be kept in suspense on urgent occasions, and to be agitated between fear and hope. Deliberation, even in things of lightest moment, is very troublesome to me; and I find my mind more put to it, to undergo the various tumbling and tossing of doubt and consultation, than to set up its rest, and to acquiesce in whatever shall happen, after the die is thrown. Few passions break my sleep; but of deliberations, the least disturbs me."

Remedy
for Precipi-
tation.

Precipitation is no incurable disease. There is for it one sure and simple remedy, if properly applied. It is only required, to speak with Confucius, manfully to restrain the wild horse of precipitancy by the curb of consideration; to weigh the reasons of decision, each and all, in the balance of cool investigation; not to allow ourselves to decide until a clear consciousness has declared these reasons to be true,—to be sufficient; and, finally, to throw out of account the suffrages of self-love, of prepossession, of passion, and to admit only those of reflection, of experience, and of evidence. This remedy is certain and effectual. In theory it is satisfactory, but its practical application requires a moral resolution for the acquisition of which no precept can be given.

2. Sloth.

In the second place, "Sloth is likewise a cause of precipitation, and it deserves the more attention as it is a cause of error extremely frequent, and one of which we are ourselves less aware, and which is less

a *Essais,* L. ii. c. 17.—ED.

notorious to others. We feel it fatiguing to continue
an investigation, therefore we do not pursue it; but
as it is mortifying to think that we have laboured in
vain, we easily admit the flattering illusion that we
have succeeded. By the influence of this disposition
it often happens, that, after having rejected what first
presented itself,—after having rejected a second time
and a third time what subsequently turned up, because not sufficiently applicable or certain, we get
tired of the investigation, and perhaps put up with
the fourth suggestion, which is not better, haply even
worse, than the preceding; and this simply because
it has come into the mind when more exhausted and
less scrupulous than it was at the commencement."[a]
"The volition of that man," says Seneca, "is often
frustrated, who undertakes not what is easy, but who
wishes what he undertakes to be easy. As often as
you attempt anything, compare together yourself, the
end which you propose, and the means by which it is
to be accomplished. For the repentance of an un-
finished work will make you rash. And here it
is of consequence whether a man be of a fervid
or of a cold, of an aspiring or of a humble, disposition."[b]

To remedy this failing it is necessary, in conformity with this advice of Seneca, to consult our forces,
and the time we can afford, and the difficulty of the
subjects on which we enter. We ought to labour only
at intervals, to avoid the tedium and disquiet consequent on unremitted application, and to adjourn the
consideration of any thought which may please us
vehemently at the moment, until the prepossession in

a Crousaz, Logique, t. iii. part ii.
ch. 7, p. 302.—Ed.   b De Ira, L. iii. c. 7. Quoted by
Crousaz, Logique, t. iii. p. 302.—Ed.

100 LECTURES ON LOGIC.

LECT.
XXIX.

3. Hope
and Fear.

its favour has subsided with the animation which gave it birth.

The two Causes of premature judgment,—the affections of Impatience and Sloth,—being considered, I pass on to the third principle of Passion, by which the intellect is turned aside from the path of truth,—I mean the disturbing influence of Hope and Fear. These passions, though reciprocally contrary, determine a similar effect upon the deliberations of the Understanding, and are equally unfavourable for the interest of truth. In forming a just conclusion upon a question of probable reasoning, that is, where the grounds of decision are not few, palpable, and of determinate effect,—and such questions may be said to be those alone on which differences of opinion may arise, and are, consequently, those alone which require for their solution any high degree of observation and ingenuity,—in such questions hope and fear exert a very strong and a very unfavourable influence. In these questions it is requisite, in the first place, to seek out the premises; and, in the second, to draw the conclusion. Of these requisites the first is the more important, and it is also by far the more difficult.

How Hope and Fear operate unfavourably on the Understanding.

Now the passions of Hope and Fear operate severally to prevent the intellect from discovering all the elements of decision, which ought to be considered in forming a correct conclusion, and cause it to take into account those only which harmonise with that conclusion to which the actuating passion is inclined. And here the passion operates in two ways. In the first place, it tends so to determine the associations of thought, that only those media of proof are suggested or called into consciousness, which support the conclusion to which the passion tends. In the second place, if the media of proof by which a counter conclusion is supported, are brought before the mind, still the mind is influenced by the passion to look on their reality with doubt, and, if such cannot be questioned, to undervalue their inferential importance; whereas it is moved to admit, without hesitation, those media of proof, which favour the conclusion in the interest of our hope or fear, and to exaggerate the cogency with which they establish this result. Either passion looks exclusively to a single end, and exclusively to the means by which that single end is accomplished. Thus the sanguine temperament, or the mind under the habitual predominance of hope, sees only and magnifies all that militates in favour of the wished-for consummation, which alone it contemplates; whereas the melancholic temperament, or the mind under the habitual predominance of fear, is wholly occupied with the dreaded issue, views only what tends to its fulfilment, while it exaggerates the possible into the probable, the probable into the certain. Thus it is that whatever conclusion we greatly hope or greatly fear, to that conclusion we are disposed to leap; and it has become almost proverbial, that men lightly believe both what they wish, and what they dread, to be true.

But the influence of Hope on our judgments, inclining us to find whatever we wish to find, in so far as this arises from the illusion of Self-love, is comprehended in this,—the fourth cause of Error,—to which I now proceed.

Self-love, under which I include the dispositions of 4. Self-love. Vanity, Pride, and, in general, all those which incline us to attribute an undue weight to those opinions in

which we feel a personal interest, is by far the most
extensive and influential impediment in the way of
reason and truth. In virtue of this principle, whatever is ours,—whatever is adopted or patronised by
us, whatever belongs to those to whom we are attached,—is either gratuitously clothed with a character of truth, or its pretensions to be accounted true
are not scrutinised with the requisite rigour and impartiality. I am a native of this country, and, therefore, not only is its history to me a matter of peculiar
interest, but the actions and character of my countrymen are viewed in a very different light from that in
which they are regarded by a foreigner. I am born
and bred a member of a religious sect, and because
they constitute my creed, I find the tenets of this
sect alone in conformity to the Word of God. I am
the partisan of a philosophical doctrine, and am, therefore, disposed to reject whatever does not harmonise
with my adopted system.

Aristotle,—
his precept.
It is the part of a philosopher, says Aristotle, inasmuch as he is a philosopher, to subjugate self-love,
and to refute, if contrary to truth, not only the opinions of his friends, but the doctrines which he himself
may have professed.[a] It is certain, however, that
philosophers,—for philosophers are men,—have been
too often found to regulate their conduct by the op-
Illustrations
of the influ-
ence of Self-
love on our
opinions.
posite principle. That man pretended to the name
of philosopher, who scrupled not to declare that he
would rather be in the wrong with Plato than in the
right with his opponents.[b] "Gisbert Voetius urged
Mersennus to refute a work of Descartes a year before
the book appeared, and before he had himself the
means of judging whether the opinions it contained

a _Eth. Nic._, i. 4 (6).— ED.   β Cicero, _Tusc. Quæst._, i. 17.

were right or wrong. A certain professor of philosophy in Padua came to Galileo, and requested that he would explain to him the meaning of the term *parallaxis*; which he wished, he said, to refute, having heard that it was opposed to Aristotle's doctrine touching the relative situation of the comets. What! answered Galileo, you wish to controvert a word the meaning of which you do not know! Redi, the naturalist, tells us that a sturdy Peripatetic of his acquaintance would never consent to look at the heavens through a telescope, lest he should be compelled to admit the existence of the new stars discovered by Galileo and others. The same Redi informs us that he knew another Peripatetic, a staunch advocate of the Aristotelian doctrine of equivocal generation, (a doctrine, by the way, which now again divides the physiologists of Europe,) and who, in particular, maintained that the green frogs which appear upon a shower come down with the rain, who would not be induced himself to select and examine one of these frogs. And why? Because he was unwilling to be convicted of his error, by Redi showing him the green matter in the stomach, and its feculæ in the intestines of the animal." [a] The spirit of the Peripatetic philosophy was, however, wholly misunderstood by these mistaken followers of Aristotle; for a true Aristotelian is one who listens rather to the voice of nature than to the precept of any master, and it is well expressed in the motto of the great French anatomist,—"Riolanus est Peripateticus; credit ea, et ea tantum, quæ vidit." From the same principle proceeds the abuse, and

LECT.
XXIX.

Self-love leads us to regard with favour the opinions of those to whom we are in any way attached.

Male-branche adduced to this effect.

This shown especially when the

sometimes even the persecution, which the discoverers of new truths encounter from those whose cherished opinions these truths subvert.

In like manner, as we are disposed to maintain our own opinion, we are inclined to regard with favour the opinions of those to whom we are attached by love, gratitude, and other conciliatory affections. "We do not limit our attachment to the persons of our friends, —we love in a certain sort all that belongs to them; and as men generally manifest sufficient ardour in support of their opinions, we are led insensibly by a kind of sympathy to credit, to approve, and to defend these also, and that even more passionately than our friends themselves. We bear affection to others for various reasons. The agreement of tempers, of inclinations, of pursuits; their appearance, their manners, their virtue, the partiality which they have shown to us, the services we have received at their hands, and many other particular causes, determine and direct our love.

"'It is observed by the great Malebranche,[a] that if any of our friends,—any even of those we are disposed to love,—advance an opinion, we forthwith lightly allow ourselves to be persuaded of its truth. This opinion we accept and support, without troubling ourselves to inquire whether it be conformable to fact, frequently even against our conscience, in conformity to the darkness and confusion of our intellect, to the corruption of our heart, and to the advantages which we hope to reap from our facility and complaisance.'"[b]

The influence of this principle is seen still more manifestly when the passion changes; for though the

---

[a] *Recherche de la Vérité*, L. iv. ch. 13.—ED.　　[b] Carr, *Nouvelle Logique*, part ii. ch. 8, p. 288.—ED.

things themselves remain unaltered, our judgments concerning them are totally reversed. How often do we behold persons who cannot, or will not, recognise a single good quality in an individual from the moment he has chanced to incur their dislike, and who are even ready to adopt opinions, merely because opposed to others maintained by the object of their aversion? The celebrated Arnauld[a] goes so far even as to assert, that men are naturally envious and jealous; that it is with pain they endure the contemplation of others in the enjoyment of advantages which they do not themselves possess; and, as the knowledge of truth and the power of enlightening mankind is of one of these, that they have a secret inclination to deprive them of that glory. This accordingly often determines them to controvert without a ground the opinions and discoveries of others. Self-love accordingly often argues thus:—' This is an opinion which I have originated, this. is an opinion, therefore, which is true;' whereas the natural malignity of man not less frequently suggests such another: ' It is another than I who has advanced this doctrine; this doctrine is, therefore, false.'

We may distinguish, however, from malignant or envious contradiction another passion, which, though more generous in its nature and not simply a mode of Self-love, tends, nevertheless, equally to divert us from the straight road of truth,—I mean Pugnacity, or the love of Disputation. Under the influence of this passion, we propose as our end victory, not truth. We insensibly become accustomed to find a reason for any opinion, and, in placing ourselves above all reasons, to surrender our belief to none. Thus it is why

L'Art de Penser (Port-Royal Logic), p. iii. ch. 20.—Ed.

two disputants so rarely ever agree, and why a question is seldom or never decided in a discussion where the combative dispositions of the reasoners have once been roused into activity. In controversy it is always easy to find wherewithal to reply; the end of the parties is not to avoid error, but to impose silence; and they are less ashamed of continuing wrong than of confessing that they are not right.[a]

These affections may be said to be the immediate causes of all error. Other causes there are, but not immediate. In so far as Logic detects the sources of our false judgments and shows their remedies, it must carefully inculcate that no precautionary precept for particular cases can avail, unless the inmost principle of the evil be discovered, and a cure applied. You must, therefore, as you would remain free from the hallucination of false opinion, be convinced of the absolute necessity of following out the investigation of every question calmly and without passion. You must learn to pursue, and to estimate, truth without distraction or bias. To this there is required, as a primary condition, unshackled freedom of thought, the equal glance which can take in the whole sphere of observation, the cool determination to pursue the truth whithersoever it may lead; and, what is still more important, the disposition to feel an interest in truth, and in truth alone. If perchance some collateral interest may first prompt us to the inquiry, in our general interest for truth we must repress,— we must forget, this interest, until the inquiry be concluded. Of what account are the most venerated opinions if they be untrue? At best they are only

Marginal notes: These affections the immediate causes of all error. Preliminary conditions for the efficiency of precepts against the motive of error.

a L'Art de Penser, p. iii. ch. 20. ch. 9, p. 311, Paris, 1820. — ED.
Cf. Caro, Nouvelle Logique, part ii.

venerable delusions. He who allows himself to be
actuated in his scientific procedure by any partial interest, can never obtain a comprehensive survey of the
whole he has to take into account, and always, therefore, remains incapable of discriminating, with accuracy, error from truth. The independent thinker must,
in all his inquiries, subject himself to the genius of
truth, — must be prepared to follow her footsteps
without faltering or hesitation. In the consciousness
that truth is the noblest of ends, and that he pursues
this end with honesty and devotion, he will dread no
consequences,—for he relies upon the truth. Does he
compass the truth, he congratulates himself on his
success ; does he fall short of its attainment, he knows
that even his present failure will ultimately advance
him to the reward he merits. Err he may, and that
perhaps frequently, but he will never deceive himself.
We cannot, indeed, rise superior to our limitary nature, we cannot, therefore, be reproached for failure ;
but we are always responsible for the calmness and
impartiality of our researches, and these alone render
us worthy of success. But though it be manifest,
that to attain the truth we must follow whithersoever
the truth may lead, still men in general are found to
yield not an absolute, but only a restricted, obedience
to the precept. They capitulate, and do not unconditionally surrender. I give up, but my cherished
dogma in religion must not be canvassed, says one ;—
my political principles are above inquiry, and must
be exempted, says a second ;—my country is the land
of lands, this cannot be disallowed, cries a third :—
my order, my vocation, is undoubtedly the noblest,
exclaim a fourth and fifth ;—only do not require that
we should confess our having erred, is the condition

which many insist on stipulating. Above all, that resolve of mind is difficult, which is ready to surrender all fond convictions, and is prepared to recommence investigation the moment that a fundamental error in the former system of belief has been detected. These are the principal grounds why, among men, opinion is so widely separated from opinion; and why the clearest demonstration is so frequently for a season frustrated of victory.

Par. XCVI.
Rules
against
Errors from
the Affec-
tions.
¶ XCVI. Against the Errors which arise from the Affections, there may be given the three following rules :—

1°, When the error has arisen from the influence of an active affection, the decisive judgment is to be annulled; the mind is then to be freed, as far as possible, from passion, and·the process of inquiry to be recommenced so soon as the requisite tranquillity has been restored.

2°, When the error has arisen from a relaxed enthusiasm for knowledge, we must reanimate this interest by a vivid representation of the paramount dignity of truth and of the lofty destination of our intellectual nature.

3°, In testing the accuracy of our judgments, we must be particularly suspicious of those results which accord with our private inclinations and predominant tendencies.

These rules require no comment.

LECTURE XXX.

MODIFIED STOICHEIOLOGY.

SECTION II.—ERROR—ITS CAUSES AND REMEDIES.

B.—AS IN THE COGNITIONS, FEELINGS, AND DESIRES.

II.—WEAKNESS AND DISPROPORTIONED STRENGTH
OF THE FACULTIES OF KNOWLEDGE.

I now go on to the Second Head of the class of Errors founded on the Natural Constitution, the Acquired Habits, and the Reciprocal Relations of our Cognitive and Affective Powers, that is, to the Causes of Error which originate in the Weakness or Disproportioned Strength of one or more of our Faculties of Knowledge themselves.

LECT.
XXX.

Weakness and Dispro-
portioned
Strength of
the Facul-
ties of
Knowledge.

Here, in the first place, I might consider the errors which have arisen from the Limited Nature of the Human Intellect in general,—or rather from the mistakes that have been made by philosophers in denying or not taking this limited nature into account.[a] The illustration of this subject is one which is relative to, and supposes an acquaintance with, some of the abstrusest speculations in Philosophy, and which belong

Neglect of
the Limited
Nature of
the Human
Intellect a
source of
error.

a [On this subject see Crusius.]
(Christian August Crusius, *Weg zur
Gewissheit und Zuverlässigkeit der* menschlichen Erkenntniss, § 443, 1st
ed. 1747.—En.]

not to Logic, but to Metaphysics. I shall not, therefore, do more than simply indicate at present what it
will be proper at another season fully to explain. It
is manifest, that, if the human mind be limited,—if it
only knows as it is conscious, and if it be only conscious as it is conscious of contrast and opposition,—
of an ego and non-ego ;—if this supposition, I say, be
correct, it is evident that those philosophers are in
error, who virtually assume that the human mind is
unlimited, that is, that the human mind is capable of
a knowledge superior to consciousness,—a cognition
in which knowledge and existence,—the Ego and
non-Ego,—God and the creature,—are identical ; that
is, of an act in which the mind is the Absolute, and
knows the Absolute. This philosophy, the statement
of which, as here given, it would require a long commentary to make you understand, is one which has
for many years been that dominant in Germany ; it
is called the *Philosophy of the Absolute*, or the *Philosophy of Absolute Identity*. This system, of which
Schelling and Hegel are the great representatives, errs
by denying the limitation of human intelligence without proof, and by boldly building its edifice on this
gratuitous negation.[a]

But there are other forms of philosophy, which err
not in actually postulating the infinity of mind, but
in taking only a one-sided view of its finitude. It is
a general fact, which seems, however, to have escaped
the observation of philosophers, that whatever we can
positively compass in thought,—whatever we can conceive as possible,—in a word, the *omne cogitabile*, lies
between two extremes or poles, contradictorily op-

[a] See *Discussions*, p. 19.—Ed.

posed, and one of which must, consequently, be true, but of neither of which repugnant opposites are we able to represent to our mind the possibility." To take one example out of many : we cannot construe to the mind as possible the absolute commencement of time ; but we are equally unable to think the possibility of the counter alternative,—its infinite or absolute non-commencement, in other words, the infinite regress of time. Now it is evident, that, if we looked merely at the one of these contradictory opposites and argued thus :—whatever is inconceivable is impossible, the absolute commencement of time is inconceivable, therefore the absolute commencement of time is impossible ; but, on the principles of Contradiction and Excluded Middle, one or other of two opposite contradictories must be true ; therefore, as the absolute commencement of time is impossible, the absolute or infinite non-commencement of time is necessary :—I say, it is evident that this reasoning would be incompetent and one-sided, because it might be converted ; for, by the same one-sided process, the opposite conclusion might be drawn in favour of the absolute commencement of time.

LECT.
XXX.

Illustrated by reference to the two contradictories,—the absolute commencement, and the infinite non-commencement, of Time.

Now, the unilateral and incompetent reasoning which I have here supposed in the case of time, is one of which the Necessitarian is guilty, in his argument to prove the impossibility of human volitions being free. He correctly lays down, as the foundation of his reasoning, two propositions which must at once be allowed : 1°, That the notion of the liberty of volition involves the supposition of an absolute commencement of volition, that is, of a volition which is a cause, but is not itself, *qua* cause, an effect ; 2°, That the absolute commencement of volition, or of aught else, cannot be conceived, that is, cannot be directly or positively thought as possible. So far he is correct ; but when he goes on to apply these principles by arguing, (and be it observed this syllogism lies at the root of all the reasonings for necessity), *Whatever is inconceivable is impossible ; but the supposition of the absolute commencement of volition is inconceivable ; therefore, the supposition of the absolute commencement of volition (the condition of free-will) is impossible,*—we may here demur to the sumption, and ask him,—Can he positively conceive the opposite contradictory of the absolute commencement, that is, an infinite series of relative non-commencements ? If he answers, as he must, that he cannot, we may again ask him,—By what right he assumed as a self-evident axiom for his sumption, the proposition,—*that whatever is inconceivable is impossible*, or by what right he could subsume his minor premise, when by his own confession he allows that the opposite contradictory of his minor premise, that is, the very proposition he is apagogically proving, is, likewise, inconceivable, and, therefore, on the principle of his sumption, likewise, impossible ?

And in the
case of the
Libertarian
Argument
in behalf of
Free-will.

The same inconsequence would equally apply to the Libertarian, who should attempt to prove that free-will must be allowed, on the ground that its contradictory opposite is impossible, because inconceivable. He cannot prove his thesis by such a process ; in fact, by all speculative reasoning from the conditions of thought, the two doctrines are in *æquili-*

brio ;—both are equally possible,—both are equally inconceivable. It is only when the Libertarian descends to arguments drawn from the fact of the Moral Law and its conditions, that he is able to throw in reasons which incline the balance in his favour.

On these matters I, however, at present only touch, in order to show you under what head of Error these reasonings would naturally fall.

Leaving, therefore, or adjourning, the consideration of the imbecility of the human intellect in general, I shall now take into view, as a source of logical error, the Weakness or Disproportioned Strength of the several Cognitive Faculties. Now, as the Cognitive Faculties in man consist partly of certain Lower Powers, which he possesses in common with other sensible existences, namely, the Presentative, the Retentive, the Representative, and the Reproductive Faculties, and partly of certain Higher Powers, in virtue of which he enters into the rank of intelligent existences, namely, the Elaborative and Regulative Faculties,—it will be proper to consider the powers of these two classes severally in succession, in so far as they may afford the causes or occasions of error.

Of the lower class, the first faculty in order is the Presentative or Acquisitive Faculty. This, as you remember, is divided into two, viz. into the faculty which presents us with the phænomena of the outer world, and the faculty which presents us with the phænomena of the inner.[a] The former is External Perception, or External Sense; the latter is Self-consciousness, Internal Perception, or Internal Sense. I commence, therefore, with the Faculty of External

a See *Lectures on Metaphysics*, vol. ii. p. 23 *et seq.*—ED.

Marginal notes:

Weakness or disproportioned strength of the several Cognitive Faculties,—a source of Error.

Cognitive Faculties of two classes, a Lower and a Higher.

1. The Lower Class,—1. The Presentative Faculty.

Perception, in relation to which I give you the following paragraph :—

Par. XCVII.
a. External
Perception,
—as a
source of
Error.

¶ XCVII. When aught is presented through the outer senses, there are two conditions necessary for its adequate perception :—1°, The relative Organs must be present, and in a condition to discharge their functions; and 2°, The Objects themselves must bear a certain relation to these organs, so that the latter shall be suitably affected, and thereby the former suitably apprehended. It is possible, therefore, that, partly through the altered condition of the organs, partly through the altered situation of the objects, dissimilar presentations of the same, and similar presentations of different, objects, may be the result.[a]

Explanation.
Conditions
of the
adequate
activity of
External
Perception.

"In the first place, without the organs specially subservient to External Perception,—without the eye, the ear, &c., sensible perceptions of a precise and determinate character, such, for example, as colour or sound, are not competent to man. In the second place, to perform their functions, these organs must be in a healthy or normal state ; for if this condition be not fulfilled, the presentations which they furnish are null, incomplete, or false. But, in the third place, even if the organs of sense are sound and perfect, the objects to be presented and perceived must stand to these organs in a certain relation,—must bear to them a certain proportion ; for, otherwise, the objects cannot be presented at all, or cannot be perceived without

a Krug, *Logik*, § 138.—Ed. [Cf. vi. p. 273. Bachmann, *Logik*, § 407, Cass, *Nouvelle Logique*, part ii. ch. p. 553.]

illusion. The sounds, for example, which we are to
hear, must neither be too high nor too low in quality ;
the bodies which we are to see, must neither be too
near nor too distant,—must neither be too feebly nor
too intensely illuminated. In relation to the second
condition, there are given, in consequence of the altered state of the organs, on the one hand, different
presentations of the same object ;—thus to a person
who has waxed purblind, his friend appears an an utter
stranger, the eye now presenting its objects with less
clearness and distinctness. On the other hand, there
are given the same, or undistinguishably similar, presentations of different objects ;—thus to a person in the
jaundice, all things are presented yellow. In relation
to the third condition, from the altered position of
objects, there are, in like manner, determined, on the
one hand, different presentations of the same objects,
as when the stick which appears straight in the air
appears crooked when partially immersed in water ;
and, on the other hand, identical presentations of
different objects, as when a man and a horse appear in the distance to be so similar, that the
one cannot be discriminated from the other. In
all these cases, these illusions are determined,—illusions which may easily become the occasions of false
judgments."*

"In regard to the detection of such illusions and
obviating the errors to which they lead, it behoves us
to take the following precautions. We must, in the
first place, examine the state of the organ. If found
defective, we must endeavour to restore it to perfection, but if this cannot be done, we must ascertain

Precautions
with a view
to the detec-
tion of Illu-
sions of the
Senses, and
obviating
the errors
to which
they lead.

* Krug, Logik, § 138. Anm.—Ed.

the extent and nature of the evil, in order to be upon
our guard in regard to the quality and degree of the
false presentation. In the second place, we must
examine the relative situation of the object, and if
this be not accommodated to the organ, we must
either obviate the disproportion and remove the
media which occasioned the illusion, or repeat the
observation under different circumstances, compare
these, and thus obtain the means of making an ideal
abstraction of the disturbing causes."[a]

In regard to the other Presentative Faculty,—the
Faculty of Self-consciousness,—Internal Perception,
or Internal Sense, as we know less of the material
conditions which modify its action, we are unable to
ascertain so precisely the nature of the illusions of
which it may be the source. In reference to this subject you may take the following paragraph:—

Par. XCVIII.
b. Self-con-
sciousness,
—as a
source of
Error.

¶ XCVIII. The faculty of Self-consciousness,
or Internal Sense, is subject to various changes,
which either modify our apprehensions of objects, or influence the manner in which we judge
concerning them. In so far, therefore, as false
judgments are thus occasioned, Self-consciousness
is a source of error.[β]

Explica-
tion.
Self-con-
sciousness
varies in
intensity.

It is a matter of ordinary observation, that the
vivacity with which we are conscious of the various
phænomena of mind, differs not only at different times,
in different states of health, and in different degrees
of mental freshness and exhaustion, but, at the same
time, differs in regard to the different kinds of these

a Krug, Logik, § 155.—ED. β Krug, Logik, § 139.—ED.

phænomena themselves. According to the greater or
less intensity of this faculty, the same thoughts of
which we are conscious are, at one time, clear and
distinct, at another, obscure and confused. At one
time we are almost wholly incapable of reflection, and
every act of self-attention is forced and irksome, and
differences the most marked pass unnoticed; while,
at another, our self-consciousness is alert, all its applications pleasing, and the most faint and fugitive
phænomena arrested and observed. On one occasion,
self-consciousness, as a reflective cognition, is strong;
on another, all reflection is extinguished in the intensity of the direct consciousness of feeling or desire. In
one state of mind our representations are feeble; in
another, they are so lively that they are mistaken for
external realities. Our self-consciousness may thus
be the occasion of frequent error; for, according to its
various modifications, we may form the most opposite
judgments concerning the same things,—pronouncing
them, for example, now to be agreeable, now to be
disagreeable, according as our Internal Sense is variously affected.

The next is the Retentive or Conservative Faculty,
—Memory strictly so called; in reference to which I
give you the following paragraph:—

¶ XCIX. Memory, or the Conservative Faculty,
is the occasion of Error, both when too weak and
when too strong. When too weak, the complement
of cognitions which it retains is small and in-
distinct, and the Understanding or Elaborative
Faculty is, consequently, unable adequately to
judge concerning the similarity and differences

of its representations and concepts. When too
strong, the Understanding is overwhelmed with
the multitude of acquired cognitions simultaneously forced upon it, so that it is unable calmly
and deliberately to compare and discriminate
these.[a]

That both these extremes—that both the insufficient and the superfluous vigour of the Conservative
Faculty—are severally the sources of error, it will not
require many observations to make apparent.

In regard to a feeble memory, it is manifest that a
multitude of false judgments must inevitably arise
from an incapacity in this faculty to preserve the
observations committed to its keeping. In consequence of this incapacity, if a cognition be not wholly
lost, it is lost at least in part, and the circumstances
of time, place, persons and things confounded with
each other. For example,—I may recollect the tenor
of a passage I have read, but from defect of memory
may attribute to one author what really belongs to
another. Thus a botanist may judge two different
plants to be identical in species, having forgotten the
differential characters by which they were discriminated ; or he may hold the same plant to be two different species, having examined it at different times
and places.[b]

Though nothing could be more erroneous than a
general and unqualified decision, that a great memory
is incompatible with a sound judgment, yet it is an
observation confirmed by the experience of all ages
and countries, not only that a great memory is no

a [Cf. Bachmann, Logik, § 408.]   β Krug, Logik, § 141. Anm.—En.

condition of high intellectual talent, but that great LECT.
memories are very frequently found in combination XXL.
with comparatively feeble powers of thought.* The
truth seems to be, that where a vigorous memory is
conjoined with a vigorous intellect, the force of the
subsidiary faculty not only does not detract from the
strength of the principal, but, on the contrary, tends
to confer on it a still higher power; whereas when
the inferior faculty is disproportionately strong, so
far from nourishing and corroborating the superior,
it tends to reduce this faculty to a lower level than
that at which it would have stood, if united with a
less overpowering subsidiary. The greater the magazine of various knowledge which the memory contains,
the better for the understanding, provided the understanding can reduce this various knowledge to
order and subjection. "A great memory is the principal condition of bringing before the mind many
different representations and notions at once, or in
rapid succession. This simultaneous or nearly simultaneous presence disturbs, however, the tranquil comparison of a small number of ideas, which, if it shall
judge aright, the intellect must contemplate with a
fixed and steady attention."* Now, where an intellect
possesses the power of concentration in a high degree,
it will not be harassed in its meditations by the offi-
cious intrusions of the subordinate faculties, however
vigorous these in themselves may be, but will control
their vigour by exhausting in its own operations the
whole applicable energy of mind. Whereas where

a Compare *Lectures on Metaphysics*, *Mute*, quoted by Stewart, *Elem.*,
vol. ii. p. 223.—Ed. part iii. ch. i. sect. vi. *Collected*
β Diderot, *Lettre sur les Sourds et Works*, vol. iv. p. 249.

the inferior is more vigorous than the superior, it will, in like manner, engross in its own function the disposable amount of activity, and overwhelm the principal faculty with materials, many even in proportion as it is able to elaborate few. This appears to me the reason, why men of strong memories are so often men of proportionally weak judgments, and why so many errors arise from the possession of a faculty, the perfection of which ought to exempt them from many mistaken judgments.

Remedies for these opposite extremes.

As to the remedy for these opposite extremes. The former,—the imbecility of Memory,—can only be alleviated by invigorating the capacity of Retention through mnemonic exercises and methods; the latter,—the inordinate vigour of Memory,—by cultivating the Understanding to the neglect of the Conservative Faculty. It will, likewise, be necessary to be upon our guard against the errors originating in these counter sources. In the one case distrusting the accuracy of the facts, in the other, the accuracy of their elaboration.[a]

2. The Reproductive Faculty.

The next faculty is the Reproductive. This, when its operation is voluntarily exerted, is called *Recollection* or *Reminiscence;* when it energises spontaneously or without volition, it is called *Suggestion.* The laws by which it is governed in either case, but especially in the latter, are called the *Laws of Mental Association.* This Reproductive Faculty, like the Retentive, is the cause of error, both if its vigour be defective, or if it be too strong. I shall consider Recollection and Suggestion severally and apart. In regard to the former I give you the following paragraph :—

a Cf. Krug, *Logik,* § 156. Anm.—Ed.

¶ C. The Reproductive Faculty, in so far as it is voluntarily exercised, or Reminiscence, becomes a source of Error as it is either too sluggish or too prompt, precisely as the Retentive Faculty, combined with which it constitutes Memory in the looser signification.

It is necessary to say very little in special reference to Reminiscence, for what was said in regard to the Conservative Faculty or Memory Proper in its higher vigour, was applicable to, and in fact supposed a corresponding degree of, the Reproductive. For, however great may be the mass of cognitions retained in the mind, that is, out of consciousness but potentially capable of being called into consciousness, these can never of themselves oppress the Understanding by their simultaneous crowding or rapid succession, if the faculty by which they are revoked into consciousness be inert ; whereas, if this revocative faculty be comparatively alert and vigorous, a smaller magazine of retained cognitions may suffice to harass the intellect with a ceaseless supply of materials too profuse for its capacity of elaboration.

On the other hand, the inactivity of our Recollection is a source of error, precisely as the weakness of our Memory proper ; for it is of the same effect in relation to our judgments, whether the cognitions requisite for a decision be not retained in the mind, or whether, being retained, they are not recalled into consciousness by Reminiscence.

In regard to Suggestion, or the Reproductive Faculty operating spontaneously, that is, not in subservience to an act of Will, I shall give you the following paragraph :—

LECT.
XXX.

Par. CL
b. Sugges-
tion,—as a
source of
Error.

¶ CI. As our Cognitions, Feelings, and Desires
are connected together by what are called the
Laws of Association, and as each link in the
chain of thought suggests or awakens into consciousness some other in conformity to these
Laws,—these Laws, as they bestow a strong
subjective connection on thoughts and objects
of a wholly arbitrary union, frequently occasion
great confusion and error in our judgments.

"Even in methodical thinking, we do not connect
all our thoughts intentionally and rationally, but
many press forward into the train, either in consequence of some external impression, or in virtue of
certain internal relations, which, however, are not of a
logical dependency. Thus, thoughts tend to suggest
each other, which have reference to things of which
we were previously cognisant as coexistent, or as immediately consequent, which have been apprehended
as bearing a resemblance to each other, or which have
stood together in reciprocal and striking contrast.
This connection, though precarious and non-logical, is
thus, however, governed by certain laws, which have
been called the *Laws of Association*." [a] These laws,
which I have just enumerated, viz. the Law of Co-
existence or Simultaneity, the Law of Continuity or
Immediate Succession, the Law of Similarity, and the
Law of Contrast, are all only special modifications of
one general law which I would call the *Law of Red-
integration;* [β] that is, the principle according to which
whatever has previously formed a part of one total
act of consciousness, tends, when itself recalled into

a Krug, *Logik*, § 144. Aen.—
Ed.

β See *Lectures on Metaphysics*, vol.
ii. p. 233 *et seq.*—Ed.

consciousness, to reproduce along with it the other parts of that original whole. But though these tendencies be denominated *laws*, the influence which they exert, though often strong and sometimes irresistible, is only contingent ; for it frequently happens that thoughts which have previously stood to each other in one or other of the four relations do not suggest each other. The Laws of Association stand, therefore, on a very different footing from the laws of logical connection. But those Laws of Association, contingent though they be, exert a great and often a very pernicious influence upon thought, inasmuch as by the involuntary intrusion of representations into the mental chain, which are wholly irrelevant to the matter in hand, there arises a perplexed and redundant tissue of thought, into which false characters may easily find admission, and in which true characters may easily be overlooked.[a] But this is not all. For, by being once blended together in our consciousness, things really distinct in their nature tend again naturally to reassociate, and at every repetition of this conjunction, this tendency is fortified, and their mutual suggestion rendered more certain and irresistible.

It is in virtue of this principle of Association and Custom, that things are clothed by us with the precarious attributes of deformity or beauty ; and some philosophers have gone so far as to maintain that our principles of Taste are exclusively dependent on the accidents of Association. But if this be an exaggeration, it is impossible to deny that Association enjoys an extensive jurisdiction in the empire of taste, and, in particular, that fashion is almost wholly subject to its control.

a Krug, *Logik*, § 144. Anm.—Ed.

On this subject I may quote a few sentences from
the first volume of Mr Stewart's *Elements.* " In matters of Taste, the effects which we consider, are produced on the mind itself, and are accompanied either
with pleasure or with pain.   Hence the tendency to
casual association is much stronger than it commonly
is with respect to physical events ; and when such
associations are once formed, as they do not lead to
any important inconvenience, similar to those which
result from physical mistakes, they are not so likely
to be corrected by mere experience, unassisted by
study.   To this it is owing that the influence of association on our judgments concerning beauty and deformity, is still more remarkable than on our speculative conclusions ; a circumstance which has led some
philosophers to suppose that association is sufficient
to account for the origin of these notions, and that
there is no such thing as a standard of taste, founded
on the principles of the human constitution.   But this
is undoubtedly pushing the theory a great deal too
far.   The association of ideas can never account for
the origin of a new notion, or of a pleasure essentially
different from all the others which we know.   It may,
indeed, enable us to conceive how a thing indifferent
in itself may become a source of pleasure, by being
connected in the mind with something else which is
naturally agreeable ; but it presupposes, in every instance, the existence of those notions and those feelings which it is its province to combine : insomuch
that, I apprehend, it will be found, wherever association produces a change in our judgments on matters
of taste, it does so by co-operating with some natural
principle of the mind, and implies the existence of
certain original sources of pleasure and uneasiness.

" A mode of dress, which at first appeared awkward, acquires, in a few weeks or months, the appearance of elegance. By being accustomed to see it
worn by those whom we consider as models of taste,
it becomes associated with the agreeable impressions
which we receive from the ease and grace and refinement of their manners. When it pleases by itself, the
effect is to be ascribed, not to the object actually before us, but to the impressions with which it has been
generally connected, and which it naturally recalls to
the mind.

" This observation points out the cause of the per-
petual vicissitudes in dress, and in everything whose
chief recommendation arises from fashion. It is evident that, as far as the agreeable effect of an ornament
arises from association, the effect will continue only
while it is confined to the higher orders. When it is
adopted by the multitude, it not only ceases to be
associated with ideas of taste and refinement, but it is
associated with ideas of affectation, absurd imitation,
and vulgarity. It is accordingly laid aside by the
higher orders, who studiously avoid every circumstance in external appearance which is debased by low
and common use; and they are led to exercise their
invention in the introduction of some new peculiarities, which first become fashionable, then common, and
last of all, are abandoned as vulgar."[a]

" Our moral judgments, too, may be modified, and
even perverted, to a certain degree, in consequence of
the operation of the same principle. In the same
manner in which a person who is regarded as a model
of taste may introduce, by his example, an absurd or
fantastical dress; so a man of splendid virtues may

a *Elements*, vol. L, part 1. chap. v. *Collected Works*, ii. p. 322 *et seq.*

attract some esteem also to his imperfections, and, if
placed in a conspicuous situation, may render his vices
and follies objects of general imitation among the
multitude.

"'In the reign of Charles II.,' says Mr Smith," 'a
degree of licentiousness was deemed the characteristic
of a liberal education.   It was connected, according
to the notions of those times, with generosity, sincerity, magnanimity, loyalty; and proved that the
person who acted in this manner was a gentleman,
and not a puritan.   Severity of manners and regu-
larity of conduct, on the other hand, were altogether
unfashionable, and were connected, in the imagination
of that age, with cant, cunning, hypocrisy, and low
manners.   To superficial minds the vices of the great
seem at all times agreeable.   They connect them not
only with the splendour of fortune, but with many
superior virtues which they ascribe to their superiors;
with the spirit of freedom and independency; with
frankness, generosity, humanity, and politeness.   The
virtues of the inferior ranks of people, on the contrary,—their parsimonious frugality, their painful in-
dustry, and rigid adherence to rules, seem to them
mean and disagreeable.   They connect them both with
the meanness of the station to which these qualities
commonly belong, and with many great vices which
they suppose usually accompany them; such as an ab-
ject, cowardly, ill-natured, lying, pilfering disposition.'"ᵝ

Condillac
quoted on
the Influ-
ence of
Association "In general," says Condillac, "the impression we
experience in the different circumstances of life, makes
us associate ideas with a force which renders them
ever after for us indissoluble.   We cannot, for exam-

α *Theory of Moral Sentiments*,   β *Elements*, vol. i. c. v. § 3.   Col-
part v. c. 2.—ED.   ·   lected *Works*, vol. ii. p. 232.

ple, frequent the society of our fellow-men without insensibly associating the notions of certain intellectual or moral qualities with certain corporeal characters. This is the reason why persons of a decided physiognomy please or displease us more than others; for a physiognomy is only an assemblage of characters, with which we have associated notions which are not suggested without an accompaniment of satisfaction or disgust. It is not, therefore, to be marvelled at that we judge men according to their physiognomy, and that we sometimes feel towards them at first sight aversion or inclination. In consequence of these associations, we are often vehemently prepossessed in favour of certain individuals, and no less violently disposed against others. It is because all that strikes us in our friends or in our enemies is associated with the agreeable or the disagreeable feeling which we severally experience; and because the faults of the former borrow always something pleasing from their amiable qualities, whereas the amiable qualities of the latter seem always to participate of their vices. Hence it is that these associations exert a powerful influence on our whole conduct. They foster our love or hatred; enhance our esteem or contempt; excite our gratitude or indignation; and produce those sympathies, those antipathies, or those capricious inclinations, for which we are sometimes sorely puzzled to render a reason. Descartes tells us that through life he had always felt a strong predilection for squint eyes; which he explains by the circumstance, that the nursery-maid by whom he had been kindly tended, and to whom as a child he was, consequently, much attached, had this defect."[a] 'S Gravesande, I think it

a *Origine des Connoissances Humaines,* sect. ii. ch. ix. § 80.—ED.

is, who tells us he knew a man, and a man otherwise
of sense, who had a severe fall from a waggon ; and
thereafter he could never enter a waggon without fear
and trembling, though he daily used, without apprehension, another and far more dangerous vehicle.[a] A
girl once and again sees her mother or maid fainting
and vociferating at the appearance of a mouse ; if she
has afterwards to escape from danger, she will rather
pass through flames than take a patent way, if ob-
structed by a *ridiculus mus.* A remarkable example
of the false judgments arising from this principle of
association, is recorded by Herodotus and Justin, in
reference to the war of the Scythians with their slaves.
The slaves, after they had repeatedly repulsed several
attacks with arms, were incontinently put to flight
when their masters came out against them with their
whips.[β]

· I shall now offer an observation in regard to the
appropriate remedy for this evil influence of Association.

Only re-
medy for
the influence
of Associa-
tion in the
Philosophy
of the
Human
Mind.

The only mean by which we can become aware of,
counteract, and overcome, this besetting weakness of
our nature, is Philosophy,—the Philosophy of the
Human Mind ; and this studied both in the consciousness of the individual, and in the history of the species. The philosophy of mind, as studied in the consciousness of the individual, exhibits to us the source
and nature of the illusion. It accustoms us to discriminate the casual, from the necessary, combinations
of thought ; it sharpens and corroborates our facul-

a *Introductio ad Philosophiam Lo-*
gica, c. 28. The example, however,
is given as a supposed case, not as a
fact. The two instances which follow are also from 'S Gravesande.—
ED.

β Herod., iv. 3. Justin, ii. 5.—
ED.

ties, encourages our reason to revolt against the blind
preformations of opinion, and finally enables us to
break through the enchanted circle within which Custom and Association had enclosed us. But in the
accomplishment of this end, we are greatly aided by
the study of man under the various circumstances
which have concurred in modifying his intellectual
and moral character. In the great spectacle of history, we behold in different ages and countries the
predominance of different systems of association, and
these ages and countries are, consequently, distinguished by the prevalence of different systems of
opinions. But all is not fluctuating ; and, amid the
ceaseless changes of accidental circumstances and precarious beliefs, we behold some principles ever active,
and some truths always commanding a recognition.
We thus obtain the means of discriminating, in so
far as our unassisted reason is conversant about mere
worldly concerns, between what is of universal and
necessary certainty, and what is only of local and
temporary acceptation ; and, in reference to the latter,
in witnessing the influence of an arbitrary association
in imposing the most irrational opinions on our fellow-men, our eyes are opened, and we are warned of
the danger from the same illusion to ourselves. And
as the philosophy of man affords us at once the indication and the remedy of this illusion, so the philosophy of man does this exclusively and alone. Our
irrational associations, our habits of groundless credulity and of arbitrary scepticism, find no medicine in
the study of aught beyond the domain of mind itself.
As Goethe has well observed, " Mathematics remove
no prejudice ; they cannot mitigate obstinacy, or

LECT.
XXI.

temper party-spirit;"[a] in a word, as to any moral
influence upon the mind they are absolutely null.
Hence we may well explain the aversion of Socrates
for these studies, if carried beyond a very limited
extent.

The next faculty in order is the Representative, or
Imagination proper, which consists in the greater or less
power of holding up an ideal object in the light of
consciousness. The energy of Representation, though
dependent on Retention and Reproduction, is not to
be identified with these operations. For though these
three functions (I mean, Retention, Reproduction, and
Representation), immediately suppose, and are immediately dependent on, each other, they are still manifestly discriminated as different qualities of mind, inasmuch as they stand to each other in no determinate
proportion. We find, for example, in some individuals the capacity of Retention strong, but the Reproductive and Representative faculties sluggish and
weak. In others, again, the Conservative tenacity is
feeble, but the Reproductive and Representative energies prompt and vivid; while in others the power of
Reproduction may be vigorous, but what is recalled is
never pictured in a clear and distinct consciousness.
It will be generally, indeed, admitted, that a strong retentive memory does not infer a prompt recollection;
and still more, that a strong memory and a prompt
recollection do not infer a vivid imagination. These,
therefore, though variously confounded by philosophers, we are warranted, I think, in viewing as elementary qualities of mind, which ought to be theoretically
distinguished. Limiting, therefore, the term *Imagination* to the mere Faculty of Representing in a more or

*The Repre-
sentative
Faculty,
or Imagina-
tion Proper.*

a *Werke*, xxil. p. 258. Quoted by Scheidler, *Psychologie*, p. 146.

less vivacious manner an ideal object,—this Faculty is the source of errors which I shall comprise in the following paragraph.

¶ CII. Imagination, or the Faculty of Representing with more or less vivacity a recalled object of cognition, is the source of Error, both when it is too languid and when it is too vigorous. In the former case, the object is represented obscurely and indistinctly; in the latter, the ideal representation affords the illusive appearance of a sensible presentation.

A strong imagination, that is, the power of holding up any ideal object to the mind in clear and steady colours, is a faculty necessary to the poet and to the artist; but not to them alone. It is almost equally requisite for the successful cultivation of every scientific pursuit; and though differently applied, and different in the character of its representations, it may well be doubted whether Aristotle did not possess as powerful an imagination as Homer. The vigour and perfection of this faculty are seen, not so much in the representation of individual objects and fragmentary sciences, as in the representation of systems. In the better ages of antiquity the perfection, —the beauty, of all works of taste, whether in Poetry, Eloquence, Sculpture, Painting, or Music, was principally estimated from the symmetry or proportion of all the parts to each other, and to the whole which they together constituted; and it was only in subservience to this general harmony that the beauty of the several parts was appreciated. In the criticism of modern times, on the contrary, the reverse is true;

and we are disposed to look more to the obtrusive qualities of details than to the keeping and unison of a whole. Our works of art are, in general, like kinds of assorted patch-work;—not systems of parts all subdued in conformity to one ideal totality, but coordinations of independent fragments, among which a "*purpureus pannus*" seldom comes amiss. The reason of this difference in taste seems to be, what at first sight may seem the reverse, that in antiquity not the Reason but the Imagination was the more vigorous;—that the Imagination was able to represent simultaneously a more comprehensive system; and thus the several parts being regarded and valued only as conducive to the general result,—these parts never obtained that individual importance, which would have fallen to them had they been created, and considered, only for themselves. Now this power of representing to the mind a complex system in all its bearings, is not less requisite to the philosopher than to the poet, though the representation be different in kind; and the nature of the philosophic representations, as not concrete and palpable like the poetical, supposes a more arduous operation, and, therefore, even a more vigorous faculty. But Imagination, in the one case and in the other, requires in proportion to its own power a powerful intellect; for imagination is not poetry nor philosophy, but only the condition of the one and of the other.

Errors
which arise
from the
dispropor-
tion be-
tween Ima-
gination and
Judgment.
But to speak now of the Errors which arise from the disproportion between the Imagination and the Judgment;—they originate either in the weakness, or in the inordinate strength, of the former.

In regard to the errors which arise from the imbecility of the Representative Faculty, it is not difficult

to conceive how this imbecility may become a cause
of erroneous judgment. The Elaborative Faculty, in
order to judge, requires an object,—requires certain
differences to be given. Now, if the imagination be
weak and languid, the objects represented by it will
be given in such confusion and obscurity, that their
differences are either null or evanescent, and judgment
thus rendered either impossible, or possible only with
the probability of error. In these circumstances, to
secure itself from failure, the intellect must not
attempt to rise above the actual presentations of
sense; it must not attempt any ideal analysis or
synthesis,—it must abandon all free and self-active
elaboration, and all hope of a successful cultivation
of knowledge.

Again, in regard to the opposite errors, those arising
from the disproportioned vivacity of imagination,—
these are equally apparent. In this case the renewed
or newly-modified representations make an equal impression on the mind as the original presentations,
and are, consequently, liable to be mistaken for these.
Even during the perception of real objects, a too lively
imagination mingles itself with the observation, which
it thus corrupts and falsifies. Thus arises what is
logically called the *vitium subreptionis*.[a] This is frequently seen in those pretended observations made by
theorists in support of their hypotheses, in which, if
even the possibility be left for imagination to interfere, imagination is sure to fill up all that the senses
may leave vacant. In this case the observers are at
once dupes and deceivers, in the words of Tacitus
"*Fingunt simul creduntque.*"[b]

a Krug, *Logik*, § 142. Anm.—
Ed.

β *Hist.* lib. II. c. 8. See *Lectures
on Metaphysics*, vol. I. p. 76.—Ed.

LECT.
III.

Remedies
for them
defects of
the Imagin-
ation.

In regard to the remedies for these defects of the Representative Faculty;—in the former case, the only alleviation that can be proposed for a feeble Imagination, is to animate it by the contemplation and study of those works of art which are the products of a strong Phantasy, and which tend to awaken in the student a corresponding energy of that faculty. On the other hand, a too powerful imagination is to be quelled and regulated by abstract thinking, and the study of philosophical, perhaps of mathematical, science.[a]

The faculty which next follows, is the Elaborative Faculty, Comparison, or the Faculty of Relations. This is the Understanding, in its three functions of Conception, Judgment, and Reasoning. On this faculty take the following paragraph.

Par. CIII.
8. Elabor-
ative Fa-
culty,—as
a source of
Error.

¶ CIII. The Affections and the Lower Cognitive Faculties afford the sources and occasions of Error; but it is the Elaborative Faculty, Understanding, Comparison, or Judgment, which truly errs. This faculty does not, however, err from strength or over-activity, but from inaction; and this inaction arises either from natural weakness, from want of exercise, or from the impotence of attention.[β]

Explica-
tion.
Error does
not lie in
the condi-
tions of our
Higher
Faculties,
but is per-

I formerly observed that error does not lie in the conditions of our higher faculties themselves, and that these faculties are not, by their own laws, determined to false judgments or conclusions:—

" Nam neque decipitur ratio, nec decipit unquam."[γ]

a Cf. Krug, Logik, § 156. Anm.   Fries, Logik, § 108.   Bachmann,
Eu.   Logik, § 411.]
β Krug, Logik, § 148—Ed.   [Cf.   γ See above, vol. ii. p. 77.—Ed.

If this were otherwise, all knowledge would be impossible,—the root of our nature would be a lie. "But
in the application of the laws of our higher faculties
to determinate cases, many errors are possible ; and
these errors may actually be occasioned by a variety
of circumstances. Thus it is a law of our intelligence,
that no event, no phænomenon, can be thought as
absolutely beginning to be ; we cannot but think that
all its constituent elements had a virtual existence
prior to their concurrence, to necessitate its manifestation to us ; we are thus unable to accord to it more
than a relative commencement, in other words, we
are constrained to look upon it as the effect of antecedent causes. Now though the law itself of our intelligence,—that a cause there is for every event,—be
altogether exempt from error, yet in the application
of this law to individual cases, that is, in the attribution of determinate causes to determinate effects, we
are easily liable to go wrong. For we do not know,
except from experience and induction, what particular
antecedents are the causes of particular consequents ;
and if our knowledge of this relation be imperfectly
generalised, or if we extend it by a false analogy to
cases not included within our observation, error is
the inevitable consequence. But in all this there is
no fault, no failure, of intelligence, there is only a deficiency,—a deficiency in the activity of intelligence,
while the Will determines us to a decision before the
Understanding has become fully conscious of certainty.
The defective action of the Understanding may arise
from three causes. In the first place, the faculty of
Judgment may by nature be too feeble. This is the
case in idiots and weak persons. In the second place,
though not by nature incompetent to judge, the in-

LECT.
XXX.

b. Want of
necessary
experience.
a. Incom-
petency of
attention.

tellect may be without the necessary experience,—
may not possess the grounds on which a correct judgment must be founded. In the third place, and this
is the most frequent cause of error, the failure of the
understanding is from the incompetency of that act of
will which is called *Attention*. Attention is the voluntary direction of the mind upon an object, with the
intention of fully apprehending it. The cognitive
energy is thus, as it were, concentrated upon a single
point. We, therefore, say that the mind collects itself,
when it begins to be attentive ; on the contrary, that
it is distracted, when its attention is not turned upon
an object as it ought to be. This fixing, this concentration, of the mind upon an object can only be
carried to a certain degree, and continued for a certain
time. This degree and this continuance are both dependent upon bodily circumstances ; and they are
also frequently interrupted or suspended by the intrusion of certain collateral objects, which are forced upon
the mind, either from without, by a strong and sudden
impression upon the senses, or from within, through
the influence of Association ; and these, when once
obtruded, gradually or at once divert the attention
from the original and principal object. If we are not
sufficiently attentive, or if the effort which accompanies
the concentration of the mind upon a single object
be irksome, there arise hurry and thoughtlessness in
judging, inasmuch as we judge either before we have
fully sought out the grounds on which our decision
ought to proceed, or have competently examined their
validity and effect. It is hence manifest that a multitude of errors is the inevitable consequence." [a]

[a] Krug. *Logik*, § 142. Anm. In some places slightly changed.—ED.

In regard to the Regulative Faculty, — Common Sense,—Intelligence,—νοῦς,—this is not in itself a source of error. Errors may, however, arise either from overlooking the laws or necessary principles which it does contain; or by attributing to it, as necessary and original data, what are only contingent generalisations from experience, and, consequently, make no part of its complement of native truths. But these errors, it is evident, are not to be attributed to the Regulating Faculty itself, which is only a place or source of principles, but to the imperfect operations of the Understanding and Self-consciousness, in not properly observing and sifting the phænomena which it reveals.

Besides these sources of Error, which immediately originate in the several powers and faculties of mind, there are others of a remoter origin arising from the different habits which are determined by the differences of sex,[a] of age,[β] of bodily constitution,[γ] of education, of rank, of fortune, of profession, of intellectual pursuit. Of these, however, it is impossible at present to attempt an analysis; and I shall only endeavour to afford you a few specimens, and to refer you for information in regard to the others to the best sources.

Intellectual pursuits or favourite studies, inasmuch as these determine the mind to a one-sided cultivation, that is, to the neglect of some, and to the disproportioned development of other, of its faculties, are among the most remarkable causes of error. This

Margin notes:
6. Regulative Faculty,—not properly a source of Error.

Remoter sources of Error in the different habits determined by sex, age, bodily constitution, education, &c.

Selected examples of these. A one-sided cultivation of the intellectual power.

a [See Stewart, *Elements*, vol. iii. part iii. sect. v. chap. i. *Works*, vol. iv. p. 238 *et seq.*]

β [Aristotle, *Rhet.*, L. ii. c. 12

Crousaz, *Logique*, t. i. part i. sect. i. ch. v. § 15, p. 104.]

γ [See Crousaz, *Logique*, t. i. p. i. sect. i. ch. v. p. 91 *et seq.*]

LECT.
XXX.

This ex-
emplified
in three
different
phases.
Exclusive
cultivation,
1. Of the
powers of
Observa-
tion.

partial or one-sided cultivation is exemplified in three different phases. The first of these is shown in the exclusive cultivation of the powers of Observation, to the neglect of the higher faculties of the Understanding. Of this type are your men of physical science. In this department of knowledge there is chiefly demanded a patient habit of attention to details, in order to detect phænomena, and, these discovered, their generalisation is usually so easy that there is little exercise afforded to the higher energies of Judgment and Reasoning. It was Bacon's boast that Induction, as applied to nature, would equalise all talents, level the aristocracy of genius, accomplish marvels by co-operation and method, and leave little to be done by the force of individual intellects. This boast has been fulfilled; Science has, by the Inductive Process, been brought down to minds, who previously would have been incompetent for its cultivation, and physical knowledge now usefully occupies many who would otherwise have been without any rational pursuit. But the exclusive devotion to such studies, if not combined with higher and graver speculations, tends to wean the student from the more vigorous efforts of mind, which, though unamusing and even irksome at the commencement, tend, however, to invigorate his nobler powers, and to prepare him for the final fruition of the highest happiness of his intellectual nature.

2. Of Meta-
physics.
3. Of Ma-
thematics.

Stewart
referred to.

A partial cultivation of the intellect, opposite to this, is given in the exclusive cultivation of Metaphysics and of Mathematics. On this subject I may refer you to some observations of Mr Stewart, in two chapters entitled *The Metaphysician* and *The Mathe-*

matician, in the third volume of his *Elements of the*
Philosophy of the Human Mind,—chapters distinguished equally by their candour and their depth of
observation. On this subject Mr Stewart's authority
is of the highest, inasmuch as he was distinguished in
both the departments of knowledge the tendency of
which he so well develops.

LECTURE XXXI.

MODIFIED STOICHEIOLOGY.

SECTION II.—ERROR,—ITS CAUSES AND REMEDIES.

C. LANGUAGE—D. OBJECTS OF KNOWLEDGE.

LECT.
XXXI.
_____

III. Lan-
guage,—as
a source of
Error.

IN my last Lecture I concluded the survey of the
Errors which have their origin in the conditions and
circumstances of the several Cognitive Faculties, and
I now proceed to that source of false judgment, which
lies in the imperfection of the Instrument of Thought
and Communication,—I mean language.

Has man
invented
Language?
Ambiguity
of the
question.

Much controversy has arisen in regard to the question,—Has man invented Language ? But the differences of opinion have in a great measure arisen from
the ambiguity or complexity of the terms, in which
the problem has been stated. By _language_ we may
mean either the power which man possesses of associating his thought with signs, or the particular systems of signs with which different portions of mankind have actually so associated their thoughts.

In what
sense Lan-
guage is
natural to
man.

Taking _language_ in the former sense, it is a natural
faculty, an original tendency of mind, and, in this
view, man has no more invented language than he
has invented thought. In fact, the power of thought
and the power of language are equally entitled to be
considered as elementary qualities of intelligence ; for

while they are so different that they cannot be identified, they are still so reciprocally necessary that the
one cannot exist without the other. It is true, indeed, that presentations and representations of given
individual objects might have taken place, although
there were no signs with which they were mentally
connected, and by which they could be overtly expressed; but all complex and factitious constructions
out of these given individual objects, in other words,
all notions, concepts, general ideas, or thoughts proper,
would have been impossible without an association to
certain signs, by which their scattered elements might
be combined in unity, and their vague and evanescent
existence obtain a kind of definite and fixed and
palpable reality. Speech and cogitation are thus the
relative conditions of each other's activity, and both
concur to the accomplishment of the same joint result.
The Faculty of Thinking,—the Faculty of forming
General Notions,—being given, this necessarily tends to
energy, but the energy of thinking depends upon the
coactivity of the Faculty of Speech, which itself tends
equally to energy. These faculties,—these tendencies,
—these energies, thus coexist and have always co-
existed; and the result of their combined action is
thought in language, and language in thought. So
much for the origin of Language, considered in general
as a faculty.

But, though the Faculty of Speech be natural and
necessary, that its manifestations are to a certain
extent contingent and artificial, is evident from the
simple fact, that there are more than a single language
actually spoken. It may, therefore, be asked,—Was the
first language actually spoken, the invention of man,
or an inspiration of the Deity? The latter hypothesis

Was the
first lan-
guage
actually
spoken,
the inven-
tion of man
or an in-
spiration of
the Deity?

LECT.
XXXI.

The latter
hypothesis
considered.

Difficulty
of the
question.

Language
has a gene-
ral and a
special
character.

cuts, but does not loose, the knot. It declares that
ordinary causes and the laws of nature are insufficient
to explain the phænomenon, but it does not prove
this insufficiency; it thus violates the rule of Parci-
mony, by postulating a second and hypothetical cause
to explain an effect, which it is not shown cannot be
accounted for without this violent assumption. The
first and greatest difficulty in the question is thus :—
It is necessary to think in order to invent a language,
and the invention of a language is necessary in order
to think : for we cannot think without notions, and
notions are only fixed by words.* This can only be
solved, as I have said, by the natural attraction between thought and speech,—by their secret affinity,
which is such that they suggest and, *pari passu*,
accompany each other. And in regard to the question,—Why, if speech be a natural faculty, it does not
manifest itself like other natural principles in a uniform manner,—it may be answered that the Faculty
of Speech is controlled and modified in its exercise
by external circumstances, in consequence of which,
though its exertion be natural and necessary, and,
therefore, identical in all men, the special forms of
its exertion are in a great degree conventional and
contingent, and, therefore, different among different
portions of mankind.

Considered on one side, languages are the results of
our intelligence and its immutable laws. In consequence of this, they exhibit in their progress and development resemblances and common characters which
allow us to compare and to recall them to certain

* See Rousseau, *Discours sur l'Origine de l'Inégalité parmi les Hommes*,
Première Partie. "Si les hommes
ont eu besoin de la parole pour ap-
prendre à penser, ils ont eu bien
plus besoin encore de savoir penser
pour trouver l'art de la parole."—
ED.

primitive and essential forms,—to evolve a system of Universal Grammar. Considered on another side, each language is the offspring of particular wants, of special circumstances, physical and moral, and of chance. Hence it is that every language has particular forms as it has peculiar words. Language thus bears the impress of human intelligence only in its general outlines. There is, therefore, to be found reason and philosophy in all languages, but we should be wrong in believing that reason and philosophy have, in any language, determined everything. No tongue, how per- fect soever it may appear, is a complete and perfect instrument of human thought. From its very conditions every language must be imperfect. The human memory can only compass a limited complement of words, but the data of sense, and still more the combinations of the understanding, are wholly unlimited in number. No language can, therefore, be adequate to the ends for which it exists; all are imperfect, but some are far less incompetent instruments than others.

From what has now been said, you will be prepared to find in Language one of the principal sources of Error ; but before I go on to consider the particular modes in which the Imperfections of Language are the causes of false judgments,—I shall comprise the general doctrine in the following paragraph :—

¶ CIV. As the human mind necessarily re- quires the aid of signs to elaborate, to fix, and to communicate its notions, and as Articulate Sounds are the species of signs which most effectually afford this aid, Speech is, therefore, an indispensable instrument in the higher functions of thought and knowledge. But as speech is a necessary, but not a perfect, instrument, its imperfection must react upon the mind. For the Multitude of Languages, the Difficulty of their Acquisition, their necessary Inadequacy, and the consequent Ambiguity of Words, both singly and in combination,—these are all copious sources of Illusion and Error.[a]

We have already sufficiently considered the reason why thought is dependent upon some system of signs or symbols, both for its internal perfection and external expression.[β] The analyses and syntheses,—the decompositions and compositions,—in a word, the elaborations, performed by the Understanding upon the objects presented by External Perception and Self-Consciousness, and represented by Imagination,— these operations are faint and fugitive, and would have no existence, even for the conscious mind, beyond the moment of present consciousness, were we not able to connect, to ratify, and to fix them, by giving to their parts, (which would otherwise immediately fall asunder), a permanent unity, by associating them with a sensible symbol, which we may always recall at pleasure, and which, when recalled, recalls along with it the characters which concur in constituting a notion or factitious object of intelligence. So far signs are necessary for the internal operation of thought itself.

But for the communication of thought from one mind to another, signs are equally indispensable. For in

a Krug, *Logik,* § 146.—Ed. [Cf. Ernesti, *Initia Doctrinæ Solidioris: Pars Altera; Dialectica,* c. 2, § 24. Wyttenbach, *Præcepta Phil. Log.* P. iii. c. ii. p. 98. Tittel, *Logik,* p. 292. Kirwan, *Logick,* i. 214. Fries, *System der Logik,* § 109. Caro, *Lo-* *gique,* Part. I. ch. I. art. 9, p. 121. Crousaz, Toussaint.] [Crousaz, *Logique,* t. iii. part I. sect. iii. c. 2, p. 68 et seq. Toussaint, *De la Pensée,* Cha. viii. x.—Ed.]

β See above, vol. I. p. 137.—Ed.

itself thought is known,—thought is knowable, only LECT.
XXXI.
to the thinking mind itself; and were we not enabled
to connect certain complements of thought to certain
sensible symbols, and by their means to suggest in
other minds those complements of thought of which we
were conscious in ourselves, we should never be able
to communicate to others what engaged our interest,
and man would remain for man, if an intelligence at
all, a mere isolated intelligence.

In regard to the question,—What may these sen- *Intonations*
sible symbols be, by which we are to compass such *of the voice the only*
memorable effects,—it is needless to show that mien *adequate sensible*
and gesture, which, to a certain extent, afford a kind *symbols of thought*
of natural expression, are altogether inadequate to the *and its communication.*
double purpose of thought and communication, which
it is here required to accomplish. This double purpose can be effected only by symbols, which express,
through intonations of the voice, what is passing in
the mind. These vocal intonations are either inarti- *These in-*
culate or articulate. The former are mere sounds or *articulate and articulate.*
cries, and, as such, an expression of the feelings of
which the lower animals also are capable. The latter *The latter*
constitute words, and these, as the expression of *constitute Language*
thoughts or notions, constitute Language Proper or *Proper.*
Speech." Speech, as we have said, as the instru- *How Lan-*
ment of elaborating, fixing, and communicating our *guage is a source of*
thoughts, is a principal mean of knowledge, and even *Error.*
the indispensable condition on which depends the exercise of our higher cognitive faculties. But, at the
same time, in consequence of this very dependence of
thought upon language, inasmuch as language is itself
not perfect, the understanding is not only restrained
in its operations, and its higher development, conse-

<hr>

a Cf. Krug. *Logik,* § 145. Anm.—ED.

quently, checked, but many occasions are given of
positive error. For to say nothing of the impediment presented to the free communication of thought
by the multitude of tongues into which human language is divided, in consequence of which all speech
beyond their mother-tongue is incomprehensible to
those who do not make a study of other languages,—
even the accurate learning of a single language is attended with such difficulties, that perhaps there never
yet has been found an individual who was thoroughly
acquainted with all the words and modes of verbal
combination in any single language,—his mother-
tongue even not excepted. But the circumstance of
principal importance is, that, how copious and expressive soever it may be, no language is competent adequately to denote all possible notions, and all possible
relations of notions, and from this necessary poverty
of language in all its different degrees, a certain inevitable ambiguity arises, both in the employment
of single words and of words in mutual connection.

*The ambi-
guity of
words the
principal
source of
error origi-
nating in
Language.*

As this is the principal source of the error originating in Language, it will be proper to be a little more
explicit. And here it is expedient to take into account two circumstances, which mutually affect each
other. The first is, that as the vocabulary of every
language is necessarily finite, it is necessarily disproportioned to the multiplicity, not to say infinity, of
thought; and the second, that the complement of
words in any given language has been always filled
up with terms significant of objects and relations of
the external world, before the want was experienced
of words to express the objects and relations of the
internal.

*Two cir-
cumstances
under this
head, which
mutually
affect each
other.*

From the first of these circumstances, considered

exclusively and by itself, it is manifest that one of two alternatives must take place. Either the words of a language must each designate only a single notion,—a single fasciculus of thought,—the multitude of notions not designated being allowed to perish, never obtaining more than a momentary existence in the mind of the individual; or the words of a language must each be employed to denote a plurality of concepts. In the former case, a small amount of thought would be expressed, but that precisely and without ambiguity; in the latter, a large amount of thought would be expressed, but that vaguely and equivocally. Of these alternatives, (each of which has thus its advantages and disadvantages,) the latter is the one which has universally been preferred; and, accordingly, all languages by the same word express a multitude of thoughts, more or less differing from each other. Now what is the consequence of this? It is plain that if a word has more than a single meaning attached to it, when it is employed it cannot of itself directly and peremptorily suggest any definite thought :—all that it can do is vaguely and hypothetically to suggest a variety of different notions; and we are obliged from a consideration of the context,— of the tenor,—of the general analogy, of the discourse, to surmise, with greater or less assurance, with greater or less precision, what particular bundle of characters it was intended to convey. Words, in fact, as languages are constituted, do nothing more than suggest,—are nothing more than hints; hints, likewise, which leave the principal part of the process of interpretation to be performed by the mind of the hearer. In this respect, the effect of words resembles the effect of an outline or shade of a countenance with which

The vocabulary of every Language necessarily finite. Consequences of this.

Words are merely hints to the mind.

we are familiar. In both cases, the mind is stimulated to fill up what is only hinted or pointed at.
Thus it is that the function of language is not so much
to infuse knowledge from one intelligence to another,
as to bring two minds into the same train of thinking,
and to confine them to the same track. In this procedure what is chiefly wonderful, is the rapidity with
which the mind compares the word with its correla-
tions, and, in general, without the slightest effort, decides which among its various meanings is the one
which it is here intended to convey. But how marvellous soever be the ease and velocity of this process
of selection, it cannot always be performed with equal
certainty. Words are often employed with a plurality of meanings; several of which may quadrate, or
be supposed to quadrate, with the general tenor of the
discourse. Error is thus possible ; and it is also probable, if we have any prepossession in favour of one
interpretation rather than of another. So copious a
source of error is the ambiguity of language, that a
very large proportion of human controversy has been
concerning the sense in which certain terms should
be understood ; and many disputes have even been
fiercely waged, in consequence of the disputants being
unaware that they agreed in opinion, and only differed
in the meaning they attached to the words in which
that opinion was expressed. On this subject I may
refer you to the very amusing and very instructive
treatise of Werenfelsius, entitled *De Logomachiis
Eruditorum.*

" In regard to a remedy for this description of error,
—this lies exclusively in a thorough study of the
language employed in the communication of knowledge, and in an acquaintance with the rules of Criti-

cism and Interpretation. The study of languages, when rationally pursued, is not so unimportant as many fondly conceive; for misconceptions most frequently arise solely from an ignorance of words; and every language may, in a certain sort, be viewed as a commentary upon Logic, inasmuch as every language, in like manner, mirrors in itself the laws of thought.

"In reference to the rules of Criticism and Interpretation,—these especially should be familiar to those who make a study of the writings of ancient authors, as these writings have descended to us often in a very mutilated state, and are composed in languages which are now dead. How many theological errors, for example, have arisen only because the divines were either ignorant of the principles of Criticism and Hermeneutic, or did not properly apply them! Doctrines originating in a corrupted lection, or in a figurative expression, have thus arisen and been keenly defended. Such errors are best combated by philological weapons; for these pull them up along with their roots.

"A thorough knowledge of languages in general accustoms the mind not to remain satisfied with the husk, but to penetrate in, even to the kernel. With this knowledge we shall not so easily imagine that we understand a system, when we only possess the language in which it is expressed; we shall not conceive that we truly reason, when we only employ certain empty words and formulæ; we shall not betray ourselves into unusual and obscure expressions, under which our meaning may be easily mistaken; finally, we shall not dispute with others about words, when we are in fact at one with them in regard to things." [a] So much for the errors which originate in Language.

[a] Krug, *Logik*, § 157. Anm.—Ed.

LECT.
XXXI.

IV. Source
of Error,—
the Objects
of our
knowledge.

As to the last source of Error which I enumerated, —the Objects themselves of our knowledge,—it is hardly necessary to say anything. It is evident that some matters are obscure and abstruse, while others are clear and palpable; and that, consequently, the probability of error is greater in some studies than it is in others. But as it is impossible to deliver any special rules for these cases, different from those which are given for the Acquisition of Knowledge in general, concerning which we are soon to speak, this source of error may be, therefore, passed over in silence.

We have now finished the consideration of the various Sources of Error, and—

Par. CV.
Rules
touching
the Causes
and Reme-
dies of our
False Judg-
ments.

¶ CV. The following rules may be given, as the results of the foregoing discussion, touching the Causes and Remedies of our False Judgments.

1°, Endeavour as far as possible to obtain a clear and thorough insight into the laws of the Understanding, and of the Mental Faculties in general. Study Logic and Psychology.

2°, Assiduously exercise your mind in the application of these laws. Learn to think methodically.

3°, Concentrate your attention in the act of Thinking; and principally employ the seasons when the Intellect is alert, the Passions slumbering, and no external causes of distraction at work.

4°, Carefully eliminate all foreign interests from the objects of your inquiry, and allow yourselves to be actuated by the interest of Truth alone.

5°, Contrast your various convictions, your

past and present judgments, with each other;
and admit no conclusion as certain, until it has been once and again thoroughly examined, and its correctness ascertained.

6°, Collate your own persuasions with those of others; attentively listen to and weigh, without prepossession, the judgments formed by others of the opinions which you yourselves maintain.[a]

a Cf. Krug, *Logik,* § 160. Bachmann, *Logik,* § 410.—ED.

LECTURE XXXII.

MODIFIED METHODOLOGY.

SECTION I.—OF THE ACQUISITION OF KNOWLEDGE.

I. EXPERIENCE.—A. PERSONAL :—OBSERVATION —INDUCTION AND ANALOGY.

LECT.
XXXII.

Means by which our knowledge obtains the character of Perfection, viz. the Acquisition and the Communication of Knowledge.

IN our last Lecture, having concluded the Second Department of Concrete Logic,—that which treats of the Causes of Error,—we now enter upon the Third part of Concrete or Modified Logic,—that which considers the Means by which our Knowledge obtains the character of Perfection. These means may, in general, be regarded as two,—the Acquisition and the Communication of knowledge,—and these two means we shall, accordingly, consider consecutively and apart.

The Acquisition of Knowledge.

In regard to the Acquisition of Knowledge,—we must consider this by reference to the different kinds of knowledge of which the human intellect is capable.

Human Knowledge of two kinds.

Now, human knowledge, viewed in its greatest universality, is of two kinds. For either it is one of which the objects are given as contingent phænomena ; or one in which the objects are given as necessary facts or laws. In the former case, the cognitions are called *empirical, experiential,* or *of experience;* in the latter, *pure, intuitive, rational,* or *of reason,* also of *common sense.* These two kinds of knowledge are, likewise, severally denominated *cognitions a posteriori* and *cognitions a priori.* The distinction of these two species of cognitions consists properly in this,—that the former are solely derived from the Presentations of Sense, External and Internal : whereas the latter, though first manifested on the occasion of such Presentations, are not, however, mere products of Sense ; on the contrary, they are laws, principles, forms, notions, or by whatever name they may be called, native and original to the mind, that is, founded in, or constituting the very nature of, Intelligence ; and, accordingly, out of the mind itself they must be developed, and not sought for and acquired as foreign and accidental acquisitions. As the Presentative Faculties inform us only of what exists and what happens, that is, only of facts and events,—such empirical knowledge constitutes no necessary and universal judgment ; all, in this case, is contingent and particular, for even our generalised knowledge has only a relative and precarious universality. The cognitions, on the other hand, which are given as Laws of Mind, are, at once and in themselves, universal and necessary. We cannot but think them, if we think at all. The doctrine, therefore, of the Acquisition of Knowledge, must consist of two parts,—the first treating of the acquisition of knowledge through the data of Experience, the second, of the acquisition of knowledge through the data of Intelligence.[a]

---

[a] See Esser, *Logik,* § 145.—Ed.
In regard to the acquisition of knowledge,—all knowledge may be called acquired, inasmuch as it is acquired either, 1°, By experience ; or, 2°, On occasion of experience.

LECT.
XXXII.

I. The Doctrine of Experience. Experience of two kinds.

In regard to the first of these sources, viz. Experience,—this is either our own experience or the experience of others, and in either case it is for us a mean of knowledge. It is manifest that the knowledge we acquire through our personal experience, is far superior in degree to that which we obtain through the experience of other men; inasmuch as our knowledge of an object, in the former case, is far clearer and more distinct, far more complete and lively, than in the latter; while at the same time the latter also affords us a far inferior conviction of the correctness and certainty of the cognition than the former. On the other hand, foreign is far superior to our proper experience in this,—that it is much more comprehensive, and that, without this, man would be deprived of those branches of knowledge which are to him of the most indispensable importance. Now, as the principal distinction of experience is thus into our own experience and into the experience of others, we must consider it more closely in this twofold relation.[a] First, then, of our Personal Experience.

I. Personal Experience.

Experience necessarily supposes, as its primary condition, certain presentations by the faculties of External or of Internal Perception, and is, therefore, of two kinds, according as it is conversant about the objects of the one of these faculties, or the objects of the other. But the presentation of a fact of the external or of the internal world is not at once an experience. To this there is required, a continued series of such presentations, a comparison of these together, a mental separation of the different, a mental combination of the similar, and it, therefore, over and above the operation of the Presentative Faculties,

a Esser, Logik, § 148.—Ed.

requires the co-operation of the Retentive, the Reproductive, the Representative, and the Elaborative
Faculties. In regard to Experience, as the first means
by which we acquire knowledge through the legitimate use and application of our Cognitive Faculties,
I give you the following paragraph :—

¶ CVI. The First Mean towards the Acquisition of Knowledge is *Experience* (*experientia*,
ἐμπειρία). Experience may be, rudely and generally, described as the apprehension of the phænomena of the outer world, presented by the
Faculty of External Perception, and of the
phænomena of the inner world, presented by the
Faculty of Self-consciousness:—these phænomena
being retained in Memory, ready for Reproduction and Representation, being also arranged in
order by the Understanding.

This paragraph, you will remark, affords only a
preliminary view of the general conditions of Experience. In the first place, it is evident, that without
the Presentative, or, as they may with equal propriety
be called, the Acquisitive, Faculties of Perception,
External and Internal, no experience would be possible. But these faculties, though affording the fundamental condition of knowledge, do not of themselves
make up experience. There is, moreover, required,
of the phænomena or appearances the accumulation
and retention, the reproduction and representation.
Memory, Reminiscence, and Imagination must, therefore, also co-operate. Finally, unless the phænomena
be compared together, and be arranged into classes,
according to their similarities and differences, it is

evident that no judgments,—no conclusions, can be formed concerning them; but without a judgment knowledge is impossible; and as experience is a knowledge, consequently experience is impossible. The Understanding or Elaborative Faculty must, therefore, likewise co-operate. Manilius has well expressed the nature of experience in the following lines :—

> " Per varios usus artem experientia fecit,
> Exemplo monstrante viam." [a]

And Afranius in the others :—

> " Usus me genuit, mater peperit Memoria ;
> Sophiam vocant me Graii, vos Sapientiam." [B]

" Our own observation, be it external or internal, is either with, or without, intention ; and it consists either of a series of Presentations alone, or Abstraction and Reflection supervene, so that the presentations obtain that completion and system which they do not of themselves possess. In the former case, the experience may be called an *Unlearned* or *Common ;* in the latter, a *Learned* or *Scientific Experience.*

Observa-
tion,—
what.
Of two
kinds,—
Observation
Proper, and
Experi-
ment.
Intentional and reflective experience is called *Observation.* Observation is of two kinds ; for either the objects which it considers remain unchanged, or, previous to its application, they are made to undergo certain arbitrary changes, or are placed in certain factitious relations. In the latter case, the observation obtains the specific name of *Experiment.* Observation and Experiment do not, therefore, constitute opposite or two different procedures,—the latter is, in propriety, only a certain subordinate modification of the former ; for, while observation may accomplish its end without

a l. 61.   *pue Puetarum Latinorum,* vol. ii. p.
B *Fragmentum e Sella.* Vide Cor-   1513, Lond. 1713.—Ed.

experiment, experiment without observation is impossible. Observation and experiment are manifestly exclusively competent upon the objects of our empirical knowledge; and they co-operate, equally and in like manner, to the progress of that knowledge, partly by establishing, partly by correcting, partly by amplifying it. Under observation, therefore, is not to be understood a common or unlearned experience, which obtrudes itself upon every one endowed with the ordinary faculties of Sense and Understanding, but an intentional and continued application of the faculties of Perception, combined with an abstractive and reflective attention to an object or class of objects, a more accurate knowledge of which, it is proposed, by the observation to accomplish. But in order that the observation should accomplish this end,—more especially when the objects are numerous, and a systematic complement of cognitions is the end proposed,—it is necessary that we should know certain præcognita,—1°, What we ought to observe; 2°, How we ought to observe; and 3°, By what means are the data of observation to be reduced to system. The first of these concerns the Object; the second, the Procedure; the third, the scientific Completion, of the observations. It is proper to make some general observations in regard to these, in their order; and first, of the Object of observation,—the *what* we ought to observe.

Præcognita of Observation.

"The Object of Observation can only be some given and determinate phænomenon, and this phænomenon either an external or an internal. Through observation, whether external or internal, there are four several cognitions which we propose to compass,—viz, to ascertain—1°, What the Phænomena themselves

First,—The Object of Observation.

This fourfold.

LECT.
XXXII.

are; 2°, What are the Conditions of their Reality; 3°, What are the Causes of their Existence; 4°, What is the Order of their Consecution.

1°, What the Phæ- nomena are.

"In regard to what the phænomena themselves are (*quid sint*), that is, in regard to what constitutes their peculiar nature,—this, it is evident, must be the primary matter of consideration, it being always supposed that the fact (the *an sit*) of the phænomenon itself has been established.[a] To this there is required,

In their individual peculiarities and con- trasts.

above all, a clear and distinct Presentation or Representation of the object. In order to obtain this, it behoves us to analyse,—to dismember, the constituent parts of the object, and to take into proximate account those characters which constitute the object, that is, which make it to be what it is, and nothing but what it is. This being performed, we must proceed to compare it with other objects, and with those especially which bear to it the strongest similarity, taking accurate note always of those points in which they reciprocally resemble, and in which they reciprocally disagree.

As under determinate genera and species.

"But it is not enough to consider the several phæ- nomena in their individual peculiarities and contrasts, —in what they are and in what they are not,—it is also requisite to bring them under determinate genera and species. To this end we must, having obtained (as previously prescribed) a clear and distinct knowledge of the several phænomena in their essential similarities and differences, look away or abstract from the latter,—the differences, and comprehend the former,—the similarities, in a compendious and characteristic notion, under an appropriate name.

a Better the Aristotelic questions,   *τ*φ*ysics*, vol. i. p. 58.—En.]
—*An Sit*, &c. [See *Lectures on Me-*

"When the distinctive peculiarities of the phænomena have been thus definitively recognised, the
second question emerges,—What are the Conditions
of their Reality. These conditions are commonly
called *Requisites,* and under *requisite* we must understand all that must have preceded, before the
phænomena could follow. In order to discover the
requisites, we take a number of analogous cases, or
cases similar in kind, and inquire what are the circumstances under which the phænomenon always
arises, if it does arise, and what are the circumstances
under which it never arises; and then, after a competent observation of individual cases, we construct
the general judgment, that the phænomenon never
occurs unless this or that other phænomenon has preceded, or at least accompanied, it. Here, however, it
must be noticed, that nothing can be viewed as a requisite which admits of any, even the smallest, exception.

"The requisite conditions being discovered, the
third question arises,—What are the Causes of the
Phænomena. According to the current doctrine, the
causes of phænomena are not to be confounded with
their *requisites;* for although a phænomenon no more
occurs without its requisite than without its cause,
still, the requisite being given, the phænomenon does
not necessarily follow, and, indeed, very frequently
does not ensue. On the contrary, if the cause occurs,
the phænomenon must occur also. In other words,
the requisite or condition is that without which the
phænomenon never is; the cause, on the other hand,
is that through which it always is. Thus an emotion
of pity never arises without a knowledge of the misfortune of another; but so little does this knowledge
necessitate that emotion, that its opposite, a feeling

of rejoicing, complacency, at such suffering may ensue;
whereas the knowledge of another's misfortune must
be followed by a sentiment of pity, if we are predis-
posed in favour of the person to whom the misfortune
has occurred. In this view, the knowledge of another's
misfortune is only a requisite; whereas our favourable predisposition constitutes the cause. It must,
however, be admitted, that in different relations one
and the same circumstance may be both requisite and
cause;"[a] and, in point of fact, it would be more correct to consider the cause as the whole sum of antecedents, without which the phænomenon never does
take place, and with which it always must. What
are commonly called *requisites*, are thus, in truth, only
partial causes; what are called *causes*, only proximate
requisites.

4°. What
the Order of
their Conse-
cution.
"In the fourth place, having ascertained the essential qualities—the Conditions and the Causes of phænomena—a final question emerges,—What is the Order
in which they are manifested? and this being ascertained, the observation has accomplished its end. This
question applies either to a phænomenon considered
in itself, or to a phænomenon considered in relation
to others. In relation to itself the question concerns
only the time of its origin, of its continuance, and of
its termination; in relation to others, it concerns the
reciprocal consecution in which the several phænomena appear."[b]

Second,—
The Manner
of Observa-
tion.
"We now go on to the Second Præcognitum,—the
Manner of Observation,—How we are to observe.
What we have hitherto spoken of,—the Object,—can
be known only in one way,—the way of Scientific
Observation. It, therefore, remains to be asked,—

a Esser, *Logik*, § 148.—Ed.　　　　　b Ibid.

How must the observation be instituted, so as to
afford us a satisfactory result in regard to all the four
sides on which it behoves an object to be observed? *1°. Proper state of the observing mind.*
In the first place, as preliminary to observation, it is
required that the observing mind be tranquil and
composed, be exempt from prejudice, partiality, and
prepossession, and be actuated by no other interest
than the discovery of truth.   Tranquillity and com-
posure of mind are of peculiar importance in our observation of the phænomena of the internal world ;
for these phænomena are not, like those of the external, perceptible by sense, enclosed in space, continuous and divisible ; and they follow each other in such
numbers, and with such a rapidity, that they are at
best observable with difficulty, often losing even their
existence by the interference of the observing, the reflective, energy itself.   But that the observation should
be always conducted in the calm and collected state
of mind required to purify this condition, we must
be careful to obtain, more and more, a mastery over
the Attention, so as to turn it with full force upon a
single aspect of the phænomena, and, consequently,
to abstract it altogether from every other.   Its proper
function is to contemplate the objects of observation
tranquilly, continuously, and without anxiety for the
result ; and this, likewise, without too intense an
activity or too vigorous an application of its forces.
But the observation and concomitant energy of atten- *2°. Conditions of the question to be determined by the observation.*
tion will be without result, unless we previously well
consider what precise object or objects we are now to
observe.   Nor will our experience obtain an answer to
the question proposed for it to solve, unless that question be of such a nature as will animate the observing
faculties by some stimulus, and give them a determi-

nate direction. Where this is not the case, attention
does not effect anything, nay, it does not operate at
all. On this account such psychological questions
as the following:—What takes place in the process
of Self-consciousness,—of Perception,—of Vision,—of
Hearing,—of Imagination, &c.?—cannot be answered,
as thus absolutely stated, that is, without reference to
some determinate object. But if I propose the problem,—What takes place when I see this or that
object, or better still, when I see this table?—the
attention is stimulated and directed, and even a child
can give responses, which, if properly illustrated and
explained, will afford a solution to the problem. If,
therefore, the question upon the object of observation
be too vague and general, so that the attention is not
suitably excited and applied,—this question must be
divided and subdivided into others more particular;
and this process must be continued until we reach a
question which affords the requisite conditions. We
should, therefore, determine as closely as possible the
object itself, and the phases in which we wish to observe it, separate from it all foreign or adventitious
parts, resolve every question into its constituent elements, enunciate each of these as specially as possible,
and never couch it in vague and general expressions.
But here we must at the same time take care, that the
object be not so torn and mangled, that the attention
feels no longer any attraction to the several parts, or
that the several parts can no longer be viewed in their
natural connection. So much it is possible to say, in
general, touching the Manner in which observation
ought to be carried on; what may further be added
under this head, depends upon the particular nature
of the objects to be observed."[a]

a Esser, Logik, § 149.—Ed.

" In this manner, then, must we proceed, until all
has been accomplished which the problem, to be answered by the observation, pointed out. When the
observation is concluded, an accurate record or notation of what has been observed is of use, in order to
enable us to supply what is found wanting in our subsequent observation. If we have accumulated a considerable apparatus of results, in relation to the object
we observe, it is proper to take a survey of these :
from what is found defective, new questions must be
evolved ; and an answer to these sought out through
new observations. When the inquiry has attained
its issue, a tabular view of all the observations made
upon the subject is convenient, to afford a conspectus
of the whole, and as an aid to the memory. But how
(and this is the Third Precognition) individual observations are to be built up into a systematic whole,
is to be sought for partly from the nature of science
in general, partly from the nature of the particular
empirical science for the constitution of which the
observation is applied. Nor is what is thus sought
difficult to find. It is at once evident, that a synthetic arrangement is least applicable in the empirical
sciences. For, anterior to observation, the object is
absolutely unknown ; and it is only through observation that it becomes a matter of science. We can,
therefore, go to work only in a problematic or interrogative manner, and it is impossible to commence by
assertory propositions, of which we afterwards lead
the demonstration. We must, therefore, determine
the object on all sides, in so far as observation is competent to this ; we must analyse every question into
its subordinate questions, and each of these must find
its answer in observation. The systematic order is
thus given naturally and of itself ; and in this pro-

Third,—
The manner
by which
the data of
Observation
are to be
reduced to
System.

cedure it is impossible that it should not be given.
But for a comprehensive and all-sided system of empirical knowledge, it is not sufficient to possess the
whole data of observation, to have collected these together, and to have arranged them according to some
external principle; it is, likewise, requisite that we
have a thorough-going principle of explanation, even
though this explanation be impossible in the way of
observation, and a power of judging of the data, according to universal laws, although these universal
laws may not be discovered by experience alone.
These two ends are accomplished by different means.
The former we compass by the aid of Hypothesis, the
latter, by the aid of Induction and Analogy." [a]   Of
these in detail.   In regard to Hypothesis, I give you
the following paragraph :—

¶ CVII. When a phænomenon is presented,
which can be explained by no principle afforded
through Experience, we feel discontented and uneasy, and there arises an effort to discover some
cause which may, at least provisorily, account
for the outstanding phænomenon; and this cause
is finally recognised as valid and true, if, through
it, the given phænomenon is found to obtain a
full and perfect explanation.   The judgment in
which a phænomenon is referred to such a problematic cause, is called an *Hypothesis.* [b]

Hypotheses have thus no other end than to satisfy
the desire of the mind to reduce the objects of its
knowledge to unity and system; and they do this in

a Esser, *Logik,* 150.—ED.   tures on *Metaphysics,* vol. i. p. 168
β Esser, *Logik,* § 181. Cf. *Lec-*   *et seq.*—ED.

recalling them, *ad interim*, to some principle, through LECT.
XXXII. which the mind is enabled to comprehend them. From this view of their nature, it is manifest how far they are permissible, and how far they are even useful and expedient; throwing altogether out of account the possibility, that what is at first assumed as hypothetical, may, subsequently, be proved true.

When our experience has revealed to us a certain correspondence among a number of objects, we are determined, by an original principle of our nature, to suppose the existence of a more extensive correspondence than our observation has already proved, or may ever be able to establish. This tendency to generalise our knowledge by the judgment,—that where much has been found accordant, all will be found accordant, —is not properly a conclusion deduced from premises, but an original principle of our nature, which we may call that of *Logical*, or perhaps better, that of *Philosophical, Presumption.* This Presumption is of two kinds; it is either Induction or Analogy, which, though usually confounded, are, however, to be carefully distinguished. I shall commence the consideration of these by the following paragraph :—

¶ CVIII. If we have uniformly observed, that Par. CVIII.
Induction
and Ana-
logy. a number of objects of the same class (genus or species) possess in common a certain attribute, we are disposed to conclude that this attribute is possessed by all the objects of that class. This conclusion is properly called an *Inference of Induction.* Again, if we have observed that two or more things agree in several internal and essential characters, we are disposed to conclude that they agree, likewise, in all other essential

LECT.
XXXII.

characters, that is, that they are constituents of the same class (genus or species). This conclusion is properly called an *Inference of Analogy.* The principle by which, in either case, we are disposed to extend our inferences beyond the limits of our experience, is a natural or ultimate principle of intelligence; and may be called the principle of *Logical,* or, more properly, of *Philosophical, Presumption.*[a]

Explica-
tion.
Induction
and Ana-
logy,—their
agreement
and differ-
ence.

"The reasoning by Induction and the reasoning by Analogy have this in common, that they both conclude from something observed to something not observed; from something within to something beyond the sphere of actual experience. They differ, however, in this, that, in Induction, that which is observed and from which the inference is drawn to that which is not observed, is a unity in plurality; whereas, in Analogy, it is a plurality in unity. In other words, in Induction, we look to the one in the many; in Analogy, we look to the many in the one: and while in both we conclude to the unity in totality, we do this, in Induction, from the recognised unity in plurality, in Analogy, from the recognised plurality in unity. Thus, as induction rests upon the principle, that what belongs, (or does not belong,) to many things of the same kind, belongs, (or does not belong,) to all things of the same kind; so analogy rests upon the principle, that things which have many observed attributes in common, have other not observed attributes in common likewise."[b] It is hardly necessary to remark that we are now speaking

a Cf. Esser, *Logik,* §§ 140, 152; Krug, *Logik,* §§ 166, 167, 168.—Ed. [Wolf, *Phil. Rationalis,* § 479; Reusch, *Systema Logicum.* §§ 572, 573; Nunnesius, *De Constitutione Artis Dialecticæ,* p. 120.]

b Esser, *Logik,* § 152.—Ed.

of Induction and Analogy, not as principles of Pure
Logic, and as necessitated by the fundamental laws
of thought, but of these as means of acquiring knowledge, and as legitimated by the conditions of objective
reality. In Pure Logic, Analogy has no place, and
only that induction is admitted, in which all the
several parts are supposed to legitimate the inference
to the whole. Applied Induction, on the contrary,
rests on the constancy,—the uniformity, of nature,
and on the instinctive expectation we have of this
stability. This constitutes what has been called the
principle of *Logical Presumption*, though perhaps it
might, with greater propriety, be called the principle
of *Philosophical Presumption*. We shall now consider these severally ; and, first, of Induction.

An Induction is the enumeration of the parts, in
order to legitimate a judgment in regard to the
whole.[a] Now, the parts may be either individuals or
particulars strictly so called. I say strictly so called,
for you are aware that the term *particular* is very
commonly employed, not only to denote the species, as
contained under a genus, but, likewise, to denote the
individual, as contained under a species. Using, however, the two terms in their proper significations, I
say, if the parts are individual or singular things, the
induction is then called *Individual*; whereas if the
parts be species or subaltern genera, the induction
then obtains the name of *Special*. An example of
the Individual Induction is given, were we to argue
thus,—*Mercury, Venus, the Earth, Mars, &c., are
bodies in themselves opaque, and which borrow their*

*Induction,
—what.*

*Of two
kinds,—
Individual
and Special.*

a [Cf. *Abu Ali (Avicenna), Viri
Docti, De Logica Forma*, l. 190. (In
Schmulders, *Documenta Philosophiæ
Arabum*, p. 38.] Bonne, 1836. Za-
barella, *Opera Logica, De Natura
Logica*, l. l. c. 18, p. 45.]

LECT. XXXII.

light from the sun. But Mercury, Venus, &c., are planets. Therefore all planets are opaque, and borrow their light from the sun. An example of the Special is given, were we to argue as follows,—*Quadrupeds, birds, fishes, the amphibia, &c., all have a nervous system. But quadrupeds, birds, &c., are animals. Therefore all animals,* (though it is not yet detected in some,) *have a nervous system.* Now, here it is manifest that Special rests upon Individual induction, and that, in the last result, all induction is individual. For we can assert nothing concerning species, unless what we assert of them has been previously observed in their constituent singulars.[a]

Two conditions of legitimate Induction,—First.

For a legitimate Induction there are requisite at least two conditions. In the first place, it is necessary, That the partial (and this word I use as including both the terms *individual* and *particular*),—I say, it is necessary that the partial judgments out of which the total or general judgment is inferred, be all of the same quality.[b] For if one even of the partial judgments had an opposite quality, the whole induction would be subverted. Hence it is that we refute universal judgments founded on an imperfect induction, by bringing what is called an instance (*instantia*), that is, by adducing a thing belonging to the same class or notion, in reference to which the opposite holds true. For example, the general assertion, *All dogs bark*, is refuted by the instance of the dogs of Labrador or California (I forget which), these do not bark. In like manner, the general assertion, *No quadruped is oviparous*, is refuted by the instance of the *Ornithorhynchus Paradoxus.* But that the universal judgment must have the same quality as the partial,

a Krug, Logik, § 187. Anm.—Ed.   b Esser, Logik, § 152.—Ed.

in self-evident ; for this judgment is simply the asser- LECT. XXXII.
tion of something to be true of all which is true of many.

The second condition required is, That a competent
number of the partial objects from which the induction departs should have been observed, for otherwise
the comprehension of other objects under the total
judgment would bo rash.[a] What is the number of
such objects, which amounts to a competent induction, it is not possible to say in general. In some
cases, the observation of a very few particular or individual examples is sufficient to warrant an assertion in
regard to the whole class; in others, the total judgment
is hardly competent, until our observation has gone
through each of its constituent parts. This distinction is founded on the difference of essential and unessential characters. If the character be essential to
the several objects, a comparatively limited observation is necessary to legitimate our general conclusion.
For example, it would require a far less induction to
prove that all animals breathe, than to prove that
the mammalia, and the mammalia alone, have lateral
lobes to the cerebellum. For the one is seen to bo a
function necessary to animal life ; the other, as far as
our present knowledge reaches, appears only as an
arbitrary concomitant. The difference of essential
and accidental is, however, one itself founded on induction, and varies according to the greater or less
perfection to which this has been carried. In the progress of science, the lateral lobes of the cerebellum
may appear to future physiologists as necessary a
condition of the function of suckling their young, as
the organs of breathing appear to us of circulation
and of life.

a Essay, Logik, § 132.—ED.

LECT.
XXXII.

Summary
of the
doctrine of
Induction.

To sum up the Doctrine of Induction,—"This is more certain, 1°, In proportion to the number and diversity of the objects observed; 2°, In proportion to the accuracy with which the observation and comparison have been conducted; 3°, In proportion as the agreement of the objects is clear and precise; and, 4°, In proportion as it has been thoroughly explored, whether there exist exceptions or not." [a]

Almost all induction is, however, necessarily imperfect; and Logic can inculcate nothing more important on the investigators of nature than that sobriety of mind, which regards all its past observations as only hypothetically true, as only relatively complete, and which, consequently, holds the mind open to every new observation, which may correct and limit its former judgments.

So much for Induction; now for Analogy. Analogy, in general, means proportion, or a similarity of relations. Thus, to judge analogically or according to analogy, is to judge things by the similarity of their relations. Thus, when we judge that as two is to four, so is eight to sixteen, we judge that they are analogically identical; that is, though the sums in other respects are different, they agree in this, that as two is the half of four, so eight is the half of sixteen.

In common language, however, this propriety of the term is not preserved. For *by analogy* is not always meant merely *by proportion*, but frequently *by comparison—by relation*, or simply *by similarity*. In so far as Analogy constitutes a particular kind of reasoning from the individual or particular to the universal, it signifies an inference from the partial

a Esser, *Logik*, § 152.—Ed.

similarity of two or more things to their complete or
total similarity. For example,—*This disease corresponds in many symptoms with those we have observed
in typhus fevers; it will, therefore, correspond in all,
that is, it is a typhus fever.*[a]

Like Induction, Analogy has two essential requisites. In the first place, it is necessary that of two
or more things a certain number of attributes should
have been observed, in order to ground the inference
that they also agree in those other attributes, which
it has not yet been ascertained that they possess. It
is evident that in proportion to the number of points
observed, in which the things compared together coincide, in the same proportion can it be with safety assumed, that there exists a common principle in these
things, on which depends the similarity in the points
known as in the points unknown.

In the second place, it is required that the predicates already observed should neither be all negative
nor all contingent; but that some at least should be
positive and necessary. Mere negative characters
denote only what the thing is not; and contingent
characters need not be present in the thing at all. In
regard to negative attributes, the inference, that two
things, to which a number of qualities do not belong,
and which are, consequently, similar to each other only
in a negative point of view,—that these things are,
therefore, absolutely and positively similar, is highly
improbable. But that the judgment in reference to
the compared things (say A and X) must be of the
same quality (i.e. either both affirmative or both nega-

a Cf. Krug, *Logik*, § 168. Anm.—
Ed. [Condillac, *L'Art de Raisonner*,
L. iv. ch. 3, p. 159. Avicenna, (in
Schmölders, *Documenta Phil. Ara-
bum*, p. 30.) Whately, *Rhetoric*, p.
74.]

tive), is self-evident. For if it be said A *is* B, X *is not* B, A *is not* C, X *is* C; their harmony or similarity is subverted, and we should rather be warranted in arguing their discord and dissimilarity in other points. And here it is to be noticed that Analogy differs from Induction in this, that it is not limited to one quality, but that it admits of a mixture of both.

In regard to contingent attributes, it is equally manifest that the analogy cannot proceed exclusively upon them. For, if two things coincide in certain accidental attributes, (for example, two men in respect of stature, age, and dress,) the supposition that there is a common principle, and a general similarity founded thereon, is very unlikely.

Summary of the doctrine of Analogy.
To conclude : Analogy is certain in proportion, 1°, To the number of congruent observations ; 2°, To the number of congruent characters observed ; 3°, To the importance of these characters and their essentiality to the objects; and, 4°, To the certainty that the characters really belong to the objects, and that a partial correspondence exists.[a] Like Induction, Analogy can only pretend at best to a high degree of probability ; it may have a high degree of certainty, but it never reaches to necessity.

Induction and Analogy compared together.
Comparing these two processes together :—" The Analogical is distinguished from the Inductive in this —that Induction regards a single predicate in many subjects as the attribute z in A, in B, in C, in D, in E, in F, &c.; and as these many belong to one class, say Q; it is inferred that z will, likewise, be met with in the other things belonging to this class, that is, in all Qs. On the other hand, Analogy re-

a *Esser, Logik*, § 152. Cf. *Krug, Logik*, § 168. Ann.—Ed.

gards many attributes in one subject (say m, n, o, p, in A); and as these many are in part found in another subject (say m, and n, in B), it is concluded that, in that second thing, there will also be found the other attributes (say o and p). Through Induction we, therefore, endeavour to prove that one character belongs, (or does not belong,) to all the things of a certain class, because it belongs, (or does not belong,) to many things of that class. Through Analogy, on the other hand, we seek to prove that all the characters of a thing belong, (or do not belong,) to another or several others, because many of these characters belong to this other or these others. In the one it is proclaimed,—*One in many, therefore one in all*—In the other it is proclaimed,—*Many in one, therefore all in one.*" [a]

" By these processes of Induction and Analogy, as observed, we are unable to attain absolute certainty; —a great probability is all that we can reach, and this for the simple reason, that it is impossible, under any condition, to infer the unobserved from the observed,—the whole from any proportion of the parts,—in the way of any rational necessity. Even from the requisites of Induction and Analogy, it is manifest that they bear the stamp of uncertainty; inasmuch as they are unable to determine how many objects or how many characters must be observed, in order to draw the conclusion that the case is the same with all the other objects, or with all the other characters. It is possible only in one way to raise Induction and Analogy from mere probability to complete certainty,—viz. to demonstrate that the principles which lie at the root of these processes, and which

a Krug, *Logik,* § 108. Anm.—ED.

we have already stated, are either necessary laws of thought, or necessary laws of nature. To demonstrate that they are necessary laws of thought is impossible; for Logic not only does not allow inference from many to all, but expressly rejects it. Again, to demonstrate that they are necessary laws of nature is equally impossible. This has indeed been attempted, from the uniformity of nature, but in vain. For it is incompetent to evince the necessity of the inference of Induction and Analogy from the fact denominated *the law of nature;* seeing that this law itself can only be discovered by the way of Induction and Analogy. In this attempted demonstration there is thus the most glaring *petitio principii.* The result which has been previously given remains, therefore, intact :— Induction and Analogy guarantee no perfect certainty, but only a high degree of probability, while all probability rests at best upon Induction and Analogy, and nothing else." [a]

a Esser, *Logik,* § 152.—Ed. [On history and doctrine of the Logic of Probabilities, see Leibnitz, *Nouveaux Essais,* L. iv. ch. xv. p. 425, ed. Raspe. Wolf, *Phil. Rat.* § 564 et seq. Platner, *Phil. Aphorismen,* § 701 (old edit.) § 504 (new edit.) Zedler, *Lexikon,* v. *Wahrscheinlich.* Walch, *Lexikon, Ibid.* Lambert, *Neues Organon,* ii. p. 318 et seq. Reusch, *Systema Logicum,* § 653 et seq. Hollmann, *Logica,* § 215 et seq.

Hoffbauer, *Anfangsgründe der Logik,* § 422 et seq. Bolzano, *Logik,* vol. ii. § 161, vol. iii. § 317. Bachmann, *Logik,* § 229 et seq. Fries, *Logik,* § 96 et seq. Prevost, *Essais de Philosophie,* ii. L. I. part iii. p. 60. Kant, *Logik,* Einleitung x. Jacob, *Grundriss der Allgemeinen Logik,* § 358, p. 131 et seq., 1800, Halle. Metz, *Institutiones Logicæ,* § 230 et seq., p. 171, 1700.]

LECTURE XXXIII.

MODIFIED METHODOLOGY.

SECTION I.—OF THE ACQUISITION OF KNOWLEDGE.

I. EXPERIENCE.—B. FOREIGN :—ORAL TESTIMONY —ITS CREDIBILITY.

HAVING, in our last Lecture, terminated the Doctrine of Empirical Knowledge, considered as obtained Immediately—that is, through the exercise of our own powers of Observation,—we are now to enter on the doctrine of Empirical Knowledge, considered as obtained Mediately—that is, through the Experience of Other Men. The following paragraph will afford you a general notion of the nature and kinds of this knowledge :—

¶ CIX. A matter of Observation or Empirical Knowledge can only be obtained Mediately, that is, by one individual from another, through an enouncement declaring it to be true. This enouncement is called, in the most extensive sense of the word, a *Witnessing* or *Testimony* (*testimonium*); and the person by whom it is made is, in the same sense, called a *Witness* or *Testifier* (*testis*). The object of the Testimony is called the *Fact* (*factum*) ; and its validity constitutes what is styled *Historical Credibility* (*credibilitas historica*). To estimate this credibility, it is requisite to consider—1°, The Subjective Trustworthiness of the Witnesses (*fides testium*), and 2°, The Objective Probability of the Fact itself. The former is founded partly on the Sincerity, and partly on the Competence, of the Witness. The latter depends on the Absolute and Relative Possibility of the Fact itself. Testimony is either Immediate or Mediate. Immediate, where the fact reported is the object of a Personal Experience ; Mediate, where the fact reported is the object of a Foreign Experience.*

"It is manifest that Foreign Experience, or the experience of other men, is astricted to the same laws, and its certainty measured by the same criteria, as the experience we carry through ourselves. But the experience of the individual is limited, when compared with the experience of the species ; and if men did not possess the means of communicating to each other the results of their several observations,—were they unable to co-operate in accumulating a stock of knowledge, and in carrying on the progress of discovery,—they would never have risen above the very lowest steps in the acquisition of science. But to this mutual communication they are competent ; and each individual is thus able to appropriate to his own benefit the experience of his fellow-men, and to confer on them in return the advantages which his own observations may supply. But it is evident that this

* Krug, *Logik*, § 172.—Ed. [Cf. Scheibler, *Topica*, c. 31.]

reciprocal communication of their respective experi-
ences among men, can only be effected inasmuch as
one is able to inform another of what he has himself
observed, and that the vehicle of this information can
only be some enouncement in conventional signs of
one character or another.   The enouncement of what
has been observed is, as stated in the paragraph,
called *a witnessing, a bearing witness, a testimony,*
&c., these terms being employed in their wider acceptation ; and he by whom this declaration is made,
and on whose veracity it rests, is called a *witness,*
voucher, or *testifier (testis)."* [*]   The term *testimony,* I
may notice, is sometimes, by an abusive metonym,
employed for *witness ;* and the word *evidence* is often
ambiguously used for *testimony,* and for *the bearer of*
testimony,—the witness.

" Such an enouncement,—such a testimony, is, however, necessary for others, only when the experience
which it communicates is beyond the compass of their
own observation.   Hence it follows, that matters of
reasoning are not proper objects of testimony, since
matters of reasoning, as such, neither can rest nor
ought to rest on the observations of others ; for a
proof of their certainty is equally competent to all,
and may by all be obtained in the manner in which it
was originally obtained by those who may bear witness to their truth.   And hence it further follows, that
matters of experience alone are proper objects of testimony ; and of matters of experience themselves, such
only as are beyond the sphere of our personal experience.   Testimony, in the strictest sense of the term,
therefore, is the communication of an experience, or,

* Esser, Logik, § 163.—Ed.

what amounts to the same thing, the report of an
observed phænomenon, made to those whose own
experience or observation has not reached so far.

"The object of testimony, as stated in the paragraph, is called the *fact;* the validity of a testimony
is called *historical credibility.* The testimony is either
immediate or mediate. Immediate, when the witness
has himself observed the fact to which he testifies;
mediate, when the witness has not himself had experience of this fact, but has received it on the testimony
of others. The former, the immediate witness, is
commonly styled an *eye-witness (testis oculatus)* ; and
the latter, the mediate witness, an *ear-witness (testis
auritus).* The superiority of immediate to mediate
testimony is expressed by Plautus, 'Pluris est oculatus
testis unus, quam auriti decem.'[a] These denominations,
eye and *ear witness,* are, however, as synonyms of *immediate* and *mediate witness,* not always either applicable or correct. The person on whose testimony a
fact is mediately reported, is called the *guarantee,* or
he on whose authority it rests; and the guarantee
himself may be again either an immediate or a mediate witness. In the latter case he is called a *second-
hand* or *intermediate witness;* and his testimony is
Testimonies
— Partial,
Complete,
Consistent,
Contradic-
tory.
commonly styled *hearsay evidence.* Further, Testimony, whether immediate or mediate, is either *partial*
or *complete;* either *consistent* or *contradictory.* These
distinctions require no comment. Finally, testimony
is either *direct* or *indirect.* Direct, when the witness
has no motive but that of making known the fact;
indirect, when he is actuated to this by other ends."[b]

The only question in reference to Testimony is that which regards its Credibility; and the question concerning the credibility of the witness may be comprehended under that touching the Credibility of Testimony. The order I shall follow in the subsequent observations is this,—I shall, in the first place, consider the Credibility of Testimony in general; and, in the second, consider the Credibility of Testimony in its particular forms of Immediate and Mediate.

First, then, in regard to the Credibility of Testimony in general:—When we inquire whether a certain testimony is, or is not, deserving of credit, there are two things to be considered: 1°, The Object of the Testimony, that is, the fact or facts for the truth of which the testimony vouches; and, 2°, The Subject of the Testimony, that is, the person or persons by whom the testimony is borne. The question, therefore, concerning the Credibility of Testimony, thus naturally subdivides itself into two. Of these questions, the first asks, — What are the conditions of the credibility of a testimony by reference to what is testified, that is, in relation to the Object of the testimony? The second asks,—What are the conditions of the credibility of a testimony by reference to him who testifies, that is, in relation to the Subject of the testimony?[a] Of these in their order.

On the first question.—"In regard to the matter testified, that is, in regard to the object of the testimony, it is, first of all, a requisite condition, that what is reported to be true should be possible, both absolutely, or as an object of the Elaborative Faculty, and relatively, or as an object of the Presentative

a Cf. Esser, Logik, § 154. —Ed.

Marginal notes:

LECT. XXXIII.

Division of the subject: I. Credibility of Testimony in general. II. Credibility of Testimony in its particular forms of Immediate and Mediate.

I. Credibility of Testimony in general. 1°, The Object of the Testimony. Its Absolute Possibility.

Faculties,—Perception, External or Internal. A thing
is possible absolutely, or in itself, when it can be construed to thought, that is, when it is not inconsistent
with the logical laws of thinking ; a thing is relatively
possible as an object of Perception, External or Internal, when it can affect Sense or Self-consciousness,
and, through such affection, determine its apprehension by one or other of these faculties. A testimony
is, therefore, to be unconditionally rejected, if the fact
which it reports be either in itself impossible, or impossible as an object of the Presentative Faculties.

But the impossibility of a thing, as an object of these
faculties, must be decided either upon physical, or
upon metaphysical, principles. A thing is physically
impossible as an object of sense, when the existence
itself, or its perception by us, is, by the laws of the
material world, impossible. It is metaphysically impossible, when the object itself, or its perception, is possible neither through a natural, nor through a super-
natural, agency. But, to establish the physical impossibility of a thing, it is not sufficient that its existence
cannot be explained by the ordinary laws of nature,
or even that its existence should appear repugnant
with these laws ; it is requisite that an universal and
immutable law of nature should have been demonstrated to exist, and that this law would be subverted
if the fact in question were admitted to be physically
possible. In like manner, to constitute the metaphysical impossibility of a thing, it is by no means enough
to show that it is not explicable on natural laws, or even
that any natural law stands opposed to it ; it is further
requisite to prove that the intervention even of super-
natural agency is incompetent to its production, that

its existence would involve the violation of some necessary principle of reason.

" To establish the credibility of a testimony, in so far as this is regulated by the nature of its object, there is, besides the proof of the absolute possibility of this object, required also a proof of its relative possibility ; that is, there must not only be no contradiction between its necessary attributes,—the attributes by which it must be thought,—but no contradiction between the attributes actually assigned to it by the testimony. A testimony, therefore, which, *qua* testimony, is self-contradictory, can lay no claim to credibility ; for what is self-contradictory is logically suicidal. And here the only question is,—Does the testimony, *qua* testimony, contradict itself ? for if the repugnancy arise from an opinion of the witness, apart from which the testimony as such would still stand undisproved, in that case the testimony is not at once to be repudiated as false. For example, it would be wrong to reject a testimony to the existence of a thing, because the witness had to his evidence of its observed reality annexed some conjecture in regard to its origin or cause. For the latter might well be shown to be absurd, and yet the former would remain unshaken. It is, therefore, always to be observed, that it is only the self-contradiction of a testimony, *qua* testimony, that is, the self-contradiction of the fact itself, which is peremptorily and irrevocably subversive of its credibility.

" We now proceed to the second question ; that is, to consider in general the Credibility of a Testimony by reference to its Subject, that is, in relation to the Personal Trustworthiness of the Witness. The trust-

LECT.
XXXIII.

of the
Witness.
This con-
sists of two
elements:—
a. Honesty
or Veracity.

worthiness of a witness consists of two elements or conditions. In the first place, he must be willing, in the second place, he must be able, to report the truth. The first of these elements is the Honesty,—the Sincerity,—the Veracity; the second is the Competency of the witness. Both are equally necessary, and if one or other be deficient, the testimony becomes altogether null. These constituents, likewise, do not infer each other; for it frequently happens that where the honesty is greatest the competency is least, and where the competency is greatest the honesty is least. But when the veracity of a witness is established, there is established also a presumption of his competency; for an honest man will not bear evidence to a point in regard to which his recollection is not precise, or to the observation of which he had not accorded the requisite attention. In truth, when a fact depends on the testimony of a single witness, the competency of that witness is solely guaranteed by his honesty. In regard to the honesty of a witness,—this, though often admitting of the highest probability, never admits of absolute certainty; for, though, in many cases, we may know enough of the general character of the witness to rely with perfect confidence on his veracity, in no case can we look into the heart, and observe the influence which motives have actually had upon his volitions. We are, however, compelled, in many of the most important concerns of our existence, to depend on the testimony, and, consequently, to confide in the sincerity, of others. But from the moral constitution of human nature, we are warranted in presuming on the honesty of a witness; and this presumption is enhanced in proportion as the following circumstances concur in its confirmation. In the first place, a witness is to be presumed veracious, in this case, in proportion as his love of truth is already established from others. In the second place, a witness is to be presumed veracious, in proportion as he has fewer and weaker motives to falsify his testimony. In the third place, a witness is to be presumed veracious, in proportion to the likelihood of contradiction which his testimony would encounter, if he deviated from the truth. So much for the Sincerity, Honesty, or Veracity of a witness.

"In regard to the Competency or Ability of a witness,—this, in general, depends on the supposition, that he has had it in his power correctly to observe the fact to which he testifies, and correctly to report it. The presumption in favour of the competence of a witness rises, in proportion as the following conditions are fulfilled :—In the first place, he must be presumed competent in reference to the case in hand, in proportion as his general ability to observe and to communicate his observation has been established in other cases. In the second place, the competency of a witness must be presumed, in proportion as in the particular case a lower and commoner amount of ability is requisite rightly to observe, and rightly to report the observation. In the third place, the competency of a witness is to be presumed, in proportion as it is not to be presumed that his observation was made or communicated at a time when he was unable correctly to make or correctly to communicate it. So much for the Competency of a witness.

"Now, when both the good will and the ability, that is, when both the Veracity and Competence, of a witness have been sufficiently established, the credibility of his testimony is not to be invalidated be-

LECT. XXXIII.

The presumption of the Honesty of a Witness now enhanced by certain circumstances.

b. Competency of a Witness.

Circumstances by which the presumption of competency is enhanced.

The credibility of Testimony not invalidated because the fact testi-

LECT.
XXXIII.

find is the
out of the
ordinary
course of
experience.

cause the fact which it goes to prove is one out of the
ordinary course of experience."[a]   Thus it would be
false to assert, with Hume, that miracles, that is, sus-
pensions of the ordinary laws of nature, are incapable of proof, because contradicted by what we have
been able to observe.   "On the contrary, where the
trustworthiness of a witness or witnesses is unimpeachable, the very circumstance that the object is
one in itself unusual and marvellous, adds greater
weight to the testimony ; for this very circumstance
would itself induce men of veracity and intelligence
to accord a more attentive scrutiny to the fact, and
secure from them a more accurate report of their observation.

Summary
regarding
the Credi-
bility of
Testimony
in general.

   " The result of what has now been stated in regard
to the credibility of Testimony in general, is :—That
a testimony is entitled to credit, when the requisite
conditions, both on the part of the object and on the
part of the subject, have been fulfilled.   On the part
of the object these are fulfilled, when the object is
absolutely possible, as an object of the higher faculty
of experience,—the Understanding,—the Elaborative
Faculty, and relatively possible, as an object of the
lower or subsidiary faculties of experience,—Sense, and
Self-consciousness.   In this case, the testimony, qua
testimony, does not contradict itself.   On the part of
the subject, the requisite conditions are fulfilled, when
the trustworthiness, that is, the veracity and competency of the witness, is beyond reasonable doubt.   In
regard to the veracity of the witness,—this cannot be
reasonably doubted, when there is no positive ground
on which to discredit the sincerity of the witness, and
when the only ground of doubt lies in the mere gen-

eral possibility of deception. And in reference to the competency of a witness,—this is exposed to no reasonable objection, when the ability of the witness to observe and to communicate the fact in testimony cannot be disallowed. Having, therefore, concluded the consideration of testimony in general, we proceed to treat of it in special, that is, in so far as it is viewed either as Immediate or as Mediate." [a] Of these in their order.

The special consideration of Testimony, when that testimony is Immediate.—" An immediate testimony, or testimony at first hand, is one in which the fact reported is an object of the proper or personal experience of the reporter. Now it is manifest, that an immediate witness is in general better entitled to credit than a witness at second hand; and his testimony rises in probability, in proportion as the requisites, already specified, both on the part of its object and on the part of its subject, are fulfilled. An immediate testimony is, therefore, entitled to credit,—1°, In proportion to the greater ability with which the observation has been made; 2°, In proportion to the less impediment in the way of the observation being perfectly accomplished; 3°, In proportion as what was observed could be fully and accurately remembered; and, 4°, In proportion as the facts observed and remembered have been communicated by intelligible and unambiguous signs.

" Now, whether all these conditions of a higher credibility be fulfilled in the case of any immediate testimony,—this cannot be directly and at once ascertained, it can only be inferred, with greater or less certainty, from the qualities of the witness; and,

a Esser, Logik, § 154.—ED.

Margin notes:
LECT. XXXIII.

II. Testimony in special, as Immediate and Mediate. 1°. Immediate Testimony.

Conditions of its Credibility.

Whether all these conditions are fulfilled in the case of any immediate testimony, cannot be directly ascertained.

consequently, the validity of a testimony can only be
accurately estimated from a critical knowledge of the
personal character of the witness, as given in his intellectual and moral qualities, and in the circumstances of his life, which have concurred to modify
and determine these. The veracity of a witness either
is, or is not, exempt from doubt; and, in the latter case,
it may not only lie open to doubt, but even be exposed to suspicion. If the sincerity of the witness be
indubitable, a direct testimony is always preferable to
an indirect; for a direct testimony being made with
the sole intent of establishing the certainty of the fact
in question, the competency of the witness is less exposed to objection. If, on the contrary, the sincerity
of the witness be not beyond a doubt, and, still more,
if it be actually suspected, in that case an indirect
testimony is of higher cogency than a direct; for
the indirect testimony being given with another view
than merely to establish the fact in question, the intention of the witness to falsify the truth of the fact
has not so strong a presumption in its favour. If both
the sincerity and the competency of the witness are
altogether indubitable, it is then of no importance
whether the truth of the fact be vouched for by a
single witness, or by a plurality of witnesses. On the
other hand, if the sincerity and competency of the
witness be at all doubtful, the credibility of a testimony will be greater, the greater the number of the
witnesses by whom the fact is corroborated. But here
it is to be considered, that when there are a plurality
of testimonies to the same fact, these testimonies are
either consistent or inconsistent. If the testimonies
be consistent, and the sincerity and competency of all
the witnesses complete, in that case the testimony

When testi-
mony at-
tains the
highest
degree of
probability.

attains the highest degree of probability of which any
testimony is capable. Again, if the witnesses be inconsistent,—on this hypothesis two cases are possible; for either their discrepancy is negative, or it
is positive. A negative discrepancy arises where one
witness passes over in silence what another witness
positively avers. A positive discrepancy arises, where
one witness explicitly affirms something, which something another witness explicitly denies. When the
difference of testimonies is merely negative, we may
suppose various causes of the silence; and, therefore,
the positive averment of one witness to a fact is not
disproved by the mere circumstance, that the same
fact is omitted by another. But if it be made out,
that the witness who omits mention of the fact, could
not have been ignorant of that fact had it taken place,
and, at the same time, that he could not have passed
it over without violating every probability of human
action,—in this case, the silence of the one witness
manifestly derogates from the credibility of the other
witness, and in certain circumstances may annihilate
it altogether. Where, again, the difference is positive,
the discrepancy is of greater importance, because,
(though there are certainly exceptions to the rule,)
an overt contradiction is, in general and in itself, of
stronger cogency than a mere non-confirmation by
simple silence. Now the positive discrepancy of testimonies either admits of conciliation, or it does not.
In the former case, the credibility of the several testimonies stands intact; and the discrepancy among the
witnesses is to be accounted for by such circumstances
as explain, without invalidating, the testimony considered in itself. In the latter case, one testimony
manifestly detracts from the credibility of another;

LECT.
XXXIII.

Negative
and Posi-
tive Dis-
crepancy.

for of incompatible testimonies, while both cannot be true, the one must be false, when reciprocally contradictory, or they may both be false, when reciprocally contrary. In this case, the whole question resolves itself into one of the greater or less trustworthiness of the opposing witnesses. Is the trustworthiness of the counter-witnesses equally great? In that case, neither of the conflictive testimonies is to be admitted. Again, is the trustworthiness of the witnesses not upon a par? In that case, the testimony of the witness whose trustworthiness is the greater, obtains the preference,—and this more especially if the credibility of the other witnesses is suspected." *

So much for the Credibility of Testimony, considered in Special, in so far as that testimony is Immediate or at First Hand ; and I now, in the second place, pass on to consider, likewise in special, the Credibility of Testimony, in so far as that testimony is Mediate, or at Second Hand.

F. Mediate Testimony. "A Mediate Testimony is one where the fact is an object not of Personal, but of Foreign, Experience. Touching the credibility of a mediate testimony, this supposes that the report of the immediate, and the report of the mediate, witness are both trustworthy. Whether the report of the immediate witness be trustworthy,—this we are either of ourselves able to determine, viz., from our personal acquaintance with his veracity and competence ; or we are unable of ourselves to do this, in which case the credibility of the immediate must be taken upon the authority of the mediate witness. Here, however, it is necessary for us to be aware, that the mediate witness is possessed of the ability requisite to estimate the credi-

* Esser, Logik, § 155.—Ed.

bility of the immediate witness, and of the honesty to
communicate the truth without retrenchment or falsification.   But if the trustworthiness both of the
mediate and of the immediate witness be sufficiently
established, it is of no consequence, in regard to the
credibility of a testimony, whether it be at first hand
or at second.   Nay, the testimony of a mediate may
even tend to confirm the testimony of an immediate
witness, when his own competence fairly to appreciate
the report of the immediate witness is indubitable.
If, however, the credibility of the immediate witness be
unimpeachable, but not so the credibility of the mediate, in that case the mediate testimony, in respect of its
authority, is inferior to the immediate, and this in the
same proportion as the credibility of the second hand
witness is inferior to that of the witness at first hand.
Further, mediate witnesses are either Proximate or
Remote ; and, in both cases, either Independent or Dependent.   The trustworthiness of proximate witnesses
is, in general, greater than the trustworthiness of remote ; and the credibility of independent witnesses
greater than the credibility of dependent.   The remote witness is unworthy of belief, when the intermediate links are wanting between him and the
original witness ; and the dependent witness deserves
no credit, when that on which his evidence depends
is recognised as false or unestablished.   Mediate testimonies are, likewise, either direct or indirect ; and,
likewise, when more than one, either reciprocally congruent or conflictive.   In both cases the credibility of
the witnesses is to be determined in the same manner
as if the testimonies were immediate.

"The testimony of a plurality of mediate witnesses,
where there is no recognised immediate witness, is

LECT.
XXXIII.

Mediate
Witnesses
are either
Proximate
or Remote,
and either
Indepen-
dent or
Dependent.

Rumour,—
what.
Tradition.

called a *rumour*, if the witnesses be contemporaneous ; and a *tradition*, if the witnesses be chronologically successive. These are both less entitled to credit, in proportion as in either case a fiction or falsification of the fact is comparatively easy, and, consequently, comparatively probable." [a]

a Esser, *Logik*, § 150.—Ed.

LECTURE XXXIV.

MODIFIED METHODOLOGY.

SECTION I.—OF THE ACQUISITION OF KNOWLEDGE.

I. EXPERIENCE.—B. FOREIGN :—RECORDED TESTIMONY AND WRITINGS IN GENERAL.

II. SPECULATION.

In our last Lecture, we were engaged in the considera- LECT.
tion of Testimony, and the Principles by which its Cre- XXXIV.
dibility is governed ; on the supposition always that Criticism of
we possess the veritable report of the witness whose Recorded
testimony it professes to be, and on the supposition Writings
that we are at no loss to understand its meaning and in general.
purport.   But questions may arise in regard to these
points, and, therefore, there is a further critical process
requisite, in order to establish the Authenticity, the
Integrity, and the Signification, of the documents in
which the testimony is conveyed.   This leads us to the
important subject,—the Criticism of Recorded Testimony, and of Writings in general.   I shall comprise
the heads of the following observations on this subject in the ensuing paragraph :—

¶ CX. The examination and judgment of Par. CX.
Writings professing to contain the testimony of and Inter-
certain witnesses, and of Writings in General, pro- pretation.

fessing to be the work of certain authors, is of
two parts. For the inquiry regards either, 1°,
The Authenticity of the document, that is,
whether it be, in whole or in part, the product
of its ostensible author ; for ancient writings in
particular are frequently supposititious or interpolated ; or, 2°, It regards the Meaning of the
words of which it is composed, for these, especially when in languages now dead, are frequently
obscure. The former of these problems is resolved by the *Art of Criticism* (*Critica*), in the
stricter sense of the term ; the latter by the *Art
of Interpretation* (*Exegetica* or *Hermeneutica*).
Criticism is of two kinds. If it be occupied with
the criteria of the authenticity of a writing in its
totality, or in its principal parts, it is called the
Higher, and sometimes the *Internal, Criticism.*
If, again, it consider only the integrity of particular words and phrases, it is called the *Lower,* and
sometimes the *External, Criticism,* The former
of these may perhaps be best styled the *Criticism
of Authenticity;* the latter, the *Criticism of
Integrity.*

The problem which Interpretation has to solve
is,—To discover and expound the meaning of a
writer, from the words in which his thoughts are
expressed. It departs from the principle, that
however manifold be the possible meanings of
the expressions, the sense of the writer is one.
Interpretation, by reference to its sources or subsidia, has been divided into the *Grammatical,* the
Historical, and the *Philosophical, Exegesis.*[a]

---

a Cf. Krug, *Logik,* § 177 *et seq.*— Kiesewetter. *Logik.* p. ii. § 185 *et*
ED. [Snell, *Logik,* p. ii. § 0, p. 195. *seq.*]

"Testimonies, especially when the ostensible witnesses themselves can no longer be interrogated, may be subjected to an examination under various forms; and this examination is in fact indispensable, seeing not only that a false testimony may be substituted for a true, and a testimony true upon the whole may yet be falsified in its parts,—a practice which prevailed to a great extent in ancient times ; while at the same time the meaning of the testimony, by reason either of the foreign character of the language in which it is expressed, or of the foreign character of thought in which it is conceived, may be obscure and undetermined. The examination of a testimony is twofold, inasmuch as it is either an examination of its Authenticity and Integrity, or an examination of its Meaning. This twofold process of examination is applicable to testimonies of every kind, but it becomes indispensable when the testimony has been recorded in writing, and when this, from its antiquity, has come down to us only in transcripts, indefinitely removed from the original, and when the witnesses are men differing greatly from ourselves in language, manners, customs, and associations of thought. The solution of the problem,—By what laws are the authenticity or spuriousness, the integrity or corruption, of a writing to be determined?—constitutes the Art of Criticism, in its stricter signification (*Critica*) ; and the solution of the problem,—By what law is the sense or meaning of writing to be determined ?—constitutes the Art of Interpretation or Exposition (*Hermeneutica, Exegetica*). In theory, Criticism ought to precede Interpretation, for the question,—Who has spoken ? naturally arises before the question,—How is what has been spoken to be understood ? But in practice, criticism and interpretation cannot be separated ; for in application they proceed hand in hand." [a]

1. Criticism. "First, then, of Criticism, and the question that presents itself in the threshold is,—What are its Definition and Divisions? Under Criticism is to be understood the complement of logical rules, by which the authenticity or spuriousness, the integrity or interpolation, of a writing is to be judged. The problems Its problems which it proposes to answer are—1°, Does a writing really proceed from the author to whom it is ascribed? and, 2°, Is a writing, as we possess it, in all its parts the same as it came from the hands of its author? Universal Criticism. The system of fundamental rules, which are supposed in judging of the authenticity and integrity of every writing, constitutes what is called the *Doctrine of* Special Criticism. *Universal Criticism*; and the system of particular rules, by which the authenticity and integrity of writings of a certain kind are judged, constitutes the Universal Criticism alone within the sphere of Logic. doctrine of what is called *Special Criticism*. It is manifest, from the nature of Logic, that the doctrine of Universal Criticism is alone within its sphere. Now Universal Criticism is conversant either with the authenticity or spuriousness of a writing considered as a whole, or with the integrity or interpolation of Its Divisions. certain parts. In the former case it is called *Higher*, in the latter *Lower*, *Criticism*; but these denominations are inappropriate. The one criticism has also been styled the *Internal*, the other the *External*; but these appellations are, likewise, exceptionable; and, perhaps, it would be preferable to call the former the *Criticism of the Authenticity*, the latter, the *Criticism of the Integrity*, of a work. I shall consider these in particular, and, first, of the Criticism of Authenticity.

[a] Esser, *Logik*, § 157.—Ed.

"A proof of the authenticity of a writing, more
especially of an ancient writing, can be rested only
upon two grounds,—an Internal and an External,—
and on these either apart or in combination. By *internal grounds*, we mean those indications of authenticity which the writing itself affords. By *external
grounds*, we denote the testimony borne by other
works of a corresponding antiquity, to the authenticity of the writing in question.

"In regard to the Internal Grounds ;—it is evident,
without entering upon details, that these cannot of
themselves, that is, apart from the external grounds,
afford evidence capable of establishing beyond a doubt
the authenticity of an ancient writing; for we can
easily conceive that an able and learned forger may
accommodate his fabrications both to all the general
circumstances of time, place, people, and language,
under which it is supposed to have been written, and
even to all the particular circumstances of the style,
habit of thought, personal relations, &c., of the author
by whom it professes to have been written, so that
everything may militate for, and nothing militate
against, its authenticity.

"But if our criticism from the internal grounds
alone be, on the one hand, impotent to establish, it is,
on the other, omnipotent to disprove. For it is sufficient to show that a writing is in essential parts, that
is, parts which cannot be separated from the whole,
in opposition to the known manners, institutions,
usages, &c., of that people with which it would, and
must, have been in harmony, were it the product of
the writer whose name it bears; that, on the contrary,
it bears upon its face indications of another country
or of a later age ; and, finally, that it is at variance

LECT.
XXXIV.

I. Criticism
of Authen-
ticity.

a. Internal
Grounds
Those of
themselves
not suffi-
cient to
establish
the authen-
ticity of a
writing.

But omni-
potent to
disprove
this.

LECT.
XXXIV.
with the personal circumstances, the turn of mind, and the pitch of intellect, of its pretended author. And here it is to be noticed, that these grounds are only relatively internal ; for we become aware of them originally only through the testimony of others, that is, through external grounds." [a]

b. External Grounds.
In regard to the External Grounds ;—they, as I said, consist in the testimony, direct or indirect, given to the authenticity of the writing in question by other works of a competent antiquity. This testimony may be contained either in other and admitted writings of the supposed author himself ; or in those of contemporary writers ; or in those of writers approximating in antiquity. This testimony may also be given either directly, by attribution of the disputed writing by title to the author ; or indirectly, by quoting as his, certain passages which are to be found in it. On this subject it is needless to go into detail, and it is hardly necessary to observe, that the proof of the authenticity is most complete when it proceeds upon the internal and external grounds together. I, therefore, pass on to the Criticism of Integrity. [b]

3. Criticism of Integrity.
"When the authenticity of an ancient work has been established on external grounds, and been confirmed on internal, the Integrity of this writing is not therewith proved ; for it is very possible, and in ancient writings indeed very probable, that particular passages are either interpolated or corrupted. The authenticity of particular passages is to be judged of precisely by the same laws, which regulate our criticism of the authenticity of the whole work. The proof most pertinent to the authenticity of particular pas-

<hr>

a Esser, Logik, § 158 · 160. —   b See Esser, Logik, §§ 101. 102.
ED.   —ED.

sages is drawn—1°, From their acknowledgment by the author himself in other, and these unsuspected, works ; 2°, From the attribution of them to the author by other writers of competent information ; and, 3°, From the evidence of the most ancient MSS. On the other hand, a passage is to be obelized as spurious,— 1°, When found to be repugnant to the general relations of time and place, and to the personal relations of the author ; 2°, When wanting in the more ancient codices, and extant only in the more modern. A passage is suspicious, when any motive for its interpolation is manifest, even should we be unable to establish it as spurious. The differences which different copies of a writing exhibit in the particular passages, are called *various readings* (*variæ lectiones* or *lectiones variantes*). Now, as of various readings one only can be the true, while they may all very easily be false, the problem which the criticism of Integrity proposes to solve is,—How is the genuine reading to be made out ?—and herein consists what is technically called the *Recension*, more properly the *Emendation*, of the text.

" The Emendation of an ancient author may be of two kinds ; the one of which may be called the *Historical*, the other the *Conjectural*. The former of these founds upon historical data for its proof ; the latter, again, proceeds on grounds which lie beyond the sphere of historical fact, and this for the very reason that historical fact is found incompetent to the restoration of the text to its original integrity. The historical emendation necessarily precedes the conjectural, because the object itself of emendation is wholly of an historical character, and because it is not permitted to attempt any other than an emendation on historical grounds,

LECT.
XXXIV.

Historical
Emendation
of two
kinds,—
External
and Inter-
nal.

until, from these very grounds themselves, it be shown
that the restitution of the text to its original integrity
cannot be historically accomplished. Historical Emendation is again of two kinds, according as its judgment
proceeds on external or on internal grounds. It founds
upon external grounds, when the reasons for the truth
or falsehood of a reading are derived from testimony ;
it founds upon internal grounds, when the reasons for
the truth or falsehood of a reading are derived from
the writing itself. Historical emendation has thus a
twofold function to perform (and in its application to
practice, these must always be performed in conjunction), viz., it has carefully to seek out and accurately
to weigh both the external and internal reasons in support of the reading in dispute. Of external grounds
the principal consists in the confirmation afforded by
MSS., by printed editions which have immediately
emanated from MSS., by ancient translations, and by
passages quoted in ancient authors. The internal
grounds are all derived either from the form, or from
the contents, of the work itself. In reference to the
form,—a reading is probable, in proportion as it corresponds to the general character of the language prevalent at the epoch when the work was written, and
to the peculiar character of the language by which the
author himself was distinguished. In reference to the
contents,—a reading is probable, when it harmonises
with the context, that is, when it concurs with the
other words of the particular passage in which it
stands, in affording a meaning reasonable in itself, and
conformable with the author's opinions, reasonings,
and general character of thought." [a]

It frequently happens, however, that, notwithstand-

a Esser, *Logik*, ¶ 183.—ED.

ing the uniformity of MSS. and other external sub- LECT.
XXXIV.
sidia, a reading cannot be recognised as genuine. In
this case, it must be scientifically shown from the Conjectural
rules of criticism itself that this lection is corrupt. tion.
If the demonstration thus attempted be satisfactory,
and if all external subsidia have been tried in vain,
the critic is permitted to consider in what manner
the corrupted passage can be restored to its integrity.
And here the conjectural or divinatory emendation
comes into play; a process in which the power and
efficiency of criticism and the genius of the critic are
principally manifested." [a]

So much for Criticism, in its applications both to
the Authenticity and to the Integrity of Writings.
We have now to consider the general rules by which
Interpretation, that is, the scientific process of ex-
pounding the Meaning of an author, is regulated.

"By the *Art of Interpretation*, called likewise techni- II. Inter-
cally *Hermeneutic* or *Exegetic*, is meant the comple- pretation.
ment of logical laws, by which the sense of an ancient
writing is to be evolved.   Hermeneutic is either Gen- General and
eral or Special.   General, when it contains those laws Special.
which apply to the interpretation of any writing
whatever; Special, when it comprises those laws by
which writings of a particular kind are to be ex-
pounded.   The former of these alone is of logical
concernment.   The problem proposed for the Art of
Interpretation to solve, is,—How are we to proceed
in order to discover from the words of a writing that
sole meaning which the author intended them to
convey?   In the interpretation of a work, it is not,
therefore, enough to show in what signification its

a *Essay, Logik*, § 106.—E.D. [*Pur-   Gocoenais, *Ars Logico-Critica*, L. iv.
chasiana*, L 359-363, 2d ed. 1701.   c. vi. *et seq.*]

LECT.
XXXIV. words may be understood ; for it is required that we show in what signification they must. To the execution of this task two conditions are absolutely necessary : 1°, That the interpreter should be thoroughly acquainted with the language itself in general, and with the language of the writer in particular ; and, 2°, That the interpreter should be familiar with the subjects of which the writing treats. But these two requisites, though indispensable, are not of themselves sufficient. It is also of importance that the expositor should have a competent acquaintance with the author's personal circumstances and character of thought, and with the history and spirit of the age and country in which he lived. In regard to the interpretation itself ;—it is to be again observed, that as a writer could employ expressions only in a single sense, so the result of the exposition ought to be not merely to show what meaning may possibly attach to the doubtful terms, but what meaning necessarily must. When, therefore, it appears that a passage is of doubtful import, the best preparative for a final determination of its meaning is, in the first place, to ascertain in how many different significations it may be construed, and then, by a process of exclusion, to arrive at the one veritable meaning. When, however, the obscurity cannot be removed, in that case it is the duty of the expositor, before abandoning his task, to evince that an interpretation of the passage is, without change, absolutely or relatively impossible.

Sources of
Interpretation. "As to the sources from whence the Interpretation is to be drawn,—these are three in all,—viz., 1°, The *Tractus literarum*, the words themselves, as they appear in MSS. ; 2°, The context, that is, the passage in immediate connection with the doubtful term ; 3°,

Parallel or analogous passages in the same, or in other, writings."[a] How the interpretation drawn from these sources is to be applied, I shall not attempt to detail; but pass on to a more generally useful and interesting subject.

So much for Experience or Observation, the first mean of scientific discovery, that, viz., by which we apprehend what are presented as contingent phænomena, and by whose processes of Induction and Analogy we carry up individual into general facts. We have now to consider the other Mean of scientific discovery, that, viz., by which, from the phænomena presented as contingent, we separate what is really necessary, and thus attain to the knowledge, not of merely generalised facts, but of universal laws. This mean may, for distinction's sake, be called *Speculation*, and its general nature I comprehend in the following paragraph :—

Speculation the Second Mean of Knowledge.

¶ CXI. When the mind does not rest contented with observing and classifying the objects of its experience, but, by a reflective analysis, sunders the concrete wholes presented to its cognition, throws out of account all that, as contingent, it can think away from, and concentrates its attention exclusively on those elements which, as necessary conditions of its own acts, it cannot but think :—by this process it obtains the knowledge of a certain order of facts,—facts of Self-consciousness, which, as essential to all Experience, are not the result of any; constituting in truth the Laws by which the possibility of our cognitive functions is determined. This process,

*Par. CXI.
Speculation, as a means of Knowledge.*

<hr>

[a] Esser, *Logik*, § 167.—Ed. [Cf. Snell, *Logik*, p. II. § 6, p. 200.]

LECT.
XXXIV.
by which we thus attain to a discriminative
knowledge of the *Necessary, Native,* and, as they
are also called, the *Noetic, Pure, a priori,* or
Transcendental, Elements of Thought, may be
styled *Speculative Analysis, Analytic Speculation,* or *Speculation* simply, and is carefully to be
distinguished from Induction, with which it is
not unusually confounded.

Explica-
tion.
"The empirical knowledge of which we have
hitherto been speaking, does not, however varied and
extensive it may be, suffice to satisfy the thinking
mind as such; for our empirical knowledge itself
points at certain higher cognitions from which it may
obtain completion, and which are of a very different
character from that by which the mere empirical cognitions themselves are distinguished. The cognitions
are styled, among other names, by those of *noetic,
pure,* or *rational,* and they are such as cannot, though
manifested in experience, be derived from experience ;
for, as the conditions under which experience is possible, they must be viewed as necessary constituents
of the nature of the thinking principle itself. Philosophers have indeed been found to deny the reality of
such cognitions native to the mind ; and to confine
the whole sphere of human knowledge to the limits of
experience. But in this case philosophers have overlooked the important circumstance, that the acts, that
is, the apprehension and judgment, of experience, are
themselves impossible, except under the supposition of
certain potential cognitions previously existent in the
thinking subject, and which become actual on occasion of an object being presented to the external or
internal sense. As an example of a noetic cognition

the following propositions may suffice :—An object
and all its attributes are convertible ;— All that is has its sufficient cause. The principal distinctions of Empirical and Rational Knowledges, or rather Empirical and Noetic Cognitions, are the following :—1°, Empirical cognitions originate exclusively in experience, whereas noetic cognitions are virtually at least before or above all experience,—all experience being only possible through them. 2°, Empirical cognitions come piecemeal and successively into existence, and may again gradually fade and disappear ; whereas noetic cognitions, like Pallas armed and immortal from the head of Jupiter, spring at once into existence, complete and indestructible. 3°, Empirical cognitions find only an application to those objects from which they were originally abstracted, and, according as things obtain a different form, they also may become differently fashioned ; noetic cognitions, on the contrary, bear the character impressed on them of necessity, universality, sameness. Whether a cognition be empirical or noetic, can only be determined by considering whether it can or cannot be presented in a sensible perception,—whether it do or do not stand forward clear, distinct, and indestructible, bearing the stamp of necessity and absolute universality. The noetic cognitions can be detected only by a critical analysis of the mental phænomena proposed for the purpose of their discovery ;" [a] and this analysis may, as I have said, be styled Speculation, for want of a more appropriate appellation.

a Esser, *Logik*, § 171.—Ed.

•

LECTURE XXXV.

MODIFIED METHODOLOGY.

SECTION I.—OF THE ACQUISITION OF KNOWLEDGE.

III. COMMUNICATION OF KNOWLEDGE.—A. INSTRUCTION —ORAL AND WRITTEN.—B. CONFERENCE— DIALOGUE AND DISPUTATION.

LECT. XXXV.

I now go on to the last Mean of Acquiring and Perfecting our knowledge; and commence with the following paragraph :—

Par. CXII. The Communication of Thought, —as a means of Acquiring and Perfecting Knowledge.

¶ CXII. An important mean for the Acquisition and Perfecting of Knowledge is the Communication of Thought. Considered in general, the Communication of thought is either One-sided or Mutual. The former is called *Instruction* (*institutio*), the latter *Conference* (*collocutio*); but these, though in theory distinct, are in practice easily combined. Instruction is again either *Oral* or *Written;* and Conference, as it is interlocutory and familiar, or controversial and solemn, may be divided into *Dialogue* (*colloquium, dialogus*), and *Disputation* (*disputatio, concertatio*). The Communication of thought in all its forms is a means of intellectual improvement, not only to him who receives, but to him who bestows, information; in both relations, therefore, it ought to be considered, and not, as is usually done, in the latter only.[a]

LECT. XXXV.

In illustrating this paragraph, I shall commence with the last sentence, and, before treating in detail of Instruction and Conference, as means of extending the limits of our knowledge by new acquisitions derived from the communication of others, I shall endeavour to show, that the Communication of thought is itself an important mean towards the perfecting of knowledge in the mind of the communicator himself. In this view the communication of knowledge is like the attribute of mercy, twice blessed,—"blessed to him that gives and to him that takes;" in teaching others we in fact teach ourselves.

This view of the reflex effect of the communication of thought on the mind, whether under the form of Instruction or of Conference, is one of high importance, but it is one which has, in modern times, unfortunately been almost wholly overlooked. To illustrate it in all its bearings would require a volume,—at present I can only contribute a few hints towards its exposition.

Man is, by an original tendency of his nature, determined to communicate to others what occupies his thoughts, and by this communication he obtains a clearer understanding of the subject of his cogitations than he could otherwise have compassed. This fact did not escape the acuteness of Plato. In the *Protagoras*,—"It has been well," says Plato (and he has

a Cf. Krug, *Logik*, § 161 *et seq.*—ED.

sundry passages to the point),—" It has been well, I think, observed by Homer—

> ' Through mutual intercourse and mutual aid,
> Great deeds are done and great discoveries made ;
> The wise new wisdom on the wise bestow,
> Whilst the lone thinker's thoughts come slight and slow.' [a]

For in company we, all of us, are more alert, in deed and word and thought. *And if a man excogitate aught by himself, forthwith he goes about to find some one to whom he may reveal it, and from whom he may obtain encouragement, aye and until his discovery be completed."* [β] The same doctrine is maintained by Aristotle, and illustrated by the same quotation; [γ] (to which, indeed, is to be referred the adage,—" Unus homo, nullus homo.")—" We rejoice," says Themistius, "in hunting truth in company, as in hunting game." [δ] Lucilius,—" Scire est nescire, nisi id me scire alius scierit;" [ε]—paraphrased in the compacter, though far inferior, verse of Persius,—" Scire tuum nihil est, nisi te scire hoc sciat alter." [ζ]—Cicero's Cato testifies to the same truth :—" Non facile est invenire, qui quod sciat ipse, non tradat alteri." [η] And Seneca :—" Sic cum hac exceptione detur sapientia, ut illam inclusam teneam nec enunciem, rejiciam. Nullius boni, sine socio, jucunda possessio est." [θ]

> " Condita tabescit, vulgata scientia crescit." [ι]

---

a Altered from Pope's *Homer*, Book x. 205.

β *Prolag.*, p. 348. Compare *Lectures on Metaphysics*, i. p. 376.

γ *Eth. Nic.*, viii. 1.

δ *Orat.*, xxi. *Explorator aut Philosophus, Orationes*, p. 254, ed. Hardouin, Paris, 1684. — Ed.

ε *Fragm.*, 28, in the Bipont edition of Persius and Juvenal, p. 176. — Ed.

ζ l. 27. — En.

η Cato apud Cicero, *De Fin.*, iii. c. 20, § 66.

θ Seneca, *Ep.*, vi.

ι Quoted also in *Discussions*, p. 778. This line appears to have been taken from a small volume, entitled, *Curmanum Proverbialium Loci Communes*, p. 17, Lond. 1583 ; but the author is not named. — Ed.

" In hoc gaudeo aliquid discere, ut doceam : nec me
ulla res delectabit, licet eximia sit et salutaris, quam
mihi uni, sciturus sim." [a]   " Ita non solum ad discen-
dum propensi sumus, verum etiam ad docendum." [β]

The modes in which the Communication of thought
is conducive to the perfecting of thought itself, are
two : for the mind may be determined to more ex-
alted energy by the sympathy of society, and by the
stimulus of opposition; or it may be necessitated
to more distinct, accurate, and orderly thinking, as
this is the condition of distinct, accurate, and orderly
communication.   Of these the former requires the
presence of others during the act of thought, and is,
therefore, only manifested in oral instruction or in
conference; whereas the latter is operative both in
our oral and in our written communications.   Of these
in their order.

In the first place, then, the influence of man on
man in reciprocally determining a higher energy of
the faculties, is a phænomenon sufficiently manifest.
By nature, a social being, man has powers which are
relative to, and, consequently, find their development
in, the company of his fellows; and this is more particularly shown in the energies of the cognitive faculties.   " As iron sharpeneth iron," says Solomon, " so
a man sharpeneth the understanding of his friend." [γ]
This, as I have said, is effected both by fellow-feeling
and by opposition.   We see the effects of fellow-feeling, in the necessity of an audience to call forth the
exertions of the orator.   Eloquence requires numbers ;
and oratory has only flourished where the condition

LKCT.
XXXV.

Modes in
which Com-
munication
is conducive
to the Per-
fecting of
Thought
are two.

1. By recip-
rocally de-
termining
a higher
energy of
the facul-
ties.

a. Through
Sympathy.

a Seneca, Epist., vi.—Ed.
β Cicero, De Fin., iii. 20.—Ed.
γ Proverbs, xxvii. 17.   The autho-
rised version is, countenance of his
friend.   Compare Lectures on Metaphysics, vol. i. p. 370.—Ed.

of large audiences has been supplied. But opposition is perhaps still more powerful than mere sympathy in calling out the resources of the intellect.

In the mental as in the material world, action

Plutarch. and reaction are ever equal ; and Plutarch [a] well observes, that as motion would cease were contention to be taken out of the physical universe, so progress in improvement would cease were contention taken out of the moral ; πόλεμος ἀπάντων πατήρ.[β]

Scaliger,
J. C.

"It is maintained," says the subtle Scaliger, " by Vives, that we profit more by silent meditation than by dispute. This is not true. For as fire is elicited by the collision of stones, so truth is elicited by the collision of minds. I myself (he adds) frequently meditate by myself long and intently ; but in vain ; unless I find an antagonist, there is no hope of a successful issue. By a master we are more excited than by a book; but an antagonist, whether by his pertinacity or his wisdom, is to me a double master." [γ]

2. By imposing the necessity of obtaining a fuller consciousness of knowledge for ourselves.

But, in the second place, the necessity of communicating a piece of knowledge to others, imposes upon us the necessity of obtaining a fuller consciousness of that knowledge for ourselves. This result is to a certain extent secured by the very process of clothing our cogitations in words. For speech is an analytic process ; and to express our thoughts in language, it is requisite to evolve them from the implicit into the explicit, from the confused into the distinct, in order to bestow on each part of the organic totality of a thought its precise and appropriate symbol. But to

a *Vita Agesilai, Opera,* 1599, vol. i. p. 596.—Ed.

β Heraclitus. Cf. Plutarch, *De Is. et Osir.,* p. 370. Brandis, *Gesch. der Philos.,* i. p. 155.—Ed.

γ *Exercit.,* f. 420. [For a criticism of Scaliger's remark as regards Vives, see *Discussions,* p. 773.—Ed.]

do this is in fact only to accomplish the first step towards the perfecting of our cognitions or thoughts.

But the communication of thought, in its higher applications, imposes on us far more than this; and in so doing it reacts with a still more beneficial influence on our habits of thinking. Suppose that we are not merely to express our thoughts as they spontaneously arise; suppose that we are not merely extemporaneously to speak, but deliberately to write, and that what we are to communicate is not a simple and easy, but a complex and difficult, matter. In this case, no man will ever fully understand his subject who has not studied it with the view of communication, while the power of communicating a subject is the only competent criterion of his fully understanding it. "When a man," says Godwin, "writes a book of methodical investigation, he does not write because he understands the subject, but he understands the subject because he has written. He was an uninstructed tyro, exposed to a thousand foolish and miserable mistakes, when he began his work, compared with the degree of proficiency to which he has attained when he has finished it. He who is now an eminent philosopher, or a sublime poet, was formerly neither the one nor the other. Many a man has been overtaken by a premature death, and left nothing behind him but compositions worthy of ridicule and contempt, who, if he had lived, would perhaps have risen to the highest literary eminence. If we could examine the school exercises of men who have afterwards done honour to mankind, we should often find them inferior to those of their ordinary competitors. If we could dive into the portfolios of their early youth, we should meet with abundant matter for laughter at their senseless incongruities, and for contemptuous astonishment." [a]

"The one exclusive sign," says Aristotle, "that a man is thoroughly cognisant of anything is that he is able to teach it ; " [β] and Ovid,—[γ]

" Quodque parum novit nemo docere potest."

In this reactive effect of the communication of knowledge in determining the perfection of the knowledge communicated, originated the scholastic maxim *Doce ut discas,*—a maxim which has unfortunately been too much overlooked in the schemes of modern education. In former ages, *teach that you may learn,* always constituted one at least of the great means of intellectual cultivation. "To teach," says Plato, "is the way for a man to learn most and best." [δ] "Homines dum docent discunt," says Seneca. [ε] "In teaching," says Clement of Alexandria, [ζ] "the instructor often learns more than his pupils." "Disce sed a doctis ; indoctos ipse doceto," is the precept of Dionysius Cato ; [η] and the two following were maxims of authority in the discipline of the middle ages. The first—

" Multa rogare, rogata tenere, retenta docere,
Haec tria, discipulum faciunt superare magistrum." [θ]

The second—

" Discere si quaeris doceas ; sic ipse doceris ;
Nam studio tali tibi proficis atque sodali."

a *Enquirer,* Part I. Essay iv. pp. 23, 24, ed. 1797.—Ed.

β *Metaphys.,* l. 1. Quoted in *Discussions,* p. 765.—Ed.

γ *Tristia,* II. 348.—Ed.

δ Pseudo-Plato, *Epinomis,* p. 989. —Ed.

ε *Epist.,* 7.—Ed.

ζ *Stromata,* lib. I. p. 275, ed. Sylb. : Διδάσκων τις μανθάνει πλέον, καὶ λέγων συνακροᾶται πολλάκις τοῖς ἐπακούουσιν αὐτοῦ.—Ed.

η IV. 29.—Ed.

θ [Crenius, p. 561.] [*Gabrielis Naudæi Syntagma de Studio Liberali.* Included in the *Consilia et Methodi Aureæ studiorum optime instituendorum,* collected by Th. Crenius, Rotterdam, 1692. The lines are quoted as from an anonymous author.— Ed.]

ι Given without author's name, in

This truth is also well enforced by the great Vives.
"Doctrina est traditio eorum quae quis novit ei qui
non novit. Disciplina est illius traditionis acceptio;
nisi quod mens accipientis impletur, dantis vero non
exhauritur,—imo communicatione augetur eruditio,
sicut ignis, motu atque agitatione. Excitatur enim
ingenium, et discurrit per ea quae ad praesens nego-
tium pertinent; ita invenit atque excudit multa, et
quae in mentem non veniebant cessanti, docenti aut
disserenti occurrunt, calore acuente vigorem ingenii.
Idcirco, nihil est ad magnam eruditionem perinde
conducens, ut docere." [a] The celebrated logician, Dr
Robert Sanderson, used to say: "I learn much from
my master, more from my equals, and most of all
from my disciples." [β]

But I have occupied perhaps too much time on the
influence of the communication of knowledge on those
by whom it is made; and shall now pass on to the
consideration of its influence on those to whom it is
addressed. And in treating of communication in
this respect, I shall, in the first place, consider it
as One-sided, and, in the second, as Reciprocal or
Bilateral.

The Unilateral Communication of knowledge, or
Instruction, is of two kinds, for it is either Oral or
Written; but as both these species of instruction propose the same end, they are both, to a certain extent,
subject to the same laws.

Oral and Written Instruction have each their peculiar advantages.

In the first place, instruction by the living voice

the *Carminum Proverbialum Loci
Communes*, Lond. 1863, p. 17. See
above, p. 206, note 1.—ED.
a *De Anima*, p. 84.

β [*Reason and Judgment, or Special Remarks of the Life of the Re-
nowned Dr Sanderson*, p. 10. London: 1663.]

has this advantage over that of books, that, as more
natural, it is more impressive. Hearing rouses the
attention and keeps it alive far more effectually than
reading. To this we have the testimony of the most
competent observers. "Hearing," says Theophrastus,[a]
"is of all the senses the most pathetic," that is, it is
the sense most intimately associated with sentiment
and passion. "Multo magis," says the younger
Pliny, "multo magis *viva vox* afficit. Nam, licet
acriora sunt quæ legas, altius tamen in animo sedent
quæ pronuntiatio, vultus, habitus, gestus etiam dicen-
tis adfigit." [β]

"Plus prodest," says Valerius Maximus, "*docentem
audire*, quam in libris studere; quia vehementior fit
impressio in mentibus audientium, ex visu doctoris et
auditu, quam ex studio et libro." [γ]

And St Jerome—"Habet nescio quid latentis ener-
giæ *viva vox*; et in aures discipuli de doctoris ore
transfusa, fortius sonat." [δ]

A second reason why our Attention (and Memory
is always in the ratio of Attention) to things spoken
is greater than to things read, is that what is written
we regard as a permanent possession to which we can
always recur at pleasure; whereas we are conscious
that the "winged words" are lost to us for ever, if we
do not catch them as they fly. As Pliny hath it:—
"Legendi semper est occasio; audiendi non semper." [ε]

A third cause of the superior efficacy of oral in-

Oral in-
struction,
—its ad-
vantages.
a. More
natural,
therefore
more im-
pressive.
Theophras-
tus.

Younger
Pliny.

Valerius
Maximus.

St Jerome.

b. Less per-
manent,
therefore
more at-
tended to.

c. Hearing
a social act.

a Ὅτι δὲ ἐμπαθὴς δ' εἶναι σε προσα-
κούεται περὶ τὴν ἀκουστικὴν αἴσθησιν,
ἦν ὁ Θεόφραστος ταῦτ᾽ εὐτύπωτε εἶναι
φησὶ καττά. Plutarch, *De Auditione*,
sub init.—Ed.

β *Epist.*, ii. 3.—Ed.

γ [Thomas Hibernicus, p. 330.]
[The above passage is quoted as from

Valerius, lib. viii., in the *Flores* of
Thomas Hibernicus, and in the *An-
thologia* of Langius, under the article
Doctrina. It is not, however, to be
found in that author.—Ed.]

δ *Epist.*, ciii. *Opera*, Antv. 1579,
tom. iii. p. 337.—Ed.

ε *Epist.*, ii. 3.—Ed.

struction is that man is a social animal. He is thus naturally disposed to find pleasure in society, and in the performance of the actions performed by those with whom he consorts. But reading is a solitary, hearing is a social, act. In reading, we are not determined to attend by any fellow-feeling with others attending; whereas in hearing, our attention is not only engaged by our sympathy with the speaker, but by our sympathy with the other attentive auditors around us.

Such are the causes which concur in rendering Oral Instruction more effectual than Written. "M. Varillas," says Menage, (and Varillas was one of the most learned of modern historians,—and Menage one of the most learned of modern scholars), "M. Varillas himself told me one day, that of every ten things he knew, he had learned nine of them in conversation. I myself might say nearly the same thing." [a]

On the other hand, Reading, though only a substitute for Oral Instruction, has likewise advantages peculiar to itself. In the first place, it is more easily accessible. In the second, it is more comprehensive in its sphere of operation. In the third, it is not transitory with the voice, but may again and again be taken up and considered, so that the object of the instruction may thus more fully be examined and brought to proof. It is thus manifest, that oral and written instruction severally supply and severally support each other; and that, where this is competent, they ought always to be employed in conjunction. Oral instruction is, however, in the earlier stages of education, of principal importance; and written ought, therefore, at first only to be brought in as a

[a] *Menagiana*, tom. iv. p. 111, ed. 1715.—ED.

subsidiary. A neglect of the oral instruction, and an
exclusive employment of the written,—the way in
which those who are self-taught (the autodidacti)
obtain their education,—for the most part betrays its
one-sided influence by a contracted cultivation of the
intellect, with a deficiency in the power of communicating knowledge to others.

Oral instruction necessarily supposes a speaker
and a hearer; written instruction a writer and a
reader. In these, the capacity of the speaker and of
the writer must equally fulfil certain common requisites. In the first place, they should be fully masters
of the subject with which their instruction is conversant; and in the second, they should be able and
willing to communicate to others the knowledge which
they themselves possess. But in reference to these
several species of instruction, there are various special
rules that ought to be attended to by those who would
reap the advantages they severally afford. I shall
commence with Written Instruction, and comprise the
rules by which it ought to be regulated, in the following paragraph.

Par.CXIII
Written
Instruction,
and its em-
ployment
as a means
of intellec-
tual Im-
provement.
¶ CXIII. In regard to Written Instruction,
and its profitable employment as a means of intellectual improvement, there are certain rules
which ought to be observed, and which together
constitute the Proper Method of Reading. These
may be reduced to three classes, as they regard,
1°, The Quantity, 2°, The Quality, of what is to be
read, or, 3°, The Mode of reading what is to be read.

I. As concerns the Quantity of what is to be
read, there is a single rule,—Read much, but
not many works (multum non multa).

II. As concerns the Quality of what is to be read,—there may be given five rules. 1°, Select the works of principal importance, estimated by relation to the several sciences themselves, or to your particular aim in reading, or to your individual disposition and wants. 2°, Read not the more detailed works upon a science, until you have obtained a rudimentary knowledge of it in general. 3°, Make yourselves familiar with a science in its actual or present state, before you proceed to study it in its chronological development. 4°, To avoid erroneous and exclusive views, read and compare together the more important works of every sect and party. 5°, To avoid a one-sided development of mind, combine with the study of works which cultivate the Understanding, the study of works which cultivate the Taste.

III. As concerns the Mode or Manner of reading itself, there are four principal rules. 1°, Read that you may accurately remember, but still more, that you may fully understand. 2°, Strive to compass the general tenor of a work, before you attempt to judge of it in detail. 3°, Accommodate the intensity of the reading to the importance of the work. Some books are, therefore, to be only dipped into; others are to be run over rapidly; and others to be studied long and sedulously. 4°, Regulate on the same principle the extracts which you make from the works you read.[a]

a Cf. Krug, *Logik*, § 180.—Ed. [Fischaber, *Logik*, p. 188, ed. 1818. Scheidler, *Grundriss der Hodegetik*, § 53, p. 106; 1832. Magirus, *Florilegium, v. Lectio.*]

LECT.
XXXV.

Explica-
tion.
I. Quantity
to be read.
Rule.

Solomon.

Quintilian.

Younger
Pliny.
Seneca.

Luther
quoted.

Sanderson.

I. In reference to the head of Quantity, the single rule is—Read much, but not many works. Though this golden rule has risen in importance, since the world, by the art of printing, has been overwhelmed by the multitude of books, it was still fully recognised by the great thinkers of antiquity. It is even hinted by Solomon, when he complains that "of making many books there is no end."[a] By Quintilian, by the younger Pliny, and by Seneca, the maxim— "multum legendum esse, non multa"—is laid down as the great rule of study.[b] "All," says Luther in his Table Talk,[γ] "who would study with advantage in any art whatsoever, ought to betake themselves to the reading of some sure and certain books oftentimes over; for to read many books produceth confusion, rather than learning, like as those who dwell everywhere, are not anywhere at home." He alludes here to the saying of Seneca, "Nusquam est qui ubique est."[b] "And like as in society, we use not daily the community of all our acquaintances, but of some few selected friends, even so likewise ought we to accustom ourselves to the best books, and to make the same familiar unto us, that is, to have them, as we use to say, at our fingers' ends." The great logician, Bishop Sanderson, to whom I formerly referred, as his friend and biographer Isaac Walton informs us, said "that he declined reading many books; but what he did read were well chosen, and read so often that he became very familiar with them. They were principally three,—Aristotle's *Rhetoric*, Aquinas's *Secunda Secundæ*, and Cicero, particularly

a *Eccles.*, xii. 12.—Ed.
β Quintilian, x. 1, 59. Pliny,
Ep., vii. 9. Seneca, *De Tranquill.
Animi*, c. 9; *Epist.*, 2, 45.—Ed.

γ No. DCCCXLIV. *Of Learned Men.*
—Ed.
b *Epist.*, H.—Ed.

his *Offices.*" [a] The great Lord Burleigh, we are told
by his biographer, carried Cicero *De Officiis,* with
Aristotle's *Rhetoric,* always in his bosom ; these being
complete pieces, "that would make both a scholar and
an honest man." "Our age," says Herder, "is the
reading age ;" and he adds, "it would have been
better, in my opinion, for the world and for science,
if, instead of the multitude of books which now overlay us, we possessed only a few works good and ster-
ling, and which, as few, would, therefore, be more
diligently and profoundly studied." [b] I might quote
to you many other testimonies to the same effect ;
but testimonies are useless in support of so manifest
a truth.

For what purpose,—with what intent, do we read ?
We read not for the sake of reading, but we read to
the end that we may think. Reading is valuable
only as it may supply to us the materials which
the mind itself elaborates. As it is not the largest
quantity of any kind of food, taken into the stomach,
that conduces to health, but such a quantity of such
a kind as can be best digested ; so it is not the
greatest complement of any kind of information that
improves the mind, but such a quantity of such a
kind as determines the intellect to most vigorous
energy. The only profitable reading is that in which
we are compelled to think, and think intensely ;
whereas that reading which serves only to dissipate
and divert our thought, is either positively hurtful,
or useful only as an occasional relaxation from severe
exertion. But the amount of vigorous thinking is

a See Walton's *Lives of Donne,*
Wotton, Hooker, Herbert, and San-
derson, vol. ii. p. 287, ed. Zouch,
York, 1817.—Ed.

β *Briefe über das Stud. der Theol.*
B. xlix., *Werke,* xlv. 287, ed. 1829.
—En.

usually in the inverse ratio of multifarious reading.
Multifarious reading is agreeable; but, as a habit, it
is, in its way, as destructive to the mental as dram-
drinking is to the bodily health.

II. Quality
of what is
to be read.
First Rule.
II. In reference to the Quality of what is to be read,
the First of the five rules is—' Select the works of
principal importance, in accommodation either to the
several sciences themselves, to your particular aim in
reading, or to your individual disposition and wants.'
This rule is too manifestly true to require any illustration of its truth. No one will deny that for the
accomplishment of an end, you ought to employ the
means best calculated for its accomplishment. This
is all that the rule inculcates. But while there is no
difficulty about the expediency of obeying the rule,
there is often considerable difficulty in obeying it. To
know what books ought to be read in order to learn
a science, is in fact frequently obtained only after the
science has been already learned. On this point no
general advice can be given. We have, on all of the
sciences, works which profess to supply the advice
which the student here requires. But in general, I
must say, they are of small assistance in pointing out
what books we should select, however useful they may
be in showing us what books exist upon a science.
In this respect, the British student also labours under
peculiar disadvantages. The libraries in this country
are, one and all of them, wretchedly imperfect; and
there are few departments of science, in which they
are not destitute even of the works of primary necessity,—works which, from their high price, but more
frequently from the difficulty of procuring them, are
beyond the reach of ordinary readers.

Under the head of Quality the Second Rule is—

'Read not the more detailed works upon a science,
until you have obtained a rudimentary knowledge of
it in general.' The expediency of this rule is sufficiently apparent. It is altogether impossible to read
with advantage an extensive work on any branch of
knowledge, if we are not previously aware of its general
bearing, and of the relations in which its several parts
stand to each other. In this case, the mind is over-
powered and oppressed by the mass of details presented to it,—details, the significance and subordination of which it is as yet unable to recognise. A conspectus,—a survey of the science as a whole, ought,
therefore, to precede the study of it in its parts; we
should be aware of its distribution, before we attend
to what is distributed,—we should possess the empty
framework, before we collect the materials with which
it is to be filled. Hence the utility of an encyclo-
pædical knowledge of the sciences in general, preliminary to a study of the several sciences in particular;
that is, a summary knowledge of their objects, their
extent, their connection with each other. By this
means the student is enabled to steer his way on the
wide ocean of science. By this means he always knows
whereabouts he is, and becomes aware of the point
towards which his author is leading him.

In entering upon the study of such authors as Plato,
Aristotle, Descartes, Spinoza, Leibnitz, Locke, Kant,
&c., it is, therefore, proper that we first obtain a preparatory acquaintance with the scope, both of their
philosophy in general, and of the particular work on
which we are about to enter. In the case of writers
of such ability this is not difficult to do; as there are
abundance of subsidiary works, affording the preliminary knowledge of which we are in quest. But in the

LECT.
XXXV.

case of treatises where similar assistance is not at hand, we may often, in some degree, prepare ourselves for a regular perusal, by examining the table of contents, and taking a cursory inspection of its several departments. In this respect and also in others, the following advice of Gibbon to young students is highly deserving of attention. " After a rapid glance (I translate from the original French)—after a rapid glance on the subject and distribution of a new book, I suspend the reading of it, which I only resume after having myself examined the subject in all its relations, —after having called up in my solitary walks all that I have read, thought, or learned in regard to the subject of the whole book, or of some chapter in particular. I thus place myself in a condition to estimate what the author may add to my general stock of knowledge ; and I am thus sometimes favourably disposed by the accordance, sometimes armed by the opposition, of our views." [a]

Gibbon quoted.

The Third Rule under the head of Quality is— ' Make yourselves familiar with a science in its present state, before you proceed to study it in its chronological development.' The propriety of this procedure is likewise manifest. Unless we be acquainted with a science in its more advanced state, it is impossible to distinguish between what is more or less important, and, consequently, impossible to determine what is or is not worthy of attention in the doctrines of its earlier cultivators. We shall thus also be overwhelmed by the infinitude of details successively presented to us ; all will be confusion and darkness, where all ought to

Third Rule.

[a] The substance of the above passage is given in English, in Gibbon's *Memoirs of my Life and Writings,* pp. 54, 55; ed. 1837. The French original is quoted by Scheidler, *Hodegetik,* § 55, p. 204.—ED.

be order and light. It is thus improper to study philosophy historically, or in its past progress, before we have studied it statistically, or in its actual results.

The Fourth Rule under the same head is—'To avoid erroneous and exclusive views, read and compare together the more important works of every party.' In proportion as different opinions may be entertained in regard to the objects of a science, the more necessary is it that we should weigh with care and impartiality the reasons on which these different opinions rest. Such a science, in particular, is philosophy, and such sciences, in general, are those which proceed out of philosophy. In the philosophical sciences, we ought, therefore, to be especially on our guard against that partiality which considers only the arguments in favour of particular opinions. It is true that in the writings of one party we find adduced the reasons of the opposite party; but frequently so distorted, so mutilated, so enervated, that their refutation occasions little effort. We must, therefore, study the arguments on both sides, if we would avoid those one-sided and contracted views which are the result of party - spirit. The precept of the Apostle, "Test all things, hold fast by that which is good," is a precept which is applicable equally in philosophy as in theology, but a precept that has not been more frequently neglected in the one study than in the other.

The Fifth Rule under the head of Quality is—' To avoid a one-sided development of mind, combine with the study of works which cultivate the Understanding, the study of works which cultivate the Taste.' The propriety of this rule requires no elucidation.

I therefore pass on to the third head—viz. the

LECT.
XXXV.

III. Manner
of Reading.
First Rule. Manner of reading itself; under which the First Rule is—'Read that you may accurately remember, but still more that you may fully understand.' This also requires no comment. Reading should not be a learning by rote, but an act of reflective thinking. Memory is only a subsidiary faculty,—is valuable merely as supplying the materials on which the understanding is to operate. We read, therefore, principally, not to remember facts but to understand relations. To commit, therefore, to memory what we read, before we elaborate it into an intellectual possession, is not only useless but detrimental; for the habit of laying up in memory what has not been digested by the understanding, is at once the cause and the effect of mental weakness.

Second
Rule. The Second Rule under this head is—'Strive to compass the general tenor of a work, before you attempt to judge of it in detail.' Nothing can be more absurd than the attempt to judge a part, before comprehending the whole; but unfortunately nothing is more common, especially among professional critics,—reviewers. This proceeding is, however, as frequently the effect of wilful misrepresentation, as of unintentional error.

Third Rule. The Third Rule under this head is—'Accommodate the intensity of the reading to the importance of the work. Some books are, therefore, to be only dipped into; others are to be run over rapidly; and others to be studied long and sedulously.' All books are not to be read with the same attention ; and, accordingly, Lectio cursoria.
Lectio stataria. an ancient distinction was taken of reading into *lectio cursoria* and *lectio stataria*. The former of these we have adopted in English, cursory reading being a familiar and correct translation of *lectio cursoria*.

But *lectio stataria* cannot be so well rendered by the expression of *stationary reading.* " Read not," says Bacon in his Fiftieth Essay—" read not to contradict and confute, nor to believe and take for granted, nor to find talk and discourse, but to weigh and consider. Some books are to be tasted, others are to be swallowed, and some few to be chewed and digested ; that is, some books are to be read only in parts ; others to be read, but not curiously ; and some few to be read wholly and with diligence and attention. Some books also may be read by deputy, and extracts made of them by others ; but that would be only in the less important arguments, and the meaner sort of books ; else distilled books are, like common distilled waters, fleshy things." "One kind of books," says the great historian, Johann von Müller," "I read with great rapidity, for in these there is much dross to throw aside, and little gold to be found ; some, however, there are all gold and diamonds, and he who, for example, in Tacitus can read more than twenty pages in four hours, certainly does not understand him."

Rapidity in reading depends, however, greatly on our acquaintance with the subject of discussion. At first, upon a science we can only read with profit few books, and laboriously. By degrees, however, our knowledge of the matters treated expands, the reasonings appear more manifest,—we advance more easily, until at length we are able, without overlooking anything of importance, to read with a velocity which appears almost incredible for those who are only commencing the study.

The Fourth Rule under this head is—' Regulate on

LECT. XXXV.

Bacon quoted.

Johann von Müller.

Fourth Rule.

a *Werke,* iv. 177. Cf. xvii. 253. 55, p. 204.—En.
Quoted by Scheidler, *Hodegetik,* §

the same principle the extracts which you make from
the works you read.'

So much for the Unilateral Communication of
thought, as a mean of knowledge. We now proceed
to the Mutual Communication of thought,—Conference.

*Conference,
—of two
kinds.* This is either mere Conversation,—mere Dialogue,
or Formal Dispute, and at present we consider both
of these exclusively, only as a means of knowledge,—
only as means for the communication of truth.

1. Dialogue. The employment of Dialogue as such a mean, requires great skill and dexterity ; for presence of mind,
confidence, tact, and pliability are necessary for this,
and these are only obtained by exercise, independently
of natural talent. This was the method which Socrates almost exclusively employed in the communication
of knowledge ; and he called it his *art of intellectual
midwifery,* because in its application truth is not given
over by the master to the disciple, but the master, by
skilful questioning, only helps the disciple to deliver
himself of the truth explicitly, which his mind had
before held implicitly. This method is not, however,
applicable to all kinds of knowledge, but only to those
which the human intellect is able to evolve out of itself, that is, only to the cognitions of Pure Reason.

*2. Disputa-
tion,—Oral
and Writ-
ten.* Disputation is of two principal kinds, inasmuch as it
is oral or written ; and in both cases, the controversy
may be conducted either by the rules of strict logical
disputation, or left to the freedom of debate. Without entering on details, it may be sufficient to state, in
*Academical
disputation.* regard to Logical Disputation, that it is here essential
that the point in question,—the *status controversia,*—
the thesis, should, in the first place, be accurately determined, in order to prevent all logomachy, or mere

verbal wrangling. This being done, that disputant who denies the thesis, and who is called the *opponent*, may either call upon the disputant who affirms the thesis, and who is called the *defendant*, to allege an argument in its support, or he may at once himself produce his counter-argument. To avoid, however, all misunderstanding, the opponent should also advance an antithesis, that is, a proposition conflictive with the thesis, and when this has been denied by the defendant the process of argumentation commences. This proceeds in regular syllogisms, and is governed by definite rules, which are all so calculated that the discussion is not allowed to wander from the point at issue, and each disputant is compelled, in reference to every syllogism of his adversary, either to admit, or to deny, or to distinguish.[a] These rules you will find in most of the older systems of Logic; in particular I may refer you to them as detailed in Heerebord's *Praxis Logica,* to be found at the end of his edition of the *Synopsis* of Burgersdicius. The practice of disputation was long and justly regarded as the most important of academical exercises; though liable to abuse, the good which it certainly insures greatly surpasses the evil which it may accidentally occasion.

a Cf. Krug, *Logik,* § 186. Anm. 138.—Ed.
2. Scheidler, *Hodegetik,* § 45, p.

APPENDIX.

APPENDIX.

I.

THE CHARACTER AND COMPREHENSION OF LOGIC. —A FRAGMENT.

(See above, Volume I., page 4.)

IN the commencement of a course of academical instruction, there are usually two primary questions which obtrude themselves; and with the answer to these questions I propose to occupy the present Lecture.

The first of these questions is,—What is the character and comprehension of the subject to be taught? The second,—What is the mode of teaching it? In regard to the former of these, the question,—What is to be taught,—in the present instance is assuredly not superfluous. The subject of our course is, indeed, professedly *Logic;* but as under that rubric it has been too often the practice, in our Scottish Universities, to comprehend almost everything except the science which that name properly denotes, it is evident that the mere intimation of a course of Lectures on Logic does not of itself definitely mark out what the professor is to teach, and what the student may rely on learning.

I shall, therefore, proceed to give you a general notion of what Logic is, and of the relation in which it stands to the other sciences, for Logic,—Logic properly so called,—is the all-important science in which it is at once my duty and my desire fully and faithfully to instruct you.

The very general,—I may call it the very vague,—conception which I can at present attempt to shadow out of the scope and nature of Logic, is of course not intended to anticipate what is hereafter to be articulately stated in regard to the peculiar character of this science.

All science, all knowledge, is divided into two great branches; for it is either, 1°, Conversant about Objects Known, or, 2°, Conversant about the Manner of knowing them, in other words, about the laws or conditions under which such objects are cognisable. The former of these is Direct Science, or Science simply; the latter, Reflex Science,—the Science of Science, or the Method of Science.

Now of these categories or great branches of knowledge, Simple Science, or Science directly conversant about Objects, is again divided into two branches; for it is either conversant about the phœnomena of the internal world, as revealed to us in consciousness, or about the phænomena of the external world, as made known to us by sense. The former of these constitutes the Science of Mind, the latter the Science of Matter; and each is again divided and subdivided into those numerous branches, which together make up nearly the whole cycle of human knowledge.

The other category,—the Science of Science, or the Methodology of Science,—falls likewise into two branches, according as the conditions which it considers are the laws which determine the possibility of the mind, or subject of science, knowing, or the laws which determine the possibility of the existence, or object of science, being known; Science, I repeat, considered as reflected upon its own conditions, is twofold, for it either considers the laws under which the human mind can know, or the laws under which what is proposed by the human mind to know, can be known. Of these two sciences of science, the former,—that which treats of those conditions of knowledge which lie in the nature of thought itself,—is Logic, properly so called; the latter,—that which treats of those conditions of knowledge which lie in the nature, not of thought itself, but of that which we think about,—this has as yet obtained no recognised appellation, no name by which it is universally and familiarly known. Various denominations have indeed been given to it in its several parts or in its special relations; thus it has been called *Heuristic*, in so far as it expounds the rules of Invention or Discovery, *Architectonic*, in so far as it treats of the method of building up our observations into system; but hitherto it has obtained, as a whole, no adequate and distinctive title. The consequence, or perhaps the cause, of this want of a peculiar name to mark out the second

science of science, as distinguished from the first, is that the two
have frequently been mixed up together, and that the name of
Logic has been stretched so as to comprehend the confused assemblage of their doctrines. Of these two sciences of the conditions
of knowledge, the one owes its systematic development principally to Aristotle, the other to Bacon; though neither of
hese philosophers has precisely marked or rigidly observed the
limits which separate them from each other; and from the circumstance, that the latter gave to his great Treatise the name of
Organum,—the name which has in later times been applied to
designate the complement of the Logical Treatises of the former,
—from this circumstance, I say, it has often been supposed, that
the aim of Bacon was to build up a Logic of his own upon the
ruins of the Aristotelic. Nothing, however, can be more erroneous, either as to Bacon's views, or as to the relation in which the
two sciences mutually stand. These are not only not inconsistent,
they are in fact, as correlative, each necessary to, each dependent
on, the other; and although they constitute two several doctrines,
which must be treated in the first instance each by and for itself,
they are, however, in the last resort only two phases,—two members, of one great doctrine of method, which considers, in the
counter relations of thought to the object, and of the object to
thought, the universal conditions by which the possibility of
human knowledge is regulated and defined.

But allowing the term *Logic* to be extended so as to denote the
genus of which these opposite doctrines of Method are the species,
it will, however, be necessary to add a difference by which these
special Logics may be distinguished from each other, and from the
generic science of which they are the constituents. The doctrine,
therefore, which expounds the laws by which our scientific procedure should be governed, in so far as these lie in the forms of
thought, or in the conditions of the mind itself, which is the subject in which knowledge inheres,—this science may be called *Formal*, or *Subjective*, or *Abstract*, or *Pure Logic*. The science, again,
which expounds the laws by which our scientific procedure should
be governed, in so far as these lie in the contents, materials, or objects, about which knowledge is conversant,—this science may be
called *Material*, or *Objective*, or *Concrete*, or *Applied Logic*.

Now it is Logic, taken in its most unexclusive acceptation,

which will constitute the object of our consideration in the following course. Of the two branches into which it falls, Formal Logic, or Logic Proper, demands the principal share of our attention, and this for various reasons. In the first place, considered in reference to the quantity of their contents, Formal Logic is a far more comprehensive and complex science than Material. For, to speak first of the latter:—if we abstract from the specialities of particular objects and sciences, and consider only the rules which ought to govern our procedure in reference to the object-matter of the sciences in general,—and this is all that a universal Logic can propose,—these rules are few in number, and their applications simple and evident. A Material or Objective Logic, except in special subordination to the circumstances of particular sciences, is, therefore, of very narrow limits, and all that it can tell us is soon told. Of the former, on the other hand, the reverse is true. For though the highest laws of thought be few in number, and though Logic proper be only an articulate exposition of the universal necessity of these, still the steps through which this exposition must be accomplished, are both many and multiform.

In the second place, the doctrines of Material Logic are not only far fewer and simpler than those of Formal Logic, they are also less independent; for the principles of the latter, once established, those of the other are either implicitly confirmed, or the foundation laid on which they can be easily rested.

In the third place, the study of Formal Logic is a more improving exercise; for, as exclusively conversant with the laws of thought, it necessitates a turning back of the intellect upon itself, which is a less easy, and, therefore, a more invigorating, energy, than the mere contemplation of the objects directly presented to our observation.

In the fourth place, the doctrines of Formal Logic are possessed of an intrinsic and necessary evidence, they shine out by their native light, and do not require any proof or corroboration beyond that which consciousness itself supplies. They do not, therefore, require, as a preliminary condition, any apparatus of acquired knowledge. Formal Logic is, therefore, better fitted than Material, for the purposes of academical instruction; for the latter, primarily conversant with the conditions of the external world, is in itself a less invigorating exercise, as determining the mind to a feebler and

more ordinary exertion, and, at the same time, cannot adequately
be understood without the previous possession of such a complement of information, as it would be unreasonable to count upon
in the case of those who are only commencing their philosophical studies.

II.

GENUS OF LOGIC.

(See above, Vol. I., p. 9.)

I.—SCIENCE.

A. Affirmative.

Stoici, (v. Alexander Aphrod. *In Topica*, Proœm.; Diogenes
Laertius, *Vita Zenonis*, L. vii., § 42). "Plato et Platonici et
Academici omnes," (v. Camerarius, *Selectæ Disput. Philos.*, Pars
i., qu. 3, p. 30).

(a)—SPECULATIVE SCIENCE.

Toletus, *In Univ. Arist. Log., De Dial. in Communi*, Qu. ii., iv.
" Communiter Thomistæ, ut Capreolus, Sotus, Masius, Flandra,
Soncinas, Javellus: Omnes fere Scotistæ cum Scoto, ut Valera,
Antonius Andreas, &c." (v. Ildephonsus de Penafiel, *Logicæ Dis-
putationes*, Disp. i. qu. 4. *Cursus*, p. 79.) For Aquinas, Durandus,
Niphus, Canariensis, see Antonius Ruvio, *Comm. in Arist. Dialect.*,
Proœm. qu. 5. For Bacchonus, Javellus, Averroes, see Conimbri-
censes, *In Arist. Dial.* Proœm. Q. iv. art. 5. Lalemandet, *Cur-
sus Phil., Logica*, Disp. iii. part iii. Derodon, *Logica Restit., De
Genere*, p. 45. Camerarius, *Disp. Phil.*, Pars i., qu. 3, 4. (That *Logica docens* a true science.) For Pseudo-Augustinus, Avicenna,
Alpharabius, see Conimbricenses, *Comm. in Arist. Dial.*, Proœm.
Qu. iv. art. 3. For Boethius, Mercado, Vera Cruce, Montanesius,
see Masius, *Comm. in Porph. et in Universam Aristotelis Lo-
gicam*, Sect. i., Proœm. qu. v. *et seq.* Poncius, *De Nat. Log.*, Disp.
ii., concl. 2. For Rapineus, Petronius, Faber, see Camerarius,
Sel. Disp. Phil., Pars i., qu. 4, p. 44.

(b)—PRACTICAL SCIENCE.

Conimbricenses, *In Universam Aristotelis Dialecticam*
Proœm. Q. iv., art. 5. Fonseca, *In Metaph.* L. ii. c. 3, qu.
1, § 7. For Venetus, Albertus Magnus, Jandonus, see Ruvio,
l. c. Schuler, *Philosophia nova Methodo Explicata*, Pars Prior,
L. v. ex. i., p. 306. (1603.) D'Abra de Raconis, *Summa Totius
Philosophiæ*, *Log. Præl.*, c. i. Isendoorn, *Cursus Logicus*, L.
i., c. 2, qu. 7. Biel, *In Sentent.*, L. ii. Prol. Occam, *Summa
Totius Logicæ*, D. xxxix. qu. 6. For Aureolus, Bern. Mirandulanus,
see *Conimbricenses*, *l. c.* For Mathisius, Murcia, Vasquez, Eckius,
see Camerarius, *Scl. Disp. Phil.* Pars i., qu. 4, p. 44. Ildephonsus de Penafiel, *Log. Disp.* D. i. qu. 4, sect. 2. Oviedo, *Cursus
Philosophicus*, *Log.*, Contr. Proœm. ii. 5. Arriaga, *Cursus Philosophicus*, Disp. iii. § 4.

(c)—SPECULATIVE AND PRACTICAL.

Suarez, *Disp. Metaph.*, Disp. l. § iv. 26; Disp. xliv. § xiii.
54. Hurtado de Mendoza, *Log. Disp.* D. ii. § 2.

B. Negative.

For almost all the Greek commentators, see Zabarella, *Opera
Logica, De Nat. Log.*, L. i. c. 5, and Smiglecius, *Logica*, D. ii. qu. 5.
See also Ildephonsus de Penafiel, *Disp. Log.* D. i. qu. i, § 1, p. 67.

II.—ART.

Scheibler, *Opera Logica*, Pars i. c. 1, p. 49. J. C. Scaliger,
Exercitationes, Exerc. i. 3. G. J. Vossius, *De Natura Artium*,
L. iv., c. 2, § 4. Balforeus, *In Org.* Q. v. § 6, Proœm., p. 31.
Burgersdicius, *Institutiones Logicæ*, Lib. i. c. 1. Pacius, *Comm.
in Org.* p. 1. Sanderson, *Log. Artis Compendium*, L. i. c. 1, p. 1.
Cf. p. 192. Aldrich, *Artis Log. Compendium*, L. i. c. 1, p. 1.
Hildenius, *Quæstiones et Commentaria in Organon*, p. 579 (1585.)
Goclenius, *Problemata Logica et Philosophica*, Pars i. qu. 3.
Ramus, *Dialectica*, L. i. c. 1. Augustinus, *De Ordine*, ii. c. 15.
Cicero, *De Claris Oratoribus*, c. 41; *De Oratore*, L. ii., c. 38.

Lovanienses, *Comm. in Arist. Dial.* Præf. p. 3. Rodolphus Agricola, *De Dialecticæ Inventione*, L. ii. p. 255. Monlorius, (Bapt.), *Comm. in Anal. Pr.* Præf. Nunnesius, *De Constitut. Dial.*, p. 43. Downam, (Ramist), *Comm. in Ram. Dial.*, L. i. c. 1, p. 3. Paracus, *Ars Logica*, p. 1, 1670. For Horatius Cornachinus, Ant. Bernardus Mirandulanus, Flamminius Nobilius, see Camerarius, *Sd. Disp. Phil.*, Pars i. q. 3, p. 30.

III.—SCIENCE AND ART.

Lalemandet, *Log.*, Disp. iii. Part iii. cl. 4. (*Logica utens*, an art; *Logica docens*, a speculative science.) Tartaretus, *In P. Hispanum*, f. 2, (Practical Science and Art.) P. Hispanus, *Copulata Omn. Tractat. Pd. Hisp. Parv. Logical.* T. i. f. 10, 1490. *Philosophia Vetus et Nova in Regia Burgundia olim Pertractata, Logica*, T. I., pp. 58, 59. 4th ed. London, 1685. Tosca, *Comp. Phil. Log.*, Tr. i. l. iv. c. 4, p. 208, (Practical Science and Art). Purchot, *Instit. Phil.*, T. I. Procem. p. 36. Eugenius, Λογική, pp. 140, 141. Dupleix, *Logique*, p. 37. Facciolati, *Rudimenta Logicæ*, p. 5. Schmier *Philosophia Quadripartita*, (v. Heumannus, *Acta Philosoph.*, iii. p. 67.) Aquinas (in Caramuel, *Phil. Realis et Rationalis*, Disp. ii. p. 3).

IV.—NEITHER SCIENCE NOR ART, BUT INSTRUMENT, ORGAN, OR HABIT, OR INSTRUMENTAL DISCIPLINE.

Philoponus, *In An. Prior., initio.* For Ammonius, (*Præf. in. Præd.*), Alexander, (*In Topica*, i. c. 4; *Metaph.* ii. t. 15). Simplicius, (*Præf. in Præd.*), Zabarella, (*De Natura Logicæ*, L. i. c. 10.), Zimara, (*In Tabula v. Absurdum*,), Averroes, see Smiglecius, *Logica*, Disp. ii. qu. 6, p. 80. Aegidius, *In An. Post.* L. i. qu. 1. For Magnesius, Niger (Petrus), Villalpandeus, see Ruvio, *In Arist. Dial.*, procem. qu. 2. F. Crellius, *Isagoge Logica*, L. i. c. 1, p. 5. P. Vallius, *Logica*, T. I. procem. c. i *et alibi*. Bartholinus, *Janitores Logici*, II. pp. 25 and 76. Bertius, *Logica Peripatetica*, pp. 6, 10. Themistius, *An. Post.* i. c. 24. Aquinas, *Opuscula*, 70, qu. *De Divisione Scientiæ Speculativæ*,—sed alibi

scientiam vocat. (See Conimbricenses, *In Arist. Dial.*, T. I. qu.
iv. art. 5, p. 42). Balduinus, *In Quæsito an Logica sit Scientia.*
Scaynus, *Paraphrasis in Organon*, Præf. p. 9.

V.—THAT, LOOSELY TAKING THE TERMS, LOGIC IS EITHER ART, OR SCIENCE, OR BOTH.

Zabarella, *Opera Logica, De Nat. Log.*, L. i. c. viii. D'Abra
de Raconis, *Summa Tot. Phil. Præl. Log.*, L. iii, c. 1, p. 8, ed.
Colon., (Practical Science). Balforeus, *In Organon*, Q. v. §§ 1, 6,
pp. 20, 32. (Art). Derodon, *Logica Restit. De Procem. Log.*,
p. 49, (Speculative Science). Crellius, *Isagoge*, pp. 1, 4. Bertius,
Logica Peripatetica, pp. 11, 13. Aldrich, *Art. Log. Comp.*, L. ii.
c. 8, T. i. (Art). Sanderson, *Log. Art. Comp. Append. Pr.*, c. 2,
p. 192. (Art). Conimbricenses, *In Arist. Dial.*, T. I., p. 33. (Practical Science). *Philosophia Burgundia*, T. I. pp. 56, 59. Eustachius,
Summa Philosophiæ, Dialectica, Quæst. Procem., i. p. 4. Nunnesius, *De Constit. Dial.*, ff. 43, 68. Scheibler, *Opera Logica*,
pp. 48, 49. Scaynus, *Par. in Org.*, pp. 11, 12. Camerarius, *Sel.
Disp. Phil.*, Pars i. qu. 3, pp. 31, 38 (Speculative Science). B.
Pereira, *De Commun. Princip. Omn. Rer. Natural.*, L. i. *De Phil.*
c. 18, p. 60, 1618.

VI.—THAT AT ONCE SCIENCE (PART OF PHILOSOPHY) AND INSTRUMENT OF PHILOSOPHY.

Boethius, *Præf. in Porphyr.* (a Victorino Transl.), *Opera*, p. 48.
Eustachius, *Summa Philosophiæ*, p. 8, (Scientia organica et practica.) For Simplicius, Alexander, Philoponus, &c., see Camerarius,
Sel. Disp. Phil., p. 30. Pacius, *Com. in. Arist. Org.*, p. 4.

VII.—THAT QUESTION, WHETHER LOGIC PART OF PHILOSOPHY OR NOT, AN IDLE QUESTION.

Pacius, *Comm. in Arist. Org.*, p. 4. Avicenna, (in Conimbricenses, *In Arist. Dial.*, Qu. iv. art. 4, T. I. p. 38.)

Buffier, *Cours des Sciences, Seconde Logique*, § 421, p. 687.

Eugenius, 'Η Λογική, p. 140, has the following :—

" From what has been said, therefore, it clearly appears of what
character are the diversities of Logic, and what its nature. For
one logic is *Natural*, another *Acquired*. And of the Natural,
there is one sort according to *Faculty*, another according to *Disposition*. And of the Acquired, there is again a kind according
to *Art*, and a kind according to *Science*. And the Natural
Logic, according to Faculty, is the rational faculty itself with
which every human individual is endowed, through which all
are qualified for the knowledge and discrimination of truth, and
which, in proportion as a man employs the less, the less is he
removed from irrationality. But the Natural Logic, according
to Disposition, is the same faculty by which some, when they
reason, are wont to exert their cogitations with care and attention, confusedly, indeed, and uncritically, still, however, in
pursuit of the truth. The Acquired, according to Art, is the
correct and corrected knowledge of the Rules, through which
the intellectual energies are, without fault or failure, accomplished. But the Acquired, according to Science, is the exact
and perfect knowledge both of the energies themselves and also
of the causes through which, and through which exclusively,
they are capable of being directed towards the truth."

Logic.
- Natural, according to — Faculty. Disposition.
- Acquired, according to — Art. Science.

"And thus Disposition adds to Faculty consuetude and a
promptness to energise. Art, again, adds to Disposition a refinement and accuracy of energy. Finally, Science adds to Art
the consciousness of cause, and the power of rendering a reason
in the case of all the Rules. And the natural logician may be
able, in his random reason, to apprehend that, so to speak, one
thing has determined another, although the nature of this deter-

mination may bo beyond his ken.   But he whose disposition is
exercised by reflection and imitation, being able easily to connect
thought with thought, is cognisant of the several steps of the
reasoning process, howbeit this otherwise may be confused and
disjointed.   But he who is disciplined in the art, knows exactly
that, in an act of inference, there are required three terms, and
that these also should be thus or thus connected.   Finally, the
scientific logician understands the reason, — why three terms
enter into every syllogism,—why they are neither more nor
fewer,—and why they behove to be combined in this, and in no
other fashion.

"Wherefore to us the inquiry appears ridiculous, which is frequently, even to nausea, clamorously agitated concerning Logic
—Whether it should be regarded as an *Art* or as a *Science*."

III.

DIVISIONS, VARIETIES, AND CONTENTS OF LOGIC.

(See above, Vol. I., p. 68.)

I. Logica,	Docens, χωρὶς πραγμάτων. Utens, ἐν χρήσει καὶ γυμνασίᾳ πραγμάτων.	v. Timpler, *Systema Logicæ*, p. 4. Isendoorn, *Effata*, Cent. i. Eff. 53. Crellius, *Isagoge*, Para Prior, L. i. a. i. p. 12. Noldius, *Logica Recognita*, Proœm. p. 13. Philoponus, *In An. Pr.*, f. 4 b. Alstedius, *Encyclopædia*, pp. 29, 406. v. Aristotle, *Metaph.*, L. vii. text 21.
II. Logica,	Doctrinalis } [Objec- Systematica } tiva]. Habitualis,[Subjectiva].	v. Timpler, *Syst. Log.*, Appendix, p. 877. Noldius, *Log. Recog.*, Proœm. p. 13.
III. Logica,	Pars Communis, Generalis. Pars Propria, Specialis.	Adopted in different significations by Timpler, *Syst. Log.*, p. 55; Theoph. Gale, *Logica*, (1681), pp. 6, 246 et seq. Crellius, *Isagoge*, Para Prior, L. i. a. i. p. 8. See also Alstedius, *Encyclop.*, pp. 29, 406.
IV. Logica,	Pura. Applicata.	N.B.—Averroes, (Pacius, *Comm.* p. 2), has "Logica appropriata seu particularis," and "Logica communis"= Universal, Abstract Logic.
V. Logica,	Abstracta. Concreta.	
VI. Logica,	Pars Communis. Pars Propria. { Apodictica, Dialectica, Sophistica.	v. Timpler, *Syst. Log.*, p. 42. Isendoorn, *Effata*, Cent. i. Eff. 54.

VII. Logica,	Εὑρετική vel τεχνική, Inventio. Κριτική, Judicium, Dispositio.	v. Timpler, *Syst. Log.*, p. 44. Rejected by Crellius, *Isagoge*, pp. 10, 11, and Isendoorn, *Effata*, Cent. i. Eff. 51. Adopted by Agricola, *De Inv. Dial.*, L. i. p. 35; Melanchthon, *Erot. Dial.*, p. 10; Ramus, *Schol. Dialect.*, L. i. c. i., and L. ii. c. i. p. 351 *et seq.*; Spencer, *Log.*, p. 11; Downam, *In Rami Dial.*, L. i. c. 2, p. 14; Perionius, *De Dialectica*, L. i. p. 6, (1544); Vossius, *De Nat. Artium*, Lib. iv., *De Logica*, c. ix. § 2 *et seq.*, p. 217.
VIII. Logica,	Pars de Propositio. Pars de Judicia.	v. Timpler, *Syst. Log.*, p. 49.
IX. Logica,	Doctrina Dividendi. Doctrina Definiendi. Doctrina Argumentandi.	v. Timpler, *Syst. Log.*, p. 51. Isendoorn, *Effata*, Cent. i. Eff. 57. Boethius, Augustin, Fonseca, &c.
X. Logica,	Simplicis Apprehensionis. Judicii. Ratiocinationis. *aliter* Noëtica, (*melius* Noëmatica.) Synthetica. Dianoetica.	v. Timpler, *Syst. Log.*, p. 52. Isendoorn, *Effata*, Cent. i. Eff. 58. Isendoorn, *Cursus Logicus*, p. 31, and *Effata*, Cent. i. Eff. 59. Noldius, *Log. Rec.*, p. 9. Aquinas.
XI. Logica,	1. Ideas (notions). 2. Judgment. 3. Reasoning. 4. Method.	*L'Art de Penser*, Part i. Clericus, *Logica*, adopts this division, but makes Method third, Reasoning fourth.
XII. Logica,	1. Doctrine of Elements. 2. Doctrine of Method.	Kant, *Logik*; Krug, *Logik*.

1st called Analytic by Metz, *Instit. Log.*, § 105, p. 71; Twesten, *Die Logik, insbesondere die Analytik*, p. lii.; Esser, *Logik*, Part i. 2d called Systematic or Architectonic by Bachmann, *Logik*, Part ii.; called Synthetic by Esser (who includes under it also Applied Logic), *Logik*, Part ii.

XIII. Logica partes,
{ Thematica—de materia operationi Logicæ subjecta.
Organica — de instrumentis sciendi. }
Mark Duncan, *Institutiones Logicæ*, Proleg. c. iii. § 2, p. 22. Burgersdicius, *Instit. Log.*, L. i. c. l. p. 5.

XIV. Logica,

Communis, Generalia,
1. De ordinibus rerum generalibus et attributis communissimis.
2. De Vocibus et Orationo.
3. De Ideis simplicibus et apprehensione simplici dirigenda.
4. De Judicio et Propositione.
5. De Discursu.
6. De Dispositione seu Methodo.

Specialia,
Genesis seu Inventio { Genesis stricta. Genesis didactica.
Analysis { Hermeneutica. Analytica and Critica.

Genetica,
In ordine ad mentem — Logica stricte dicta.
In ordine ad alios—Interpretativa vel Hermeneutica genetica.

Analytica,
Hermeneutica analytica.
Analytica stricte vel in specie.

Theophilus Gale (*Logica*, 1681), follows, (besides Keckermann and Burgersdyk), principally Clauberg and *L'Art de Penser* of Port Royal.

XV. Logica,
{ Theoretica pars.
Practica pars—(this including the Methodology and Applied Logic of Kant.) }
Wolf, *Philos. Rationalis*, Pars I. and ii.

XVI.
{ On Adrastean order, &c. of the books of the Organon, *vide* Ramus, *Scholæ Dial.*, L. ii. c. 8, p. 354. Piccartus, *In Organum*, Prolegomena, p. 1 *et seq*. }

XVII. Logica partes,
1. Περὶ προσλήψεως.
2. Περὶ σκέψεως.
3. Περὶ κρίσεως.
4. Περὶ βουλίας.
5. Περὶ μεθόδου.
} Eugenius Diaconus, Λογική, p. 144.

XVIII. Logica,	1. Emendatrice. 2. Inventrice. 3. Giudicatrice. 4. Ragionatrice. 5. Ordinatrice.	Genovesi. A division different in some respects is given in his Latin Logic, Prolog. § 51, p. 22. The fourth part of the division in the Latin Logic is omitted in the Italian, or rather reduced to the second, and the fifth divided into two.
XIX. Logica,	Vetus, { Porphyrii Isag. Præd. Interpret. } Nova, { Analyt. Pr. Analyt. Post. Top. Elench. }	Isendoorn, *Effata*, Cent. i. Eff. 52. Reason of terms, Pacius, *Comment. in Org., in Porph. Isag.* p. 3.
XX. Logica,	Στοιχειολογική. Συλλογιστική. { Apodictica. Topica. Sophistica. }	Isendoorn, *Effata*, Cent. i. Eff. 56. (From John Hospinian, *De Controversiis Dialecticis*.)
XXI. Logica,	Στοιχειολογική. Συλλογιστική. { Analytica { Priora. Posteriora. } Dialectica { Topica. Sophistica. } }	Vossius, *De Nat. Art.*, L. iv. c. ix. § 12, p. 220.
XXII. Logica,	Analytica { Prodromus de Interpretatione. Universe de Syllogismo. Speciatim de Demonstratione. } Dialectica { Prodromus de Categoriis. De Syll. Verisimili. De Syll. Sophistico sive Pirastico. }	Vossius, *De Nat. Art.*, L. iv. c. ix. §§ 13, 14, p. 220.
XXIII. Logica,	Dialectica. Analytica.	Aristotle, in Laertius v. Vossius, *De Nat. Art.*, L. iv. c. ix. § 11, p. 219.
XXIV. Logica,	De Rebus quæ significantur. De Vocibus quæ significant.	Stoicorum, see Vossius, *De Nat. Art.*, L. iv. c. ix. § 7, p. 218.
XXV. Logicæ partes,	De Loquendo. De Eloquendo. De Proloquendo. Proloquiorum summa.	Varro, vide Vossius, *De Nat. Art.*, L. iv. c. ix. § 8, p. 219.

XXVI. Logica,	Πρὸς εὕρεσιν. Πρὸς κρίσιν. Πρὸς χρῆσιν.	Aristotle (?) in Laertium, L. v. § 28, p. 281., Alexander Aphrod. ibi, in nota Aldobrandini.
Logica,	Νοητικὴ, Apprehensiva. Ἐπιστημονικὴ vel Κριτικὴ, Judicativa. Διαλεκτικὴ, Argumentativa.	Caramuel Lobkowitz, *Rationalis et Realis Philosophia Logica seu Phil. Rat.*, Disp. ii. p. 2.
Logicæ partes,	Divisio. Definitio. Argumentatio.	v. Crellius, *Isagoge*, Pars prior, L. i. c. i. p. 10.
Logicæ partes,	Apodictica. Dialectica. Sophistica.	v. Crellius, *Isagoge*, Pars prior, L. i. c. i. p. 10. Isendoorn, *Effata*, Cent. i. Eff. 54.
Logicæ partes,	Analytica. Topica.	Crellius, *Isagoge*, Pars prior, L. i. c. i. p. 10.

Stoicheiology (pure) should contain the doctrine of Syllogism, without distinction of Deduction or Induction. Deduction, Induction, Definition, Division, from the laws of thought, should come under pure Methodology. All are processes. (v. Cæsalpinus, *Quæst. Perip.*, sub init.)

Perhaps, 1°, Formal Logic, (from the laws of thought proper), should be distinguished from, 2°, Abstract Logic, (material, but of abstract general matter); and then, 3°, A Psychological Logic might be added as a third part, considering how Reasoning, &c., is affected by the constitution of our minds. Applied Logic is properly the several sciences.

Or may not Induction and Deduction come under abstract Material Logic?

IV.

NOTE OF LOGICAL TREATISES RECOMMENDED BY SIR WILLIAM HAMILTON TO HIS CLASS.

(See above, Vol. I., p. 71.)

I. Editions of Aristotle's Organon, and works immediately relative to the Aristotelic Treatises.

1. Organon Pacii.
2. Organon Waitzii.
   Both of these have the original texts.
3. A French Translation of the Organon by St Hilaire.
4. Synopsis Organi by Trendelenburg, in Latin.

II. Modern Systems of Logic—Foreign and British.

1. The *Port Royal Logic*, L'Art de Penser—of which a stereo-typed edition was issued in Paris by Hachette. This work has been well translated into English by Mr Baynes, and published in this city.

2. The Italian and the Latin Logics of *Genovesi* are worthy of your attention. The Italian has been once and again reprinted, and, with the valuable additions of Romagnosi, can easily be obtained. The Latin Logic (the Genuensis Ars Logica Critica) is comparatively rare.

3. In Latin there is a very elegant compend by the late illustrious *Daniel Wyttenbach* of Leyden; and, besides the Dutch editions, there is a cheap reprint of this work published by Professor Maass of Halle, who has, however, ventured on the un-warrantable liberty of silently altering the text, besides omitting what he considered as not absolutely indispensable for a text-book.

4. The Logic of *Facciolati*, of Padua, is also written with classical elegance, and on every account merits a careful study; but though frequently reprinted in Italy, the various treatises of which it is composed are by no means easily procured in a collected form.

5. The Latin Logic of *Burgersdicius* has been repeatedly re-printed at the English universities, where for a long time it deserv-

edly enjoyed a high reputation, and has been latterly superseded in them from no want of relative merit in itself.

6. There are two Logical works of the celebrated *Christian Wolf.*

(1.) The vernacular Logic of this author, originally published in German in 1712, 8vo. Of this there is a good English translation, entitled "Logic, or Rational Thoughts," &c., published in London, 1770, also in 8vo.

(2.) Logica sive Philosophia Rationalis, first published at Leipsic, 1728, 4to. It has been frequently reprinted.

(3.) There is likewise an excellent compend of Wolf's Logic by *Frobesius,* 1746, in 4to, entitled *Christiani Wolfii Philosophia Rationalis sive Logica.* This book, however, is not merely an abridgment; it contains many original views of great value.

7. Of English logical works I may mention those published in Oxford by *Dr Wooley, Mr Mansel,* and *Mr Thomson.* Among the books of Mr Mansel is an abridgment of the Compendium of Aldrich with a commentary; the entire work, however, is well deserving of your study.

8. I need hardly name the Logic of *Archbishop Whately; Mr Mill's Logic,* although he is not a member of either English University, I may also recommend to your attention; and there are published by North American authors several Logic treatises of great merit, as the works of Dr Hickock and Dr Wayland.

V.

LAWS OF THOUGHT.

(See Vol. I, p. 79.)

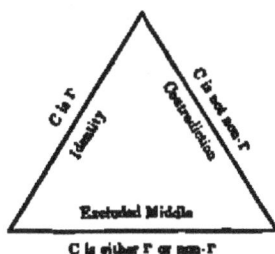

The laws of Identity and Contradiction, each infers the other, but only through the principle of Excluded Middle; and the principle of Excluded Middle only exists through the supposition of the two others. Thus, the principles of Identity and Contradiction cannot move,—cannot be applied, except through supposing the principle of Excluded Middle; and this last cannot be conceived existent, except through the supposition of the two former. They are thus co-ordinate but inseparable. Begin with any one, the other two follow as corollaries.

(a)—PRIMARY LAWS OF THOUGHT,—IN GENERAL.

See the following authors on :—Dreier, *Disput. ad Philosophiam Primam*, Disp. v. Aristotle, *Analyt. Post.* i. c. 11, §§ 2, 3, 4, 5, 6, 7. Schramm, *Philosophia Aristotelica*, p. 36. Lippius, *Metaphysica Magna*, L. i. c. i., p. 71 *et seq.* Stahl, *Regulæ Philosophicæ*, Tit. i., reg. i p. 2 *et seq.*, reg. ii., p. 8 *et seq.*, Tit. xix. reg. viii., p. 520 *et seq.* Chauvin, *Lexicon Philosophicum, v. Metaphysica.* Bisterfeld, evolves all out of *ens*,—*ens est.* See *Philosophia Prima*, c. ii., p. 24 *et seq.* Dobrik, *System der Logik*, § 70, p. 247 *et seq.*

Laws of Thought are of two kinds :—1°. The laws of the Thinkable,—Identity, Contradiction, &c. 2°. The laws of Thinking in a strict sense—viz. laws of Conception, Judgment, and Reasoning. See Scheidler, *Psychologie*, p. 15, ed. 1833.

That they belong to Logic :—Ramus *Schol. Dial.*, L. ix. p. 549.

Is Affirmation or Negation prior in order of thought ? and thus on order and mutual relation of the Laws among themselves, as co-ordinate or derived ; (see separate Laws). Fracastorius, *Opera, De Intellectione*, L. i. f. 125 b., makes negation an act prior to affirmation ; therefore principle of Contradiction prior to principle of Identity.—Esser, *Logik*, § 28, p. 57. Sigwart, *Handbuch zu Vorlesungen über die Logik*, § 38 *et seq.* Piccolomineus, *De Mente Humana*, L. iii. c. 4, p. 1301, on question—Is affirmative or negative prior? Schulz, *Prüf. der Kant. Krit. der reinen Vernunft*, l. p. 78, 2d ed. Weiss, *Lehrbuch der Logik*, § 81 *et seq.* pp. 61, 62, 1805. Castillon, *Mémoires de l'Académie de Berlin* (1803) p. 8, (Contradiction and Identity co-ordinate). A. Andreas, *In Arist. Metaph.* iv. Qu. 5, p. 21. (Affirmative prior to negative.) Leibnitz, *Œuvres Philosophiques, Nouv. Essais*, L. iv. ch. 2, § 1, p. 327, ed. Raspe. (Identity prior to Contradiction.) Wolf, *Ontologia*, §§ 55, 268—(Contradiction first, Identity second.) Derodon, *Metaphysica*, c. iii. p. 75 *et seq.* 1669. (Contradiction first, Excluded Middle second, Identity third). Fonseca, *In Metaph.*, I. 849. Biunde, *Psychologie*, Vol. I., part. ii. § 151, p. 159. (That principle of Contradiction, and principle of Reason and Consequent not identical, as Wolf and Reimarus hold.) Nic. Taurellus, *Philosophia Triumphus*, &c., p. 124. Arnheim, 1617. "Cum simplex aliqua sit affirmatio, negatio non Item, hanc illam sequi concludimus," &c. Chauvin, *Lexicon Philosophicum, v. Metaphysica.*

By whom introduced into Logic :—Eberstein, (*Über die Beschaffenheit der Logik und Metaphysik der reinen Peripatetiker*, p. 21, Halle, 1800), says that Darjes, in 1737, was the first to introduce Principle of Contradiction into Logic. That Buffier, and not Reimarus, first introduced principle of Identity into Logic, see Bobrik, *Logik*, § 70, p. 249.

(b)—PRIMARY LAWS OF THOUGHT,—IN PARTICULAR.

1. Principle of Identity. " Omne ens est ens." Held good by

Antonius Andreas, *In Metaph.* iv. qu. 5. (apud Fonsecam, *In Metaph.* I. p. 849; melius apud Suarez, *Select. Disp. Metaph.* Disp. iii. sect. iii. n. 4.) Derodon, *Metaphysica*, c. iii. p. 77. J. Sergeant, *Method to Science*, p. 133—136 and after. (Splits it absurdly.) Boethius—"Nulla propositio est verior illa in qua idem prædicatur do seipso." (Versor, *In P. Hispani Summulas Logicales*, Tr. vii., p. 441 (1st ed. 1487); et Buridanus, *In Sophism.*) "Propositiones illas oportet esse notissimas per se in quibus idem do se ipso prædicatur, ut ' Homo est Homo,' vel quarum prædicata in definitionibus subjectarum includuntur, ut 'Homo est animal.'" Aquinas, *Contra Gentiles*, L. i. c. 10. *Opera*, T. xviii. p. 7, Venet. 1786. Prior to principle of Contradiction—Leibnitz, *Nouveaux Essais*, p. 377. Buffier, *Principes du Raisonnement*, II. art. 21, p. 204. Rejected as identical and nugatory by Fonseca, *loc. cit.* Suarez, *loc. cit.* Wolf, *Ontologia*, §§ 55, 288, calls it Principium Certitudinis, and derives it from Principium Contradictionis.

2. Principle of Contradiction—ἀξίωμα τῆς ἀντιφάσεως.
Aristotle, *Metaph.*, L. iii. 3; x. 5. (Fonseca, *In Metaph.*, T. I., p. 850, L. iv. (iii.) c. 3.) *Anal. Post.* L. i. c. 11 c. 2, § 13. (On Aristotle and Plato, see Mansel's *Prolegomena*, pp. 283, 284.) Stahl, *Regulæ Philosophicæ*, Tit. i. reg. i. Suarez, *Select. Disp. Phil.* Disp. iii § 3. Timpler, *Metaph.* L. i. c. 8, qu. 14. Derodon, *Metaphysica*, p. 75 etc. Lippius, *Metaphysica*, L. i. c. i., p. 73. Bernardi, *Thes. Aristot., vv. Principium, Contradictio.* Leibnitz, *Œuvres Philosophiques, Nouv. Ess.*, L. iv. c. 2. Ramus, "Axioma Contradictionis," *Scholæ Dial.* L. ix. c. i., L. iv. c. 2, § 1, p. 548. Gul. Xylander, *Institutiones Aphoristicæ Logices Aristot.*, p. 24, (1577), "Principium principiorum, hoc est, lex Contradictionis." Philoponus, ἀξίωμα τῆς ἀντιφάσεως, v. *In Post. An.* f. 30 b. et seq. Ammonius, ἀξίωμα τῆς ἀντιφάσεως, *In De Interpret.* f. 94, Ald. 1503. Scheibler, *Topica*, c. 19.

On Definition of Contradictories, v. Scheibler, *Ibid.*
On Two Principles of Contradiction,—Negative and Positive, v. Zabarella, *Opera Logica, In An. Post.* i. t. 83, p. 807.

Conditions of.—Aristotle, *Metaph.*, L. iv. c. 6. Bernardi, *Thesaurus Arist.*, v. *Contrad.*, p. 300.

Proof attempted by—Clauberg, *Ontosophia*, § 26, (Degerando, *Histoire de Philosophie*, T. II. p. 57), through Excluded Middle.

3. Principle of Excluded Middle—ἀξίωμά διαιρετικόν.

"'Αξίωμα διαιρετικόν, divisivum, dicitur a Graecis *principium contradictionis affirmativum*; 'Oportet de omni re affirmare aut negare,'" Goclenius, *Lexicon Philosophicum*, *Lat.* p. 136. Zabarella, *In An. Post.*, L. i., text 83, *Opera Logica*, p. 807. Conimbricenses, *In Org.*, ii., 125. Lucian, *Opera*, ii. p. 44, (ed. Hemsterhuis). Aristotle, *Metaph.*, L. iv. (iii.) c. 7; *An. Post.*, L. i. 2; ii. 13, (Mansel's *Prolegomena*, p. 283). Joannes Philoponus, (v. Bernardi, *Thes. v. Contrad.*, p. 300). Piccartus, *Isagoge*, pp. 290, 291. Javellus, *In Metaph.*, L. iv. qu. 9. Suarez, *Disp. Metaph.*, Disp. iii. sect. 3, § 5. Stahl, *Regulæ Philos.*, Tit. i. reg. 2. Wolf, *Ontologia*, §§ 27, 29, 56, 71, 498. Fonseca, *In Metaph.*, L. iv. c. iii. qu. 1 *et seq.*, T. I. p. 850. (This principle not first). Timpler, *Metaphysica*, L. ii. c. 8, qu. 15. Derodon, *Metaph.*, p. 76 (Secundum principium). Lippius, *Metaphysica*, L. i. c. 1, pp. 72, 75. Chauvin, *Lexicon Philosophicum*, v. *Metaphysica.* Scheibler, *Topica*, c. 19. Hurtado de Mendoza, *Disp. Metaph.*, Disp. iii. § 3, (Caramuel, *Rat. et Real. Phil.*, § 452, p. 68).

Whether identical with Principle of Contradiction.

Affirmative,—

Javellus, *l. c.* Mendoza, *Disp. Metaph.*, D. iii. § 3. Leibnitz, *Œuvres Philosophiques*, *Nouv. Ess.*, L. iv. c. 2, p. 327.

Negative,—

Fonseca, *Disp. Met.* Disp. iv. c. 3, 9. Suarez, *Disp. Metaph.*, Disp. iii. § 3. Stahl, *Reg. Phil.* Tit. i. reg. 2.

Whether a valid and legitimate Law.

Fischer, *Logik*, § 64 *et seq.* (Negative).—Made first of all principles by Alexander de Ales, *Metaph.*, xiv. text 9 : "Conceptus omnes simplices, ut resolvuntur ad ens, ita omnes conceptus compositi resolvuntur ad hoc principium—*De quolibet affirmatio vel negatio.*" J. Picus Mirandulanus, (after Aristotle), *Conclusiones*, *Opera*, p. 90. Philoponus, *In An. Post.* i. f. 9 b, (Brandis, *Scholia*, p. 199.) Τὸ δ' ἅπαν φάναι ἢ ἀποφάναι, ἢ εἰς τὸ

ἀδύνατον ἀπόδειξις λαμβάνει. Aristotle, *An. Post.* i. c. 11, § 3.
'Αντίφασις δὲ ἀντίθεσις ἧς οὐκ ἔστι μεταξὺ καθ' αὑτήν.
An. Post. i. c. 2, § 13. Μεταξὺ ἀντιφάσεως οὐκ ἐνδέχεται
οὐθέν. *Metaph.*, L. iii. c. 7. 'Επεὶ ἀντιφάσεως οὐδὲν ἀνὰ
μέσον, φανερὸν ὅτι ἐν τοῖς ἐναντίοις ἔσται τὸ μεταξύ.
Physica, L. v. c. 3, § 5. See also *Post. An.* L. i. c. i. § 4, p. 414;
c. 2, § 13, p. 417; c. 11, § 3, p. 440, (vide Scheibler, *Topica*, c.
19; and Mansel's *Prolegomena*, p. 283, on Aristotle.)

4. Principle of Reason and Consequent.
    That can be deduced from Principle of Contradiction.
    Wolf, *Ontologia*, § 70. Baumgarten, *Metaphysik*, § 18. Jakob,
*Grundriss der allgemeinen Logik und kritische Anfangsgründe
der allgemeinen Metaphysik*, p. 38, 3d ed., 1794. (See Kiesewetter, as below.)

    That not to be deduced from Principle of Contradiction.
    Kiesewetter, *Allgemeine Logik*; *Weitere Auseinandersetzung*,
P. I. *ad* §§ 20, 21, p. 57 *et seq.* Hume *On Human Nature*, Book
i. part iii. § 8. Schulze, *Logik*, § 16, 5th ed., 1831.

VI.

NEW ANALYTIC OF LOGICAL FORMS—GENERAL RESULTS—FRAGMENTS.

(a)—EXTRACT FROM PROSPECTUS OF "ESSAY TOWARDS A NEW ANALYTIC OF LOGICAL FORMS."

(First published in 1846.[a] See above, Vol. I., pp. 144, 244.—ED.)

" Now, what has been the source of all these evils, I proceed to relate, and shall clearly convince those who have an intellect and a will to attend,—that a trivial slip in the elementary precepts of a Logical Theory, becomes the cause of mightiest errors in that Theory itself."—GALEN. (*De Temperamentis*, l. i. c. 5.)

" THIS New Analytic is intended to complete and simplify the old ;—to place the keystone in the Aristotelic arch. Of Abstract Logic, the theory in particular of Syllogism, (bating some improvements, and some errors of detail), remains where it was left by the genius of the Stagirite ; if it have not receded, still less has it advanced. It contains the truth ; but the truth, partially, and not always correctly, developed,—in complexity,—even in confusion. And why? Because Aristotle, by an oversight, marvellous certainly in him, was prematurely arrested in his analysis ; began his synthesis before he had fully sifted the elements to be recomposed and, thus, the system which, almost spontaneously, would have evolved itself into unity and order, he laboriously, and yet imperfectly, constructed by sheer intellectual force under a load of limitations and corrections and rules, which, deforming the symmetry,

[a] An extract corresponding in part with that now given from the Prospectus of "Essay towards a New Analytic of Logical Forms," is republished in the *Discussions on Philosophy*, p. 650. To this extract the Author has prefixed the following notice regarding the date of his doctrine of the Quantification of the Predicate :—" Touching the principle of an explicitly *Quantified Predicate*, I had by 1833 become convinced of the necessity to extend and correct the logical doctrine upon this point. In the article on Logic (in the *Edinburgh Review*) first published in 1833, the theory of Induction there maintained proceeds on a thoroughgoing quantification of the predicate, in affirmative propositions. Before 1840, I had, however, become convinced that it was necessary to extend the principle equally to negatives ; for I find, by academical documents, that in that year, at latest, I had publicly taught the unexclusive doctrine." —*Discussions*, p. 650, 2d edition.—ED.

has seriously impeded the usefulness, of the science. This imperfection, as I said, it is the purpose of the New Analytic to supply.

"In the *first* place, in the Essay there will be shown, that the Syllogism proceeds, not as has hitherto, virtually at least, been taught, in one, but in the *two* correlative and counter *wholes* (Metaphysical) of *Comprehension*, and (Logical) of *Extension;* the major premise in the one whole, being the minor premise in the other, &c.—Thus is relieved, a radical defect and vital inconsistency in the present logical system.

" In the *second* place, the self-evident truth,—That we can only rationally deal with what we already understand, determines the simple logical postulate,—*To state explicitly what is thought implicitly.* From the consistent application of this postulate, on which Logic ever insists, but which Logicians have never fairly obeyed, it follows:—that, logically, we ought to take into account the *quantity*, always understood in thought, but usually, and for manifest reasons, elided in its expression, not only of the *subject*, but also of the *predicate*, of a judgment. This being done, and the necessity of doing it will be proved against Aristotle and his repeaters, we obtain *inter alia*, the ensuing results:—

" 1°. That the *preindesignate terms* of a proposition, whether subject or predicate, are never, on that account, thought as *indefinite* (or indeterminate) in quantity. The only indefinite, is *particular*, as opposed to *definite*, quantity ; and this last, as it is either of an extensive *maximum* undivided, or of an extensive *minimum* indivisible, constitutes quantity *universal*, (general), and quantity *singular*, (individual). In fact, *definite* and *indefinite* are the only quantities of which we ought to hear in Logic ; for it is only as indefinite that particular, it is only as definite that individual and general, quantities have any (and the same) logical avail.

" 2°. The revocation of the *two terms of a proposition* to their *true relation ;* a proposition being always an *equation* of its subject and its predicate.

" 3°. The consequent reduction of the *Conversion of Propositions* from three species to *one*,—that of Simple Conversion.

" 4°. The reduction of all the *General Laws of Categorical Syllogisms* to a *Single Canon*.

" 5°. The evolution from that *one canon* of all the *Species and varieties of Syllogism.*

" 6°. The *abrogation* of all the *Special Laws of Syllogism.*

" 7°. A demonstration of the *exclusive possibility of Three syllogistic Figures;* and (on new grounds) the scientific and final *abolition of the Fourth.*

" 8°. A manifestation that *Figure* is an *unessential variation* in syllogistic form ; and the consequent *absurdity of Reducing* the syllogisms of the other figures to the first.

" 9°. An enouncement of *one Organic Principle* for each *Figure.*

" 10°..A determination of the true *number* of the legitimate *Moods;* with

" 11°. Their *amplification* in number (*thirty-six*);

" 12°. Their numerical *equality* under all the figures ; and,

" 13°. Their *relative equivalence,* or virtual identity, throughout every schematic difference.

" 14°. That, in the *second* and *third* figures, the extremes holding both the same relation to the middle term, there *is not,* as in the first, an *opposition and subordination between a term major and a term minor, mutually containing and contained, in the counter wholes of Extension and Comprehension.*

" 15°. Consequently, in the *second* and *third* figures, there is *no determinate major and minor premise,* and there are *two indifferent conclusions;* whereas, in the *first* the *premises are determinate,* and there is a *single proximate conclusion.*

" 16°. That the *third,* as the figure in which *Comprehension* is predominant, is more appropriate to *Induction.*

" 17°. That the *second,* as the figure in which *Extension* is predominant, is more appropriate to *Deduction.*

" 18°. That the *first,* as the figure in which *Comprehension* and *Extension* are in *equilibrium,* is common to *Induction* and *Deduction,* indifferently.

" In the *third* place, a scheme of Symbolical Notation will be given, wholly different in principle and perfection from those which have been previously proposed ; and showing out, in all their old and new applications, the propositional and syllogistic forms, with even a mechanical simplicity.

" This Essay falls naturally into two parts. There will be contained—in the *first,* a systematic exposition of the new doctrine itself ; in the *second,* an historical notice of any occasional anticipations of its several parts which break out in the writings of previous philosophers.

"Thus, on the new theory, many valid *forms* of judgment and reasoning, in ordinary use, but which the ancient logic continued to ignore, are now openly recognised as legitimate; and many *relations*, which heretofore lay hid, now come forward into the light. On the one hand, therefore, Logic certainly becomes more complex. But on the other, this increased complexity proves only to be a higher development. The developed Syllogism is, in effect, recalled, from multitude and confusion, to order and system. Its laws, erewhile many, are now few,—we might say one alone,—but thoroughgoing. The exceptions, formerly so perplexing, have fallen away; and the once formidable array of limitary rules has vanished. The science now shines out in the true character of beauty,—*as One at once and Various.* Logic thus accomplishes its final destination; for as 'Thrice-greatest Hermes,' speaking in the mind of Plato, has expressed it,—' *The end of Philosophy is the intuition of Unity.'* "

(δ)—LOGIC,—ITS POSTULATES.

(November 1848—See above, Vol. L, p. 114.)

I. To state explicitly what is thought implicitly. In other words, to determine what is meant before proceeding to deal with the meaning. Thus in the proposition "Men are Animals," we should be allowed to determine whether the term "Men" means *all* or *some* Men,—whether the term "Animals" means *all* or *some* Animals; in short, to quantify both the subject and predicate of the proposition. This postulate applies both to Propositions and to Syllogisms.[a]

----

[a] See (quoted by Wallis, *Logica,* p. 291) Aristotle, *An. Prior.,* L. i. c. 33 (Pacio, c. 32, §§ 2, 3, 4, p. 261), and Ramus (from Downam, *In P. Rami Dialect.,* L. ii. c. 9, p. 410). What is understood to be supplied; *Ramus Dial.,* L. ii. c. 9: "Si qua (de argumentationis consequentia propter crypsin) dubitatio fuerit, explenda quæ desunt; amputanda quæ supersunt; et pars quælibet in locum redigenda est." [Cf. Ploucquet, *Elementa Philosophiæ*

II. Throughout the same proposition, or immediate (not mediate) reasoning, to use the same words, and combinations of words, to express the same thought,* (that is, in the same Extension and Comprehension,) and this identity is to be presumed.

Thus a particular in one (prejacent) proposition of an immediate reasoning, though indefinite, should denote the *same part* in the other. This postulate applies to immediate inference, *e.g.* Conversion.

Predesignates in same logical unity, (proposition or syllogism,) in same sense, both Collective or both Distributive.

III. And, *e contra*, throughout the same logical unity, (immediate reasoning), to denote and presume denoted the same sense (notion or judgment) by the same term or terms.β

This does not apply to the different propositions of a Mediate Inference.

IV. (or V.) To leave, if necessary, the thought undetermined, as subjectively uncertain, but to deal with it only as far as certain or determinable. Thus a whole may be truly predicable, though we know only the truth of it as a part. Therefore we ought to be able to say "some at least" when we do not know, and cannot, therefore, say determinately, either that "some only" or that "all" is true.

(January 1850.)

III. (or IV.) To be allowed in an immediate reasoning, to de-

Contemplatica, (Statgardiæ, 1778), § 29, p. 5: " Secundum sensum logicum cum omni termino jungendum est signum quantitatis."—Ed.]

a That words must be used in the same sense. See Aristotle, *Anal. Pr.*, L. i., cc. 33, 34, 35, 36, 37, &c.

β If these postulates (II. and III.) were not cogent, we could not convert, at least not use the converted proposition, (unless 1. were cogent, the *convertenda* would be false). "All man is (some) animal," is converted into "Some animal is (all) man." But if the "some animal" here were not thought to and limited to the sense of the con-

vertend, it would be false. So in the Hypothetical proposition, "If the Chinese are Mahometans, they are (some) infidels," the word "infidel," unless thought in a meaning limited to and true of "Mahometans," is inept. But if it be so limited, we can (contrary to the doctrine of the logicians) argue back from the position of the consequent to the position of the antecedent, and from the sublation of the antecedent to the sublation of the consequent, though false. If not granted, Logic is a mere childish play with the vagueness and ambiguities of language. [Cf. Titius, *Ars Cogitandi*, c. xli., § 28.—Ed.]

note, that *another* part, or other "some," is used in the conclusion
from what was in the antecedent. Inference of *Sub-contrariety.*

That the "some," if not otherwise qualified, means "some
only,"—this by presumption.

That the term (Subject, or Predicate) of a Proposition shall
be converted *with its quantity unchanged, i.e.* in the same *extension*. This violated, and violation cause of error and confusion.
No conversion *per accidens*. For the real terms compared are
the *quantified* terms, and we convert only the *terms compared*
in the prejacent or convertenda.

That the *same terms*, apart from the quantity, *i.e.* in the same
comprehension, should be converted. As before stated, such
terms are new and different. No Contraposition. For contraposition is only true in some cases, and even in these it is true
accidentally, not by conversion, but through contradiction; *i.e.*
same Comprehension.

That we may see the truth from the necessary validity of the
logical process, and not infer the validity of the logical process
from its accidental truth. Conversion *per accidens*, and Contraposition, being thus accidentally true in some cases only, are
logically inept, as not true in all.

To translate out of the complexity, redundance, deficiency, of
common language into logical simplicity, precision, and integrity."

(December 1849.)
As Logic considers the form and not the matter, but as the
form is only manifested in application to some matter, Logic
postulates to employ any matter in its examples.

(January 1850.)
That we may be allowed to translate into logical language the
rhetorical expressions of ordinary speech. Thus the Exceptive
and Limitative propositions in which the predicate and subject
are predesignated, are to be rendered into logical simplicity.

(May 1850.)

As Logic is a formal science, and professes to demonstrate by abstract formulæ, we should know, therefore, nothing of the notions and their relations except *ex facie* of the propositions. This implies the necessity of overtly quantifying the predicate.

(c)—QUANTIFICATION OF PREDICATE,—IMMEDIATE INFERENCE,—CONVERSION,—OPPOSITION.[a]

(See above, Vol. I., pp. 244, 262.)

We now proceed to what has been usually treated under the relation of Propositions, and previously to the matter of Inference altogether; but which I think it would be more correct to consider as a species of Inference, or Reasoning, or Argumentation, than as merely a preparatory doctrine. For in so far as these relations of Propositions warrant us, one being given, to educe from it another, this is manifestly an inference or reasoning. Why it has not always been considered in this light, is evident. The inference is immediate; that is, the conclusion or second proposition is necessitated directly and without a medium, by the first. There are only two propositions and two notions in this species of argumentation; and the logicians have in general limited reasoning or inference to a mediate eduction of one proposition out of the correlation of two others, and have thus always supposed the necessity of three terms or collated notions.

But they have not only been, with few exceptions, unsystematic in their procedure, they have all of them, (if I am not myself mistaken,) been fundamentally erroneous in their relative doctrine.

There are various immediate inferences of one proposition from another. Of these some have been wholly overlooked by the logicians; whilst what they teach in regard to those which they do consider, appears to me at variance with the truth.

---

[a] Appendix (c), from p. 257 to p. 276, was usually delivered by the Author as a Lecture supplementary to the doctrine of Conversion as given above, vol. i. p. 262.—ED.

I shall make no previous enumeration of all the possible species
of Immediate Inference, but shall take them up in this order :—
I shall consider, 1°, Those which have been considered by the
logicians ; and, 2°, Those which have not. And in treating of
the first group, I shall preface what I think the true doctrine
by a view of that which you will find in logical books.

The first of these is *Conversion.* When, in a categorical proposition, (for to this we now limit our consideration,) the Subject and Predicate are transposed, that is, the notion which was
previously the subject becomes the predicate, and the notion
which was previously the predicate becomes the subject, the
proposition is said to be converted.[a]

The proposition given, and its product, are together called the
judicia conversa, or *propositiones conversæ,* which I shall not attempt to render into English. The relation itself in which the
two judgments stand, is called *conversion, reciprocation, transposition,* and sometimes *obversion,* (*conversio, reciprocatio, trans-
positio, obversio.*) The original or given proposition is called the
converse, or *converted,* sometimes the *præjacent, judgment,* (*judi-
cium,* or *propositio, conversum, conversa, præjacens*) ; the other,
that into which the first is converted, is called the *converting,*
and sometimes the *subjacent, judgment,* (*prop.* or *jud. convertens,
subjacens*). It would be better to call the former the *convertend,*
(*pr. convertenda*), the latter the *converse,* (*pr. conversa*). This
language I shall use.[β]

a [Definitions of conversion in general. 'Ἀντιστροφή ἐστιν ἀντιστροφή τις,
Philoponus (or Ammonius), *In An. Pr.,*
l. c. 2, f. 11 b. So Magentinus, *In An.
Pr.* l. c. 2, f. 3 b. Anonymus, *De
Syllogismo,* l. 42 b. Πρὸς δύσιν ἀντιστροφή ἐστι κοινωνία δύο προτασίων κατὰ
τοὺς ὅρους ἀνάπαλιν τιθεμένους, μετὰ τοῦ
συναληθεύειν. Alexander, *In An. Pr.*
l. c. 4, f. 15 b. See the same in different words, by Philoponus (or Ammonius), *ut supra,* and copied from him by
Magentinus, *ut supra.* Cf. Boethius,
Opera, Introductio ad Syllogismos, p.
574 ; Wegelin, in *Gregorii Aneponymi
Phil. Syntag.* (circa 1260), L. v. c. 12,
p. 621 ; Nicephorus Blemmidas, *Epit.
Log.,* c. 31, p. 221.]

β See above, vol. i. p. 262.—F.D.
[I. Names for the two correlative
propositions—'Ἀντιστρεφόμεναι προτά-
σεις, Philoponus, (quoted by Wegelin,
ut supra, p. 622.) *Conversa, Contrapo-
sita,* Twesten, *Logik,* § 87.

II. Original or Given proposition.
a) ἡ προηγουμένη, προσιμένη, ἀντιστρε-
φομένη πρότασις—Cf. Strigelius, *In
Melanchth. Erot. Dial.,* L. ii. p. 581.
b) Conversa (= *Convertenda*) vulgo,
Scotus, *Quæstiones in An. Prior.,* l.
q. 12. Corvinus, *Instit. Phil.,* §510.
Richter, *De Conversione,* (1740. Hala
Magdeb.) Baumgarten, *Logica,* §
278. Ulrich, *Instit. Log. et Met.,* §
162, p. 188.
c) Convertibilis (raro).

Such is the doctrine touching Conversion, taught even to the present day. This in my view is beset with errors; but all these errors originate in two, as these two are either the cause or the occasion of every other.

The First cardinal error is,—That the quantities are not converted with the quantified terms. For the real terms compared in the Convertend, and which, of course, ought to reappear without change, except of place, in the Converse, are not the naked, but the quantified terms. This is evident from the following considerations:

1°, The Terms of a Proposition are only terms as they are terms of relation; and the relation here is the relation of comparison.

2°, As the Propositional Terms are terms of comparison, so they are only compared as Quantities,—quantities relative to each other. An Affirmative Proposition is simply the declaration of an equation, a Negative Proposition is simply the declaration of a non-equation, of its terms. To change, therefore, the quantity of either, or of both Subject and Predicate, is to change their correlation,—the point of comparison; and to exchange their quantities, if different, would be to invert the terminal interdependence, that is, to make the less the greater, and the greater the less.

3°, The Quantity of the Proposition in Conversion remains always the same; that is, the absolute quantity of the Converse

d) *Convertens*, Micraelius, *Lex. Phil.* s. *Conversio.* Twesten, *Logik,* § 87. *Antecedens*, Scotus, l. c. Strigelius, l. c.

e) *Projacens*, Scheibler, *Opera Logica, De Propositionibus*, Pars lil. c. x. p. 478.

f) *Exposita*, Aldrich, *Comp.*, L., l. c. 2. Whately, *Logic*, p. 69. *Propositio exposita*, or *exponens*, quite different as used by Logicians, v. Scheghium, *In Arist. Org.* 162 (and above, vol. I. p. 263.)

g) *Convertenda*, Corvinus, l. c. Richter, l. c.

h) *Contraponens*, Twesten, l. c.

i) Prior, Boethius, *De Syllog. Categ.* L. I. *Opera*, p. 588.

k) *Principium*, Darjes, *Via ad Veritatem*, § 234.

III. Product of Conversion.

a) ἡ ἀντιστρέφουσα. See Strigelius, loc. cit.

b) *Convertens, Subjacens, Consequens*, Scotus, *Quaestiones, In. An. Prior.*, l. 9, 24, f. 276 et passim. Krug, *Logik*, § 65, p. 205, and logicians in general.

c) *Conversa*, Boethius, *Opera, Introd. ad Syll.*, pp. 575 et seq., 587 et seq.; Melanchthon, *Erotemata*, L. il. p. 581, and Strigelius, ad loc. Micraelius, *Lex. Phil.* s. *Conversio.* Noldius, *Logica Recognita*, p. 263, says that the first should more properly be called *Convertibilia*, or *Convertenda*, and the second *Conversa.*

d) *Conversum*, Twesten, *Logik*, § 87.

e) *Contrapositum*, Twesten, l. c.

f) *Conclusio*, Darjes, *Via ad Veritatem*, § 234.)

must be exactly equal to that of the Convertend. It was only from overlooking the quantity of the predicate, (the second error to which we shall immediately advert,) that two propositions, exactly equal in quantity, in fact the same proposition, perhaps, transposed, were called the one *universal*, the other *particular*, by exclusive reference to the quantity of the subject.

4°, Yet was it of no consequence, in a logical point of view, which of the notions collated were Subject or Predicate; and their comparison, with the consequent declaration of their mutual inconclusion or exclusion, that is, of affirmation or negation, of no more real difference than the assertions,—" London is four hundred miles distant from Edinburgh,"—" Edinburgh is four hundred miles distant from London." In fact, though logicians have been in use to place the subject first, the predicate last, in their examples of propositions, this is by no means the case in ordinary language, where, indeed, it is frequently even difficult to ascertain which is the determining, and which the determined, notion. Out of logical books the predicate is found almost as frequently before as after the subject, and this in all languages. You recollect the first words of the *First Olympiad* of Pindar, Ἄριστον μὲν ὕδωρ, " Best is water;" and the Vulgate, (I forget how it is rendered in our English translation), has, " Magna est veritas, et prævalebit."ᵃ Alluding to the Bible, let us turn up any Concordance under any adjective title, and we shall obtain abundant proof of the fact. As the adjective *great*, (*magnus*,) has last occurred, let us refer to Cruden under that simple title. Here, in glancing it over, I find— " Great is the wrath of the Lord "—" Great is the Lord and greatly to be praised "—" Great is our God "—" Great are thy works "— " Great is the Holy One of Israel "—" Great shall be the peace of thy children "—" Great is thy faithfulness "—" Great is Diana of the Ephesians "—" Great is my boldness "—" Great is my glorying "—" Great is the mystery of godliness," &c.

The Second cardinal error of the logicians is, the not considering that the Predicate has always a quantity in thought, as much as the Subject; although this quantity be frequently not explicitly enounced, as unnecessary in the common employment of language;

---

a III. Esdras iv. 41: " Magna est veritas, et prævalet." In the English version: "Great is truth, and mighty above all things."—ED.

for the determining notion or predicate being always thought as at least adequate to, or coextensive with, the subject or determined notion, it is seldom necessary to express this, and language tends ever to elide what may safely be omitted. But this necessity recurs, the moment that, by conversion, the predicate becomes the subject of the proposition; and to omit its formal statement is to degrade Logic from the science of the necessities of thought, to an idle subsidiary of the ambiguities of speech. An unbiassed consideration of the subject will, I am confident, convince you that this view is correct.

1°, That the predicate is as extensive as the subject is easily shown. Take the proposition,—"All animal is man," or, "All animals are men." This we are conscious is absurd, though we make the notion *man* and *men* as wide as possible; for it does not mend the matter to say,—"All animal is all man," or, "All animals are all men." We feel it to be equally absurd as if we said,—"All man is all animal," or "All men are all animals." Here we are aware that the subject and predicate cannot be made coextensive. If we would get rid of the absurdity, we must bring the two notions into coextension, by restricting the wider. If we say—"Man is animal," (*Homo est animal*), we think, though we do not overtly enounce it, "*All* man is animal." And what do we mean here by *animal?* We do not think, *all*, but *some*, animal. And then we can make this indifferently either subject or predicate. We can think,—we can say, "Some animal is man," that is, *some* or *all* man; and, *e converso*, "Man (*some* or *all*) is animal," viz. *some* animal.

It thus appears that there is a necessity in all cases for thinking the predicate, at least, as extensive as the subject. Whether it be absolutely, that is, out of relation, more extensive, is generally of no consequence; and hence the common reticence of common language, which never expresses more than can be understood,—which always, in fact, for the sake of brevity, strains at ellipsis.

2°, But, in fact, ordinary language quantifies the Predicate so often as this determination becomes of the smallest import. This it does either directly, by adding *all, some*, or their equivalent predesignations to the predicate; or it accomplishes the same end indirectly, in an exceptive or limitative form.

*) Directly,—as "Peter, John, James, &c., are *all* the Apostles" —"Mercury, Venus, &c., are *all* the planets."

*) But this is more frequently accomplished indirectly, by the equipollent forms of *Limitation* or *Inclusion*, and *Exception*.ᵃ

For example, by the limitative designations, *alone* or *only*, we say,—"God alone is good," which is equivalent to saying, *God is all good*, that is, *God is all that is good;* "Virtue is the only nobility," that is, *Virtue is all noble*, that is, *all that is noble*.β

The symbols of the Catholic and Protestant divisions of Christianity may afford us a logical illustration of the point. The Catholics say,—"Faith, Hope, and Charity alone justify;" that is, *the three heavenly virtues together are all justifying*, that is, *all that*

ᵃ By the logicians this is called simply *Exclusion*, and the particles, *tantum, &c., particulæ exclusivæ.* This, I think, is inaccurate; for it is inclusion, limited by an exclusion, that is meant. —(See Schcibler, *Opera Logica*, P. iii., c. vii., tit. 3, p. 457 et seq.]

β (February 1850.) On the Indirect Predesignation of the Predicate by what are called *Exclusive* and *Exceptive particles.*

Names of the particles.

Latin,—*Unus, unicus, unice; solus, solum, solummodo; tantum, tantummodo; dumtaxat; præcise; adæquate.*—*Nihil præter—præterquam—ni, nisi, non.*

English,—*One, only, alone, exclusively, precisely, just, sole, solely. Nothing but —not—except—beyond.*

1. These particles annexed to the Subject predesignate the Predicate universally, or to its whole extent, denying its particularity or indefinitude, and definitely limiting it to the Subject alone. As, *Man alone philosophises*, (though not all do). *The dog alone barks*, or, *dogs alone bark*, (though some do not). *Man only is rational*, or *No animal but man is rational. Nothing but rational is risible. Of material things there is nothing living* (but) *not organised, and nothing organised not living. God alone is to be worshipped. God is the single,—sole object of worship.*

Some men only are elect.

II. Annexed to the Predicate, they limit the Subject to the Predicate, but do not define its quantity, or exclude from it other Subjects. As, *Peter only plays. The sacraments are only two. The categories are only ten. John drinks only water.*

III. Sometimes the particles *sole, solely, single, alone, only, &c.,* are annexed to the predicate as a predesignation tantamount to *all.* As, *God is the single,—one,—alone,—only,—exclusive, —adequate, object of worship.*

On the relation of Exclusive propositions to those in which the predicate is predesignated, see Titius, *Ars Cogitandi*, c. vi. §§ 66, 67. Hollman, *Philosophia Rationalis*, § 475. Kreil, *Handbuch der Logik*, § 62. Derodon, *Logica Restituta, De Enunciations*, c. v. p. 569 et seq. Kerkerman, *Systema Logicæ*, lib. iii. c. 11. *Opera*, t. i. p. 703.

The doctrine held by the logicians as to the *exclusum prædicatum, exclusum subjectum*, and *exclusum signum*, is erroneous.—See Scheibler, *Opera Logica*, P. iii., c. vii. tit. 3, p. 457 et seq. Jac. Thomasius, *System. Log.*, c. xxx. p. 67 et seq. [Cf. Fonseca, *Instit. Dial.*, L. III. c. 23. For a detailed exposition of this doctrine by Scheibler, see below, p. 253, note a.—Ed.]

justifies; omne justificans, justum faciens. The Protestants say,—"Faith alone justifies;" that is, *Faith*, which they hold to comprise the other two virtues, *is all justifying*, that is, *all that justifies; omne justificans.* In either case, if we translate the watchwords into logical simplicity, the predicate appears predesignated.

"Of animals man alone is rational;" that is, *Man is all rational animal.* "What is rational is alone or only risible;" that is, *All rational is all risible,* &c.

I now pass on to the Exceptive Form. To take the motto overhead,—"On earth there is nothing great but man." What does this mean? It means, *Man—is—all earthly great. (Homo —est—omne magnum terrestre.)* And the second clause—"In man there is nothing great but mind,"—in like manner gives as its logical equipollent—*Mind—is—all humanly great,* that is, *all that is great in man. (Mens—est—omne magnum humanum.)*[a]

---

[a] Vide Scheibler, *Opera Logica,* P. lii. c. vii. pp. 458, 460, where his examples, with the exposition of the Logicians, may be well contrasted with value.

[Scheibler, after referring to the *Parva Logicalia* of the schoolmen, as containing a proposed supplement of the doctrines of Aristotle, proceeds to expound the *Propositiones Exponibiles* of those treatises : "Exclusiva enunciatio est, quæ habet particulam exclusivam, ut : *Solus* homo est rationalis.

. . . . . Porro exclusivæ enunciationes sunt duplicis generis. Aliæ sunt exclusivæ prædicati, aliæ exclusivæ subjecti ; hoc est, in aliis particula exclusiva excludit a subjecto, in aliis excludit a prædicato. Valuti hæc propositio exclusiva est : Deus tantum est immortalis. Estque exclusiva a subjecto, hoc sensu, Deus tantum, et non homo vel lapis, &c. Si intelligatur ut exclusiva a prædicato, falsa est, (hoc sensu, Deus tantum est immortalis, immortalis, inquam, *et non omnipotens vel omnisciens.*) Atque ad hunc modum omnes propositiones exclusivæ ambiguæ sunt, si habeant particulam ex-

clusivam post subjectum propositionis, ante vinculum, ut erat in proposito exemplo. Carent autem propositiones exclusivæ illa ambiguitate, si vel exclusiva particula ponatur ante subjectum propositionis, vel etiam sequatur copulam. Ibi enim indicatur esse propositio exclusiva subjecti, ut : Solus homo discurrit. Hic autem indicatur, esse propositio exclusiva prædicati, ut : Sacramenta Novi Testamenti sunt tantum duo ; Prædicamenta sunt tantum decem."

Scheibler then proceeds to give the following general and special rules of Exclusion :—

"I. Generaliter tenendum est, *quod aliter sint exponendæ exclusivæ a prædicato, et aliter exclusivæ a subjecto.*

"II. *Exclusiva propositio non excludit concomitantia.* . . . . .

"III. *Omnis exclusiva resolvitur in duas simplices, alteram affirmatam, alteram negatam.* Atque hoc est quod vulgo dicitur, quod omnis exclusiva sit hypothetica. Hypothetica enim propositio est quæ includit duas alias in virtute, vel dispositione sua. Veluti hæc : *Solus homo est rationalis,* æquiva

We ought, indeed, as a corollary of the postulate already
stated, to require to be allowed to translate into equivalent logical terms the rhetorical enouncements of common speech. We
should not do as the logicians have been wont,—introduce and
deal with these in their grammatical integrity; for this would
be to swell out and deform our science with mere grammatical
accidents; and to such fortuitous accrescences the formidable
volume, especially of the older Logics, is mainly owing. In
fact, a large proportion of the scholastic system is merely
grammatical.

3°, The whole doctrine of the non-quantification of the predicate is only another example of the passive sequacity of the
logicians. They follow obediently in the footsteps of their great

let his duabus : *Homo est rationalis, et
Quod non est homo, non est rationale.
Et in specie : Bestia non est rationalis.
Planta non est rationalis.* . . . .
Atque hæ duæ propositiones vocantur
expositæ, sicut propositio exclusiva
dicitur *exponibilis.* . . . .

"Specialem autem regulam explicandi
exclusivas sunt octo : sicut et octo sunt
genera locutionum exclusivarum.

"I. *Propositio exclusiva universalis
affirmativa, cujus signum non negatur,
ut, Tantum omnis homo currit,* exponitur sic, *Omnis homo currit, et nihil
aliud ab homine currit.* Vocari solet
hæc expositio PATER, quia prior ejus
pars est universalis affirmativa, quod
notat A. Et, altera pars est universalis negativa, quod indicat in posteriori syllaba litera E.

"II. *Propositio particularis, vel indefinita affirmativa, in qua signum non
negatur, ut Tantum homo currit,* exponitur sic, *Homo currit, et nihil aliud ab
homine currit.* Vocatur hæc expositio,
NINE.

"III. *Propositio exclusiva, in qua
signum non negatur, universalis negativa, ut, Tantum nullus homo currit,*
exponitur sic, *Nullus homo currit, et
quodlibet aliud ab homine currit,* vocatur, TENAX.

"IV. *Exclusiva cujus signum non negatur particularis vel indefinita negativa,*

ut, *Tantum homo non currit,* exponitur
sic, *Homo non currit, et quodlibet aliud
ab homine currit,* vocatur, STORAX.

"V. *Exclusiva, in qua signum negatur, affirmativa et universalis, ut, Non
tantum omnis homo currit,* exponitur
sic, *Omnis homo currit, et aliquod aliud
ab homine currit,* vocatur, CANOR.

"VI. *In qua signum negatur, existens
universalis affirmativa, ut, Non tantum
nullus homo currit,* sic exponitur, *Nullus
homo currit, et aliquid aliud ab homine
non currit,* vocatur, FROIT.

"VII. *Exclusiva, in qua signum negatur, existens particularis affirmativa,
ut, Non tantum aliquis homo currit,* exponitur sic, *Aliquis homo currit, aliquid
aliud ab homine currit,* vocatur, PILOR.

"VIII. *Negativa particularis exclusiva propositiones, cujus signum negatur, ut, Non tantum aliquis homo non
currit,* exponitur sic, *Aliquis homo non
currit, et aliquid aliud ab homine non
currit,* vocatur, NOBIS.

"Differentia autem propositionis exclusivæ et exceptivæ est evidens. Nempe
exclusiva prædicatum vendicat uni subjecto, aut a subjecto excludit alia prædicata, ut, *Solus Deus bonus est.* Ex-
ceptiva autem statuit universale subjectum, indicatque aliquid contineri
sub illo universali, de quo non dicatur
prædicatum, ut, *Omne animal est irra-
tionale, præter hominem.*"— ED.]

master. We owe this doctrine and its prevalence to the precept and authority of Aristotle. He prohibits once and again the annexation of the universal predesignation to the predicate. For why, he says, such predesignation would render the proposition absurd; giving as his only example and proof of this, the judgment—"All man is all animal." This, however, is only valid as a refutation of the ridiculous doctrine, held by no one, that any predicate may be universally quantified; for, to employ his own example, what absurdity is there in saying that "some animal is all man"? Yet this nonsense, (be it spoken with all reverence of the Stagirite,) has imposed the precept on the systems of Logic down to the present day. Nevertheless, it could be shown by a cloud of instances from the Aristotelic writings themselves, that this rule is invalid; nay, Aristotle's own doctrine of Induction, which is far more correct than that usually taught, proceeds upon the silent abolition of the erroneous canon. The doctrine of the logicians is, therefore, founded on a blunder; which is only doubled by the usual averment that the predicate, in what are technically called *reciprocal propositions*, is taken universally *vi materiæ* and not *vi formæ*.

But, 4°, The non-quantification of the predicate in thought is given up by the logicians themselves, but only in certain cases where they were forced to admit, and to the amount which they could not possibly deny. The predicate, they confess, is quantified by particularity in affirmative, by universality in negative, propositions. But why the quantification, formal quantification, should be thus restricted in thought, they furnish us with no valid reason.

To these two errors I might perhaps add as a third, the confusion and perplexity arising from the attempt of Aristotle and the logicians to deal with *indefinite* (or, as I would call them, *indesignate*) terms, instead of treating them merely as verbal ellipses, to be filled up in the expression before being logically considered; and I might also add as a fourth, the additional complexity and perplexity introduced into the science by viewing propositions, likewise, as affected by the four or six modalities. But to these I shall not advert.

These are the two principal errors which have involved our systems of Logic in confusion, and prevented their evolution in simplicity, harmony, and completeness;—which have condemned them to bits and fragments of the science, and for these bits and fragments have made a load of rules and exceptions indispensable, to avoid falling into frequent and manifest absurdity. It was in reference to these two errors chiefly, that I formerly gave you as a self-evident Postulate of Logic—"Explicitly to state what has been implicitly thought;" in other words, that before dealing logically with a proposition, we are entitled to understand it, that is, to ascertain and to enounce its meaning. This quantification of the predicate of a judgment, is, indeed, only the beginning of the application of the Postulate; but we shall find that at every step it enables us to cast away, as useless, a multitude of canons, which at once disgust the student, and, if not the causes, are at least the signs, of imperfection in the science.

I venture then to assert, that there is only one species of Conversion, and that one thorough-going and self-sufficient. I mean Pure, or Simple Conversion. The other species,—all are admitted to be neither thorough-going nor self-sufficient,—they are in fact only other logical processes, accidentally combined with a transposition of the subject and predicate. The *conversio per accidens* of Boethius, as an Ampliative operation, has no logical existence; it is material and precarious, and has righteously been allowed to drop out of science. It is now merely an historical curiosity. As a Restrictive operation, in which relation alone it still stands in our systems, it is either merely fortuitous, or merely possible through a logical process quite distinct from Conversion, I mean that of Restriction or Subalternation, which will be soon explained. *Conversio per contrapositionem* is a change of terms,—a substitution of new elements, and only holds through contradiction,[a]

---

a [See Aristotle, *Topica*, L. ii. c. 8. Scotus, Bannes, Mendoza, silently following each other, have held that contraposition is only mediate, infinitation requiring *Constantia*, &c. Wholly wrong. See Arriaga.—*Cursus Philosophicus*, D. II. a. 4, p. 18.] ["Observandum est prædictas consequentias (per contrapositionem) malas esse et instabiles, nisi accesserit alia propositio in antecedenti quæ impartit existentiam subjecti consequentis. Tunc enim fir-

ma erit consequentia, e. g. *Omnis homo est albus et non album est, ergo omne non album est non homo*. Alioquin si *constantiam* illam non posueris in antecedenti, instabitur illi consequentiæ in eventu, in quo nihil sit non album, et omnis homo sit albus." Bannes, *Instit. Min. Dial.*, L. vi. c. 2, p. 530.—Ed.]

[Rule for Finite Prejacents given. With the single exception of E n E (A n A) the other seven propositions

being just as good without as with conversion. The Contingent Conversion of the lower Greeks* is not a conversion,—is not a logical process at all, and has been worthily ignored by the Latin

may be converted by Counterposition under the following rule :—" Let the terms be infinitated and transposed, the predesignations remaining as before."

With the two additional exceptions of the two convertible propositions, A f I, and I f A, the infinitated propositions hold good without the transposition of the terms.

Rule for Infinite Prejacents given.

With the single exception of n I f n I, (nE-n-nE being impossible,) the other six propositions may be converted by Counterposition under the following rule :—" Let the terms be uninfinitated and transposed, the predesignations remaining as before."

Contraposition is not explicitly evolved by Aristotle in *Prior Analytics*, but is evolved from his *Topics*, L. ii. cc. 1, 8, *alibi*. *De Interpretatione*, c. 14. See Conimbricenses, *In An. Prior.*, L. i. q. i. p. 271. Bannes, *Instit. Minoris Dialecticæ*, L. v. c. 2, p. 532. Burgersdicius, *Instit. Log.*, L. l. c. 32.

First explicitly enounced by Averroes according to Molinæus, (*Elementa Logica*, L. i. c. 4, p. 54). I cannot find any notice of it in Averroes. He ignores it, name and thing. It is in Anonymus, *De Syllogismo*, f. 42 h., in Nicephorus Blemmidas, *Epit. Log.*, c. xxxi. p. 222; but long before him Boethius has all the kinds of Conversion, —*Simplex, Per Accidens*, et *Per Oppositionem*, (*Introductio ad Syllogismos*, p. 576), what he calls *Per Contrapositionem*, (*De Syllogismo Categorico*, L. l. 589.) Is he the inventor of the name? It seems so. Long before Boethius, Apuleius, (in second century), has it as one of the five species of Conversion, but gives it no name—only descriptive, see *De Habitud. Doct. Plat.*, L. iii. p. 33. Alexander, *In An. Pr.*, i. c. 2, f. 10 a, has it as of propositions, not of

terms, which is conversion absolutely. Vide Philoponus, *In An. Pr.*, L. f. 12 a. By them called ἀντιστροφή σὺν ἀντιθέσει. So Magentinus, *In An. Pr.*, L. 2, f. 3 b.

That Contraposition is not properly Conversion—(this being a species of consequence)—an æquipollence of propositions, not a conversion of their terms.

Noldius, *Logica Recognita*, c. xii. p. 292. Crakanthorpe, *Logica*, L. iii. c. 10, p. 180. Bannes, *Instit. Min. Dial.*, L. v. c. 2, p. 530. Eustachius, *Summa Philosophiæ*, *Logica*, P. ii. tract. i. q. 3, p. 104. Herbart, *Lehrbuch der Logik*, p. 78. Scotus, *Quæstiones, In An. Prior.*, L. i. q. 15, f. 258 b. Chauvin, v. *Conversio*. Lsndown, *Cursus Logicus*, p. 308.

That Contraposition is useless and perplexing. See Chauvin, v. *Conversio*. Arriaga, *Cursus Philosophicus*, p. 18. Titius, *Ars Cogitandi*, c. viii. § 19 et seq. D'Abra de Raconis, *Tot. Phil. Tract.*, *Logica*, it. q. 4, p. 315. Bannes, *Instit. Min. Dial.*, p. 529.]

a [Blemmidas.] [*Epitome Logica*, c. 31, p. 222. The following extract will explain the nature of this conversion: ʽΗ δʼ ἐν ἀπορδάσει γενομένη ἀντιστροφή, ἡ τὴν μὲν τάξιν τῶν ὅρων φυλάττει, τὸν αὐτὸν τηροῦσα κατηγορούμενον καὶ τὸν αὐτὸν ὑποκείμενον μόνον δὲ τὴν ποιότητα μεταβάλλει, ποιοῦσα τὴν ἀποφατικὴν ὑπόφασιν καταφατικήν, καὶ καταφατικὴν ἀποφατικήν. Καὶ λέγεται αὕτη ἐνδεχομένη ἀντιστροφή, ὡς ἐπὶ μόνης τῆς ἐνδεχομένης ὕλης συνισταμένη οἷον, τις ἀνθρωπος λευκαι, τις ανθρωπος οὐ λευκαι· αὕτη δʼ οὐκ ἐν τῇ τυχίει ἀντιστροφή. This so-called *contingent conversion* is in fact nothing more than the assertion, repeated by many Latin logicians, that in contingent matter subcontrary propositions are both true. —Ed.]

world. But let us now proceed to see that Simple Conversion, as I have asserted, is thorough-going and all-sufficient. Let us try it in all the eight varieties of categorical propositions. But I shall leave this explication to yourselves, and in the examination will call for a statement of the simple conversion, as applied to all the eight propositional forms.

It thus appears, that this one method of conversion has every advantage over those of the logicians. 1°, It is Natural; 2°, It is Imperative; 3°, It is Simple; 4°, It is Direct; 5°, It is Precise; 6°, It is Thorough-going: Whereas their processes are—1°, Unnatural; 2°, Precarious; 3°, Complex; 4°, Circuitous; 5°, Confused; 6°, Inadequate: breaking down in each and all of their species. The Greek Logicians, subsequent to Aristotle, have well and truly said, ἀντιστροφή ἐστιν ἰσοστροφή τις "omnis conversio est æquiversio,"* that is, all conversion is a conversion of equal into equal; and had they attended to this principle, they would have developed conversion in its true unity and simplicity. They would have considered, 1°, That the absolute quantity of the proposition, be it convertend or converse, remains always identical; 2°, That the several quantities of the collated notions remain always identical, the whole change being the transposition of the quantified notion, which was in the subject place, into the place of predicate, and *vice versâ.*

Aristotle and the logicians were, therefore, wrong; 1°, In not considering the proposition simply as the complement, that is, as the equation or non-equation, of two compared notions, but, on the contrary, considering it as determined in its quantity by one of these notions more than by the other. 2°, They were wrong, in according too great an importance to the notions considered as propositional terms, that is, as subject and predicate, independently of the import of these notions in themselves. 3°, They were wrong, in according too preponderant a weight to one of these terms over the other; but differently in different parts of the system. For they were wrong, in the doctrine of Judgment, in allowing the quantity of the proposition to be determined exclusively by the quantity of the subject term; whereas they were wrong, as we shall see, in the doctrine of Reasoning, in considering a syllogism as exclusively relative to the quantity of the pre-

* See above, p. 258.—En.

dicate (extension). So much for the theory of Conversion. Before concluding, I have, however, to observe, as a correction of the prevalent ambiguity and vacillation, that the two propositions of the process together might be called the *convertent* or *converting*, (*propositiones convertentes*); and whilst of these the original proposition is named the *convertend*, (*propositio convertenda*), its product would obtain the title of *converse, converted,* (*propositio conversa*).[a]

The other species of Immediate Inference will not detain us long. Of these, there are two noticed by the logicians.

The first of these, *Equipollence*, (*æquipollentia*), or, as I would term it, *Double Negation*, is only deserving of bare mention. It is of mere grammatical relevancy. The negation of a negation is tantamount to an affirmation. "*B is not not-A*," is manifestly only a roundabout way of saying "*B is A ;*" and, *vice versâ*, we may express a position, if we perversely choose, by sublating a sublation. The immediate inference of Equipollence is thus merely the grammatical translation of an affirmation into a double negation, or of a double negation into an affirmation. *Non-nullus* and *Non-nemo*, for example, are merely other grammatical expressions for *aliquis* or *quidam*. So *Nonnihil, Nonnunquam, Nonnusquam*, &c.

The Latin tongue is almost peculiar among languages for such double negatives to express an affirmative. Of course the few which have found their place in Logic, instead of being despised or relegated to Grammar, have been fondly commented on by the ingenuity of the scholastic logicians. In English, some authors are fond of this indirect and idle way of speaking; they prefer saying—" I entertain a not unfavourable opinion of such a one," to saying directly, I entertain of him a favourable opinion. Neglecting this, I pass on to

The third species of Immediate Inference, noticed by the logicians. This they call *Subalternation*, but it may be more unambiguously styled *Restriction*. If I have £100 at my credit in the bank, it is evident that I may draw for £5 or £10. In like manner, if I can say unexclusively, that *all men are animals*, I can say restrictively, that *negroes or any other fraction of mankind are animals*. This restriction is Bilateral, when we restrict both subject and predicate, as—

a See above, vol. l. p. 262.—Ed.

All Triangle is all trilateral.   *All rational is all risible.*
∴ *Some triangle is some trilateral.*   ∴ *Some rational is some risible.*

It is Unilateral, by restricting the omnitude or universality
either of the Subject or of the Predicate.

Of the Subject—

All man is some animal ;
∴ *Some man is some animal.*

Of the Predicate, as—

Some animal is all risible ;
∴ *Some animal is some risible.*

It has not been noticed by the logicians, that there is only an
inference by this process, if the *some* in the inferred proposition
means *some at least*, that is, *some not exclusive of all ;* for if we
think by the *some, some only*, that is, *some, not all*, so far from
there being any competent inference, there is in fact a real opposition. The logicians, therefore, to vindicate their doctrine of
the Opposition of Subalternation, ought to have declared, that
the *some* was here in the sense of *some only ;* and to vindicate
their doctrine of the Inference of Subalternation, they ought, in
like manner, to have declared, that the *some* was here taken in
the counter sense of *some at least.* It could easily be shown,
that the errors of the logicians in regard to Opposition, are not
to be attributed to Aristotle.

Before leaving this process, it may be proper to observe that
we might well call its two propositions together the *restringent*
or *restrictive* (*propositiones restringentes* vel *restrictivæ*); the
given proposition might be called the *restringend*, (*propositio
restringenda*), and the product the *restrict* or *restricted*, (*propositio restricta*).

So much for the species of Immediate Inference recognised by
the logicians.

There is, however, a kind of immediate inference overlooked by
logical writers. I have formerly noticed, that they enumerate,
(among the species of Opposition), *Subcontrariety*, (*subcontra-
rietas*, ὑπεναντιότης), to wit,—*some is, some is not ;* but that this
is not in fact an opposition at all, (as in truth neither is Subalternation in a certain sense.) Subcontrariety, in like manner, is

with them not an opposition between two partial *sames*, but between different and different; in fact, no opposition at all. But if they are thus all wrong by commission, they are doubly wrong by omission, for they overlook the immediate inference which the relation of propositions in Subcontrariety affords. This, however, is sufficiently manifest. If I can say, "*All* men are *some* animals," or, "*Some* animals are *all* men," I am thereby entitled to say,—"*All* men are *not some* animals," or "*Some* animals are *not some* men." Of course here the *some* in the inferred propositions means *some other*, as in the original proposition, *some only;* but the inference is perfectly legitimate, being merely a necessary explication of the thought: for inasmuch as I think and say that *all men are some animals*, I can think and say that they are some animals only, which implies that they are a certain some, and not any other animals.[a] This inference is thus not only to some others indefinitely, but to all others definitely. It is further either affirmative from a negative antecedent, or negative from an affirmative. Finally, it is not bilateral, as not of subject and predicate at once; but it is unilateral, either of the subject or of the predicate. This inference of Subcontrariety, I would call *Integration*, because the mind here tends to determine all the parts of a whole, whereof a part only has been given. The two propositions together might be called the *integral* or *integrant*, (*propositiones integrales vel integrantes*). The given proposition would be styled the *integrand*, (*propositio integranda*); and the product, the *integrate* (*propositio integrata*).[β]

I may refer you for various observations on the Quantification of the Predicate, to the collection published under the title, *Discussions on Philosophy and Literature.*

The grand general or dominant result of the doctrine on which I have already partially touched, but which I will now explain

[a] If we say *some animal is all men*, and *some animal is not any man*; in that case, we must hold *some* as meaning *some only*. We may have a mediate syllogism on it, as:—

  Some animals are all men;
  Some animals are not any man;

Therefore, some animals are not some animals.

[β] Mem. Immediate inference of Contradiction omitted. Also of Relation, which would come under Equipollence. [For Tabular Schemes of Propositional Forms, and of their Mutual Relations, see below, pp. 279-80, 288.—Ed.]

consecutively and more in detail, is as follows:—Touching Propositions,—Subject and Predicate;—touching Syllogisms,—in Categoricals, Major and Minor Terms, Major and Minor Premises,
Figures First, Second, Third, Fourth, and even what I call *No
Figure*, are all made convertible with each other, and all conversion reduced to a simple equation; whilst in Hypotheticals, both
the species, (viz. Conjunctive and Disjunctive reasonings), are
shown to be forms not of mediate argumentation at all, but merely
complex varieties of the immediate inference of Restriction or
Subalternation, and are relieved of a load of perversions, limitations, exceptions, and rules. The differences of Quantity and
Quality, &c., thus alone remain; and by these exclusively are
Terms, Propositions, and Syllogisms formally distinguished. Quantity and Quality combined constitute the only real discrimination
of Syllogistic Mood. Syllogistic Figure vanishes, with its perplexing apparatus of special rules; and even the General Laws of
Syllogism proper are reduced to a single compendious canon.

This doctrine is founded on the postulate of Logic :—" To state
in language, what is efficient in thought;" in other words, " Before proceeding to deal logically with any proposition or syllogism,
we must be allowed to determine and express what it means."

First, then, in regard to Propositions.—In a proposition, the
two terms, the Subject and Predicate, have each their quantity in
thought. This quantity is not always expressed in language, for
language tends always to abbreviation; but it is always understood. For example, in the proposition, *Men are animals*, what
do we mean ? We do not mean that *some men*, to the exclusion of
others, are animals, but we use the abbreviated expression *men*
for the thought *all men*. Logic, therefore, in virtue of its postulate, warrants, nay requires, us to state this explicitly. Let us,
therefore, overtly quantify the subject, and say, *All men are animals*. So far we have dealt with the proposition,—we have quantified in language the subject, as it was quantified in thought.

But the predicate still remains. We have said—*All men are
animals*. But what do we mean by *animals?* Do we mean *all
animals*, or *some animals?* Not the former; for dogs, horses,
oxen, &c., are animals as well as men, and dogs, horses, oxen,
&c., are not men. Men, therefore, are animals, but exclusively
of dogs, horses, oxen, &c. *All men*, therefore, are not equivalent

to *all animals ;* that is, we cannot say, as we cannot think, that *all men are all animals.* But we can say, for in thought we do affirm, that *all men are some animals.*

But if we can say, as we do think, that *all men are some animals,* we can, on the other hand, likewise say, as we do think, that *some animals are all men.*

If this be true, it is a matter of indifference, in a logical point of view, (whatever it may be in a rhetorical), which of the two terms be made the subject or predicate of the proposition; and whichsoever term is made the subject in the first instance, may, in the second, be converted into the predicate, and whichsoever term is made the predicate in the first instance, may, in the second, be converted into the subject.

From this it follows :—

1°, That a proposition is simply an equation, an identification, a bringing into congruence, of two notions in respect to their Extension. I say, in respect to their Extension, for it is this quantity alone which admits of ampliation or restriction, the Comprehension of a notion remaining always the same, being always taken at its full amount.

2°, The total quantity of the proposition to be converted, and the total quantity of the proposition the product of the conversion, is always one and the same. In this unexclusive point of view, all conversion is merely *simple conversion ;* and the distinction of a conversion, as it is called, *by accident,* arises only from the partial view of the logicians, who have looked merely to the quantity of the subject. They, accordingly, denominated a proposition *universal* or *particular,* as its subject merely was quantified by the predesignation *some* or *all ;* and where a proposition like, *All men are animals,* (in thought, *some animals*), was converted into the proposition, *Some animals are men,* (in thought, *all men*), they erroneously supposed that it lost quantity, was restricted, and became a particular proposition.

It can hardly be said that the logicians contemplated the reconversion of such a proposition as the preceding ; for they did not (or rarely) give the name of *conversio per accidens* to the case in which the proposition, on their theory, was turned from a particular into a universal, as when we reconvert the proposition, *Some animals are men*, into the proposition, *All men are animals.*[a] They likewise neglected such affirmative propositions as had in thought both subject and predicate quantified to their whole extent; as, *All triangular figure is trilateral*, that is, if expressed as understood, *All triangular is all trilateral figure,—All rational is risible*, that is, if explicitly enounced, *All rational is all risible animal.* Aristotle, and subsequent logicians, had indeed frequently to do with propositions in which the predicate was taken in its full extension. In these the logicians,—but, be it observed, not Aristotle,—attempted to remedy the imperfection of the Aristotelic doctrine, which did not allow the quantification of the predicate to be taken logically or formally into account in affirmative propositions, by asserting that in the obnoxious cases the predicate was distributed, that is, fully quantified, in virtue of the matter, and not in virtue of the form, (*vi materiæ non ratione formæ*). But this is altogether erroneous. For in thought we generally do, nay often must, fully quantify the predicate. In our logical conversion, in fact, of a proposition like *All men are animals,—some animals*, we must formally retain in thought, for we cannot formally abolish, the universal quantification of the predicate. We, accordingly, must formally allow the proposition thus obtained,—*Some animals are all men.*

The error of the logicians is further shown by our most naked logical notation; for it is quite as easy and quite as natural to

---

a See above, vol. i. p. 264.—Ed. [A mistake by logicians in general, that partial conversion, *ἐν μέρει*, is a more synonym of *per accidens*, and that the former is so used by Aristotle. See Vallius *Logica*, t. ii. l. i. q. i. c. 2, p. 32. For Aristotle uses the terms *universal* and *partial conversion*, simply to express whether the *convertens* is an universal or particular proposition. See § 4 of the chapter on Conversion, (*An. Prior.*, i. 2), where particular affirmatives are said to be necessarily converted, *ἐν μέρει*.

Conversio per accidens is in two forms differently defined by different logicians. The first by Boethius, by whom the name was originally given, is that in which the quantity of the proposition is contingently changed either from greater to less, or from less to greater, *salvo veritate*, the quality of the terms and propositions remaining always the same. So Ridiger, *De Sensu Veri et Falsi*, p. 303. The second is that of logicians in general, where the quantity of the proposition is diminished, the quality of the propositions and terms remaining the same, *salva veritate*.]

quantify A, B, or C, as predicate, as to quantify A, B, or C, as subject. Thus, *All* B *is some* A ; *Some* A *is all* B.

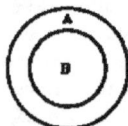

I may here also animadvert on the counter defect, the counter error, of the logicians, in their doctrine of Negative Propositions. In negative propositions they say the predicate is always distributed,—always taken in its full extension. Now this is altogether untenable. For we always can, and frequently do, think the predicate of negative propositions as only partially excluded from the sphere of the subject. For example, we can think, as our naked diagrams can show,—*All men are not some animals*, that is, not irrational animals. In point of fact, so often as we think a subject as partially included within the sphere of a predicate, *eo ipso* we think it as partially, that is, particularly, excluded therefrom. Logicians are, therefore, altogether at fault in their doctrine, that the predicate is always distributed, *i.e.* always universal, in negative propositions.[a]

. a [Melanchthon, (*Erotemata*, L. II. De *Conversione*, p. 129), followed by his pupil and commentator Strigelius, (*In Erotemata*, p. 576-81), and by Keckermann, (*Syst. Log. Minus*, L. ii. c. 3, *Op.* p. 222), and others, thinks that "there is a greater force of the particle *none*, (*nullus, non any*), than of the particle *all*, (*omnis*). For, in a universal negative, the force of the negation is so spread over the whole proposition, that in its conversion the same sign is retained, (as—*No star is consumed; therefore, no flame which is consumed is a star*) : whereas such conversion does not take place in a universal affirmative." This Strigelius compares to the diffusion of a ferment or acute poison ; adding that "the affirmative particle is limited to the subject, whilst the negative extends to both subject and predicate, in other words, to the whole proposition."

This doctrine is altogether erroneous. It is an erroneous theory devised to explain an erroneous practice. In the first place, we have here a commutation of negation with quantification ; and, at the same time, conversion, direct conversion at least, will not be said to change the quality either of a negative or affirmative proposition. In the second place, it cannot be pretended that negation has an exclusive or even greater affinity to universal than to particular quantification. We can equally well say *not some, not all, not any;* and the reason why one of these forms is preferred, lies certainly not in any attraction or affinity to the negative particle.]

But, 3°, If the preceding theory be true,—if it be true that subject and predicate are, as quantified, always simply convertible, the proposition being in fact only an enouncement of their equation, it follows, (and this also is an adequate test), that we may at will identify the two terms by making them both the subject or both the predicate of the same propositions. And this we can do. For we can not only say—as A *is* B, so conversely B *is* A, or as *All men are some animals*, so, conversely, *Some animals are all men*; but equally say—A and B are *convertible*, or, *Convertible are* B and A; *All men and some animals are convertible*, (that is, *some convertible things*), or, *Convertible*, (that is, *some convertible things*), *are some animals and all men*. By *convertible*, I mean the same, the identical, the congruent, &c.[a]

---

a [With the doctrine of Conversion taught in the text, compare the following authorities:—Laurentius Valla, *Dialectica*, L. ii. c. 24, f. 37. Titius, *Ars Cogitandi*, (v. Ridiger, *De Sensu Veri et Falsi*, L. ii. c. i. p. 232). Reusch, *Systema Logicum*, ¶ 380, p. 413 *et seq.*, ed. 1741. Hollmann, *Logica*, ¶ 89, p. 172. Ploncquet, *Fries, Logik*, ¶ 33, p. 146. E. Reinhold, *Logik*, ¶ 117, p. 286. Ancients referred to by Ammonius, *In De Interp.*, c. vii., ¶ 4, f. . . . Paulus Vallius, *Logica*, t. ii., *In An. Prior.*, L. i. q. ii. c. iv.] [Valla *l. c.* says:—"Non amplius ac latius accipitur praedicatum quam subjectum. Ideoque cum illo converti potest, ut *omnis homo est animal;* non atique totum genus *animal*, sed aliqua pars hujus generis . . . ergo, *Aliqua pars animalis est in omni homine.* Item, *Quidam homo est animal*, scilicet est *quaedam pars animalis*, ergo, *Quaedam pars animalis est quidam homo*, &c." Gottlieb Gerhard Titius, *Ars Cogitandi*, c. vii. ¶ 3 *et seq.*, p. 123. Lipsiae, 1723 (first ed. 1701). "Nihil autem aliud agit Conversio, quam ut simpliciter praedicatum et subjectum transponat, hinc neo qualitatem nec quantitatem iis largitur, aut eam mutat, sed prout reperit, ita convertit. Ex quo necessario sequitur conversionem esse uniformem, ac omnes propositiones eodem plane modo converti. Per exempla, (1), *Nullus homo est lapis (universaliter)*, ergo, *Nullus lapis est homo*, (2) *Quidam homo non est medicus (omnis)*, ergo, *Omnis medicus non est homo quidam*, seu *Nullus medicus est homo quidam* . . . (3), *Hic Petrus non est doctus (omnis)*, ergo, *Omnis doctus non est hic Petrus* . . . (4), *Omnis homo est animal (quoddam)*, ergo, *Quoddam animal est homo*, (5), *Quidam homo currit (particulariter)*, ergo, *Quidam currens est homo*, (6), *Hic Paulus est doctus (quidam)*, ergo, *Quidam doctus est hic Paulus.* In omnibus his exemplis subjectum cum sua quantitate in locum praedicati, et hoc, eodem modo, in illius sedem transponitur, ut nulla penitus ratio solida appareat, quare conversionem in diversas species divellere debeamus. Vulgo tamen aliter sentiunt, quando triplicem conversionem, nempe simplicem, per accidens, ac per contrapositionem, adstruunt . . . Enimvero conversio per accidens et per contrapositionem gratis asseritur, nam conversio propositionis affirmantis universalis perinde simplex est ac ea qua universalis negans convertitur, licet post eam subjectum sit particulare; conversionis enim hic nulla culpa est, quae quantitatem, quae non adest, largiri nec potest nec debet . . . Error vulgaris doctrinae, nisi fallor, inde

The general errors in regard to Conversion,—the errors from which all the rest proceed,—are

1°, The omission to quantify the predicate throughout.

2°, The conceit that the quantities did not belong to the terms.

3°, The conceit that the quantities were not to be transposed with their relative terms.

4°, The one-sided view that the proposition was not equally composed of the two terms, but was more dependent on the subject than on the predicate.

5°, The consequent error that the quantity of the subject term determines the quantity of the proposition absolutely.

6°, The consequent error that there was any increase or diminution of the total quantity of the proposition.

7°, That thoroughgoing conversion could not take place by one, and that the simple, form.

8°, That all called in at least the form of Accidental Conversion; all admitting at the same time that certain moods remain inconvertible.

---

ent, quod existimaverint ad conversionem simplicem requiri, *ut praedicatum assumat signum et quantitatem subjecti* . . . Conversionem *per contrapositionem* quod attinet, facile intendi potest (1) exempla huic jactari solita posse converti simpliciter; (2) conversionem per contrapositionem revera non esse conversionem; interim (3) putativam istam conversionem non in universali affirmante, et particulari negante solum, sed in omnibus potius propositionibus locum habere . . . *e.g., Quoddam animal non est quadrupes, ergo, Nullus quadrupes est animal quoddam.*" See the criticism of the doctrine of Titius by Ridiger, quoted below, p. 318. Ploucquet, *Methodus Calculandi in Logicis,* p. 49 (1763) : "Intellectio *identitatis* subjecti et praedicati est *affirmatio.* . . . *Omnis circulus est linea curva.* Quae propositio logice expressa haec est :—*Omnis circulus est quaedam linea curva.* Quo pacto id, quod intelligitur in praedicato identificatur cum eo quod intelligitur in subjecto. Sive norim, sive non norim

praeter circulum dari quoque alias curvarum species, verum tamen est quandam lineam curvam esse comprehensivo sumtam, eme omnem circulum, est omnem circulum esse quandam lineam curvam." Vallius, *l.c.* "Negativae vero convertuntur et in particulares et in universalem negativas; ut si dicamus, *Socrates non est lapis,* convertes illius erit, *Aliquis lapis non est Socrates,* et *Nullus lapis est Socrates,* et idem dicendum erit de omni alia simili propositione."—ED.]

[That Universal Affirmative Propositions may be converted simply, if their predicates are reciprocating, see Corvinus, *Instit. Phil. Rat.,* § 514, (Icum, 1742). Baumgarten, *Logica,* § 280, (1765). Scotus, *In An. Pr.,* L. l. qu. 14. Ulrich, *Instit. Log. et Met.,* § l. 2, 177, (1785). Kreil, *Logik,* §§ 46, 62, (1789). Isendoorn, *Logica Peripatetica,* L. iii. c. 8, pp. 430, 431. Wallis, *Logica,* L. ii. c. 7. Zabarella, *In An. Prior. Tabula,* p. 148. Lambert, *De Universaliori Calculi Idea,* § 24 et seq.]

9°. That the majority of logicians resorted to Contraposition, (which is not a conversion at all) ; some of them, however, as Burgersdyk, admitting that certain moods still remained obstinately inconvertible.

10°. That they thus introduced a form which was at best indirect, vague, and useless, in fact not a conversion at all.

11°. That even admitting that all the moods were convertible by one or other of the three forms, the same mood was convertible by more than one.

12°. That all this mass of error and confusion was from their overlooking the necessity of one simple and direct mode of conversion ; missing the one straight road.

We have shown that a judgment (or proposition) is only a comparison resulting in a congruence, an equation, or non-equation of two notions in the quantity of Extension ; and that these compared notions may stand to each other, as the one subject and the other predicate, as both the subject, or as both the predicate of the judgment. If this be true, the transposition of the terms of a proposition sinks in a very easy and a very simple process ; whilst the whole doctrine of logical Conversion is superseded as operose and imperfect, as useless and erroneous. The systems, new and old, must stand or fall with their doctrines of the Conversion of propositions.

Thus, according to the doctrine of the logicians, conversion applies only to the naked terms themselves :—the subject and predicate of the prejacent interchange places, but the quantity by which each was therein affected is excluded from the movement ; remaining to affect its correlative in the subjacent proposition. This is altogether erroneous. In conversion we transpose the compared notions,—the correlated terms. If we do not, *eversion*, not conversion, is the result.

If, (as the Logicians suppose), in the *convertens* the subject and predicate took each other's quantity, the proposition would

be not the *same relation* of the *same notions*. It makes no
difference that the converse only takes place when the subject
chances to have an equal amount or a less than the predicate.
There must be at any rate a reasoning, (concealed indeed), to
warrant it: in the former case—that the predicate is entitled to
take all the quantity of the subject, being itself of equivalent
amount; in the second, (a reasoning of subalternation), that it is
entitled to take the quantity of the subject, being less than its
own. All this is false. Subject and predicate have a right to
their own, and only to their own, which they carry with them,
when they become each other.

(d)—APPLICATION OF DOCTRINE OF QUANTIFIED PREDICATE TO PROPOSITIONS.

(1.) New Propositional Forms—Notation.

Instead of four species of Proposition determined by the Quantity and Quality taken together, the Quantity of the Subject being alone considered, there are double that number, the Quantity of the Predicate being also taken into account.

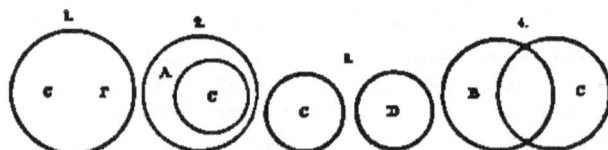

Affirmative.

(1) [A f A] C :———: Γ All Triangle is all Trilateral [fig. 1].
(ii) [A f I] C :———, A All Triangle is some Figure (A) [fig. 2].
(3) [I f A] A,———: C Some figure is all Triangle [fig. 2].
(iv) [I f I] C,———, B Some Triangle is some Equilateral (I) [fig. 4].

Negative.

(v) [E n E]   C : ⊶ : D Any Triangle is not any Square (E)
(A) (A)   [fig. 3].

(6) [E n O]   C : ⊶ , B Any Triangle is not some Equilateral
(A) (I)   [fig. 4].

(vii) [O n E] B , ⊶ : C Some Equilateral is not any Triangle
(I) (A)   (O) [fig. 4].

(8) [O n O]   C , ⊶ , B Some Triangle is not some Equila-
(I) (I)   teral [fig. 4].*

* [In this table the Roman numerals distinguish such propositional forms as are recognised in the Aristotelic or common doctrine, whereas the Arabic ciphers mark those (half of the whole) which I think ought likewise to be recognised. In the literal symbols, I simplify and disintricate the scholastic notation; taking A and I for universal and particular, but, extending them to either quality, marking affirmation by f, negation by n, the two first consonants of the verbs *affirmo* and *nego*,—verbs from which I have no doubt that Petrus Hispanus drew, respectively, the two first vowels, to denote his four complications of quantity and quality.] —*Discussions*, p. 686.

[In the notation employed above, the comma (,) denotes *some*; the colon (:) *all*, *any*; the line ⸻ denotes the affirmative copula, and negation is expressed by drawing a line through the affirmative copula ⊶; the thick end of the line denotes the subject, the thin end the predicate, of Extension. In Intension the thin end denotes the subject, the thick end the predicate. Thus :—O : ⸻ , A is read, *All C is some A.* C : ⊶ : D is read, *No C is any D.* The Table given in the text is from a copy of an early scheme of the Author's new Propositional Forms. For some time after his discovery of

the doctrine of a quantified predicate, Sir W. Hamilton seems to have used the vowels E and O in the formulæ of Negative Propositions; and the full period (.) as the symbol of *some* (indefinite quantity). In the college session of 1845-46, he had adopted the comma (,) as the symbol of indefinite quantity. As the period appears in the original copy of this table as the symbol of *some*, its date cannot be later than 1845. The comma (,) has been substituted by the Editors, to adapt the table to the Author's latest form of notation. The translation of its symbols into concrete propositions, affords decisive evidence of the meaning which the Author attached to them on the new doctrine. That this, moreover, was the uniform import of Sir W. Hamilton's propositional notation, from the earliest development of the theory of a quantified predicate, is placed beyond doubt by numerous passages in papers (not printed), and by marginal notes on books, written at various periods between 1839-40, and the date of his illness, July 1844, when he was compelled to employ an amanuensis. The letters in round brackets, (A) and (I), are the vowels finally adopted by the Author, in place of E and O. See below, p. 287.—Ed.]

(2.) QUANTITY OF PROPOSITIONS—DEFINITUDE AND INDEFINITUDE.

Nothing can exceed the ambiguity, vacillation, and uncertainty of logicians concerning the Quantity of Propositions.

I. As regards what are called *indefinite* (ἀδιόριστοι), more properly *indesignate* or *preindesignate*, *propositions*. The absence of overt quantification applies only to the subject; for the predicate was supposed always in affirmatives to be particular, in negatives to be universal. Referring, therefore, only to the indesignation of the subject:—indefinites were by some logicians (as the Greek commentators on Aristotle (?), Apuleius *apud* Waitz, *In Org.* i. p. 338, but see Wegelin, *In Anepunymi Phil. Syn.*, p. 588), made tantamount to particulars: by others, (as Valla, *Dialectica*, L. ii. c. 24, f. 37), made tantamount to universals. They ought to have been considered as merely elliptical, and to be definitely referable either to particulars or universals. [a]

II. A remarkable uncertainty prevails in regard to the meaning of particularity and its signs,—*some*, &c. Here *some* may mean *some only—some, not all*. Here *some*, though always in a certain degree indefinite, is definite so far as it excludes omnitude,—is used in opposition to *all*. This I would call its *Semi-definite* meaning. On the other hand, *some* may mean *some at least*,—

---

a [That indefinite propositions are to be referred to universals, see Purchot, *Instit. Phil. Logica*, L. § ii. c. 2, pp. 124, 125, 125. Rottenbeccius, *Logica Contracta*, c. vi. p. 92, (1560). Baumeister, *Inst. Phil. Rat.*, § 213. J. C. Scaliger, *Exercitationes*, Ex. 212, § 2. Drobisch, *Logik*, § 39. Neomagus, *Ad Trapezuntium*, f. 10. To be referred to particular; see Lovanienses, *Com. In Arist. Dial.*, p. 101. Molinaeus, *Elementa Logica*, L. i. c. 2. Alex. Aphrod, *In An. Prior.*, c. ii. p. 19. Denzinger, *Logica*, § 71. Either universal or particular, Keckermann, *Opera*, p. 230. Aristotle doubts: see *An. Prior.*, L. i. c. 27, § 7, and *De Interp.*, c. 7.

That Indefinitude is no separate species of quantity, see Scheibler, *Opera Logica*, p. iii. c. 6, p. 443. Grateus Anonymus, *De Syllogismo*, L. i. c. 4, f. 42. Leibnitz, *Opera*, t. iv. p. iii. p. 123. Fries, *System der Logik*, § 30, p. 137. Ramus, *Schol. Dial.*, L. vii. c. 2, p. 467. Downam, *In Rami Dialect.*, L. ii. c. 4, p. 359. Facciolati, *Rud. Log.*, p. ii. c. iii., p. 67. Delariviere, *Nouvelle Logique Classique*, L. ii. s. ii. c. 3, s. 580, p. 334.

That Indefinitude has sometimes a logical import, when we do not know whether *all*, or *some*, of the one be to be affirmed or denied of the other. E. Reinhold, *Logik*, § 88, Anm. 2, pp. 193, 194. Ploucquet, *Methodus Calculandi*, pp. 48, 53, ed. 1773. Lambert, *Neues Organon*, I., § 235, p. 143.]

some, perhaps all. In this signification *some* is thoroughly indefinite, as it does not exclude omnitude or totality. This meaning I would call the *Indefinite.*

Now of these two meanings there is no doubt that Aristotle used particularity only in the second, or thoroughly Indefinite, meaning. For 1°, He does not recognise the incompossibility of the superordinate and subordinate. 2°, He makes *all* and οὐ πᾶς or particular negative to be contradictories; that is, one necessarily true, the other necessarily false. But this is not the case in the Semi-definite meaning. The same holds good in the universal negative, and particular affirmative.

The particularity,—the *some,*—is held to be a definite *some* when the other term is definite, as in ii and 3, in 6 and vii. On the other hand, when both terms are indefinite or particular, as in iv and 8, the *some* of each is left wholly indefinite.

The quantification of *definitude* or *non-particularity* (:) may designate ambiguously or indifferently one or other of three concepts. 1°, It may designate explicit omnitude or totality; which, when expressed articulately, may be denoted by (: :). Thus—*All triangles are all trilaterals.* 2°, It may designate a class considered as undivided, though not positively thought as taken in its whole extent; and this may be articulately denoted by (: .). Thus—*The triangle is the trilateral;—The dog is the latrant.*— (Here note the use of the definite article in English, Greek, French, German,ª &c.) 3°, It may designate not what is merely

---

ª [On effect of the definite article and its absence in different languages, in reducing the definite to the indefinite, see Delarivière, *Logique*, §§ 580, 581.

On the Greek article, see Ammonius, *In De Interp.*, c. vii. l. 57 b.

On use of the article in quantification, see Averroes, *De Interp.*, p. 39, ed. 1552; "*Al* in the Arabic tongue, and *Ha* in the Hebrew, and in like manner the articles in other languages, sometimes have the power of universal predesignations, sometimes of particular. If the former, then they have the force of contraries; if the latter, then the force of sub-contraries. For it is true to say, *al*, that is, *ipse homo is white,* and *al,* that is, *ipse homo is not white*; that is, when the article *al* or *ha,* that is, *ipse,* denotes the designation of particularity. They may, however, be at once false, when the article *al* or *ha* has the force of the universal predesignation." (See also p. 52 of the same book.]

In English the definite article always defines,—renders definite,—but sometimes individualises, and sometimes generalises. If we would use mon

undivided, though divisible,—a class, but what is indivisible,—an individual; and this may be marked by the small letter or by (:). Thus—*Socrates is the husband of Xanthippe;—This horse is Bucephalus.*

In like manner particularity or indefinitude (,), when we wish to mark it as thoroughly indefinite, may be designated by (',), whereas when we would mark it as definitely indefinite, as excluding *all* or *not any*, may be marked by (").

The indefinites (ἀόριστα) of Aristotle correspond sometimes to the particular, sometimes to one or other, of the two kinds of universals.[a]

The designation of *indefinitude* or *particularity, some* (, or ,) may mean one or other of two very different things.

1°, It may mean *some and some only*, being neither *all* nor *none*, and, in this sense, it will be both affirmative and negative, (, ,).

2°, It may mean, negatively, *not all, perhaps none,—some at most;* affirmatively, *not none, perhaps all,—some at least,* (, ,).[β]

Aristotle and the logicians contemplate only the second mean-

<hr>

generally, we must not prefix the article, as in Greek, German, French, &c., no *wealth, government, &c.* But in definition of *horse, &c.*, the reverse, as *the dog, (le chien, ò xiwv, &c.) A* in English is often equivalent to *any*.]

a [Logicians who have marked the Quantities by *Definite, Indefinite, &c.*

Aristotle, *An. Pr.*, I. I. c. iv. § 21, and there Alexander, Pacius. Theophrastus, (Facciolati, *Bud. Log.*, p. 1. c. 4, p. 39.) Ammonius, *In De Inter.*, f. 72 b. (Brandis, *Scholia*, p. 113.) Stoics and Non-peripatetic Logicians in general, see Sext. Empiricus, *Adv. Log.*, § 98 et sq., p. 476, ed. Fabricii; Diog. Laert., Lib. vii. seg. 71, *ubi* Menagius. Downam, *In Rami Dialecticam*, L. ii. c. 4, p. 363, notices that a particular proposition " was called by the Stoics *indefinite, (ἀόριστον*); by some Latins, and sometimes by Ramus himself, *infinite;* because it does not designate some certain species, but leaves it uncertain and indefinite." Hurtado

de Mendoza, *Disp. Log. et Met.*, disp. iv. § 2, t. i. p. 114. Lovanienses, *In Arist. Dial.*, p. 161. Hollmann, *Logica*, p. 173. Boethius, *Opera*, p. 345. Reusch, *Syst. Log.*, p. 424. Esser, *Logik*, § 58. Weiss, *Logik*, § 149, 150. So Kiesewetter, *Logik*, § 102, 103.]

β The indefinite *some at least* is here and elsewhere very clearly stated by the Author to mean, in affirmatives, *some, perhaps all;* in negatives, *some, possibly none. Some at least are, perhaps all are; Some at least are not, possibly none are.* These meanings are stated and distinguished with the greatest clearness in the text and footnote to the *Discussions*, p. 690, 2d edition. By an extraordinary misconception, however, of the Author's meaning in that passage, the expression "*some at least, possibly, therefore, all or none,*" has been understood as implying that *some at least* in affirmative propositions may include " possibly none."—ED.

ing. The reason of this perhaps is, that this distinction only emerges in the consideration of Opposition and Immediate Inference, which were less elaborated in the former theories of Logic; and does not obtrude itself in the consideration of Mediate Inference, which is there principally developed.ª On the doctrine of the logicians, there is no opposition of subalternation; and by Aristotle no opposition of subalternation is mentioned. By other logicians it was erroneously introduced. The opposition of Sub-contraries is, likewise, improper, being precarious and not between the same things. Aristotle, though he enumerates this opposition, was quite aware of its impropriety, and declares it to be merely verbal, not real.ᵝ

By the introduction of the first meaning of *some*, we obtain a veritable opposition in Subalternation; and an inference in Subcontrariety, which I would call *Integration*.

(3.) OPPOSITION OF PROPOSITIONS.

Propositions may be considered under two views; according as their particularity, or indefinitude, is supposed to be thoroughly indefinite, unexclusive even of the definite; *some*, meaning *some at least*,—*some, perhaps all*,—*some, perhaps not any;* or definite indefinitude, and so exclusive of the definite; *some*, meaning *some at most*,—*some only*,—*some not all*, &c. The latter thus excludes omnitude or totality, positive or negative; the former does not. The former is the view promulgated as alone contemplated by

ᵇ The distinction between the more and less definite senses of *some* (*some only* and *some at least*) does not obtrude itself so as to be available for mediate or syllogistic inference. In other words, a more definite sense given to *some* in the premiss of a syllogism will not warrant a corresponding definiteness in the conclusion; but the latter remains just the same as if the *some* of the premiss were thoroughly indefinite. If, e. g. in a syllogism in Darii, we give a more definite meaning to *some* in the minor premiss, "All B is C; *some* (and some only) A is B;" we cannot infer, "Some (and some only) A is C;" but only "*some* A is C;" just as if the premiss were thoroughly indefinite.

That this is obviously Sir W. Hamilton's meaning may be seen by comparing Postulates II., III., and IV., p. 255. Indeed, no explanation would have been thought necessary, had not a misapprehension of the Author's language, as stated without comment in the first edition, given rise to some captious and wholly groundless criticisms of his theory as a whole.—ED.

ᵝ On both forms of Opposition, see Scheibler [*Opera Logica*, ∫ iii. *de Propositionibus*, c. xi. p. 487, and above, vol. i. p. 263.—ED.]

Aristotle; and has been inherited from him by the Logicians, without thought of increase or of change. The latter is the view which I would introduce; and, though it may not supersede, ought, I think, to have been placed alongside of the other.

Causes of the introduction of the Aristotelic system alone :—

1°, To allow a harmony of Logic with common language; for language eliding all that is not of immediate interest, and the determination of the subject-notion being generally that alone intended, the predicate is only considered in so far as it is thought to cover the subject, that is, to be at least coextensive with it. But if we should convert the terms, the inadequacy would be brought to light.

2°, A great number of notions are used principally, if not exclusively, as attributes, and not as subjects. Men are, consequently, very commonly ignorant of the proportion of the extension between the subjects and predicates, which they are in the habit of combining into propositions.

3°, In regard to negatives, men naturally preferred to attribute positively a part of one notion to another, than to deny a part. Hence the unfrequency of negatives with a particular predicate.

On the doctrine of Semi-definite Particularity, I would thus evolve the Opposition or Incompossibility of propositions, neglecting or throwing aside (with Aristotle) those of *Subalternation* and *Sub-contrariety*, but introducing that of *Inconsistency*.

Incompossibility is either of propositions of the same, or of different, quality. Incompossible propositions differing in quality are either Contradictories without a mean,—no third,—that is, if one be true the other must be false, and if one be false the other must be true; or Contraries with a mean,—a third,—that is, both may be false, but both cannot be true. Incompossible propositions of the same quality are Inconsistents, and, like Contraries, they have a mean, that is, both may be false, but both cannot be true.

Contradictories are either simple or complex. The simple are either, 1°, Of Universals, as undivided wholes; or, 2°, of Individuals, as indivisible parts.[a] The complex are of universals divided, as 4—5.

[a] General terms, used as individual terms, when opposed to each other, may be contradictories, as *Man is* mortal, *Man is not mortal*. So that there are three kinds of contradictories.

Contraries, again, which are only of divided universals, are 1°, Bilateral, as 1—5; or, 2°, Unilateral, as 1—6, 1—7, 2—5, 3—5; or, 3°, Cross, as 2—7, 3—6.

Inconsistents are either, 1°, Affirmatives; or, 2°, Negatives. Affirmatives, as 1—2, 1—3, 2—3. Negatives, as 5—6, 5—7. The propositions 6—7 are sometimes Inconsistents, sometimes Consistents.

All the other propositional forms, whether of the same or of different qualities, are Composible or Unopposed.

The differences in Composibility of the two schemes of indefinite and definite particularity lies, 1°, in the whole Inconsistents; 2°, in two Contraries for Contradictories. 1°, According to the former, all affirmative and all negative propositions are consistent, whereas in the latter these are inconsistent, 1—2, 1—3, 2—3 among the affirmatives; and among the negatives, 5—6, 5—7. (As said before, 6—7 is in both schemes sometimes composible, and sometimes incomposible). 2°, Two incomposibles, to wit, 2—7, 3—6, which, on the Aristotelic doctrine, are Contradictories, are in mine Contraries.

The propositional form 4 is consistent with all the affirmatives; 8 is not only consistent with all the negatives, but is composible with every other form in universals. It is useful only to divide a class, and is opposed only by the negation of divisibility.

By adopting exclusively the indefinite particularity, logicians threw away some important Immediate Inferences; those, to wit, 1°, from the affirmation of one *some* to the negation of another, and *vice versa;* and 2°, from the affirmation of one inconsistent to the negation of another.

1°, Thus, on our system, but not on theirs, affirming *all man to be some animal*, we have a right to infer that *no man is some* (*other*) *animal;* affirming that *some animal is all man*, we have a right to infer that *some* (*other*) *animal is not any man;* affirming *some men are some blacks*, (*Negroes*), we are entitled to say that (*same*) *some men are not some* (*other*) *blacks* (*Hindoos*), and also that (*other*) *some men are not* (*the same*) *some blacks.* And so backwards from negation to affirmation. This inference I would call that of [Integration].

2°, Affirming *all men are some animals*, we are entitled to infer the denial of the propositions, *all men are all animals, some men are all animals.* And so in the negative inconsistents.

AFFIRMATIVES.

1.) Toto-total   — AFA = All   — is all   —.
ii.) Toto-partial = AFI = All   — is some —.   (A)
3.) Parti-total   = IFA = Some — is all   —.
iv.) Parti-partial = IFI   = Some — is some —.   (I)

NEGATIVES.

v.) Toto-total   — ANA = Any — is not any —. (E)
6.) Toto-partial = ANI = Any — is not some —.
vii.) Parti-total = INA = Some — is not any —. (O)
8.) Parti-partial = INI   = Some — is not some—.

TABLE of the Mutual Relations of the Eight Propositional Forms on Either System of Particularity. (For Generals only.)

ABBREVIATIONS:—bl. = bilateral; cr. = cross; Contrar. = Contrarius; di. = direct; Incom. = Incompatible; Int. or Integr. = Integration; Repugn. = Repugnant, Contradictories; Rev. or Restr. = Restriction, Subalternation; un. = unilateral.—Blanks: in I. = Compatibles; in II. = No inference.—(Unilateral, bilateral, cross, direct, refer to the Extremes.)a

The preceding Table may not be quite accurate in details.

a The terms unilateral, bilateral, direct, cross, may perhaps need a few additional words of explanation. A relation, whether of incompatibility or of inference, between two propositions, is unilateral when the two propositions are equally affected by it in one term, and unequally in the other; thus the relation between A ls and

(4.) NOTE RELATIVE TO TABLE OF MUTUAL RELATIONS OF PROPOSITIONAL FORMS.

Every proposition may be contradicted; for every proposition is either true or false; and between affirmation and negation of the same there is no medium. But it by no means follows that this one general contradiction should be supported on a single contradictory alternative; for in this case, on my system of definite particularity, only one propositional form—to wit, No. iv., the weakest of affirmatives, and between which and absolute negation there is no medium—this form alone can be contradictorily denied, to wit, by No. v. To show how each propositional form is contradictorily affirmed and denied, take the following:—

No. 1. Toto-total affirmation.—*All triangles* (Γ) *are all trilaterals* (C);—the class triangle (Γ) is identical—convertible with the class trilateral (C). " *This is true* "—affirmation ; " *This is not true* "—negation or contradiction. And the contradictory negation is supported if any one of the five following cases be true :—1°, *All* Γ *is* (*only*) *some* C; 2°, (*only*) *some* Γ *is all* C—these are inconsistents ; 3°, *no* Γ *is any* C; *no* Γ *is some* C; *some* Γ *is no* C—these are contraries.

In like manner the other forms affirmative and negative.

No. 8. Parti-partial negation.—*Some is not some*—which has been wholly neglected by logicians, though a most useful and important form, can be contradicted, like its pendant parti-partial affirmation, only on a single incompossible case, which, as having

Ani is unilateral, the subject in each being universal, and therefore affected in its whole extent, while the predicate is universal in the one and particular in the other. When both terms are equally affected, being equally definite or equally indefinite, the relation is *bilateral* and *direct*; as in Afa-Ana, and Ifi-Ana. When the amount of quantity is equal in each proposition as a whole, but unequal in the separate terms, greater definiteness of subject in the one being balanced by greater definiteness of predicate in the other

as in Afi-Ina, the relation is cross. A cross relation is necessarily bilateral.

In a bilateral direct relation, both propositions remain the same after simple conversion. In a bilateral cross relation, both are reversed; thus Afi-Ina becomes Ifa-Ani. In an unilateral relation, one proposition remains the same, and the other is reversed; thus Afa-Ani becomes Afa-Ina. This explanation seems to suit all the cases in the Table, except the fourth (Afi-Ifa) where *un* is probably an erratum for *bi*.—ED.

no medium, is itself a contradictory. This form is necessarily supposed in the division of a whole of any kind ; for the division itself is virtually the declaration, *some is not some*. Now it is evident that, applied to aught divisible, be it a universal whole or aught divisible, this proposition is not incompatible with any of the other seven propositions. We can say *some A is not some A* if A be divisible ; and the only possible way of contradicting is the following :—This proposition is not true ; for A is indivisible ; has no parts—no some and some. Thus, when I say, *Some of the individual Socrates is not some of the individual Socrates (mens cujusque, &c.); some of an atom is not some of an atom*—I assert *Socrates* and *an atom* to be divisible, and the assertion can only be contradicted by declaring it false, from the indivisibility of one and other.

EXAMPLES OF 1 AND 8.

" My mind is myself ; " " Some of my mind is not some of myself."

This bad in universal whole and any other wholes.

Remarks on Propositional Form 8.—An individual subject and individual predicate form a proposition in which the indivisible subject is indivisibly determined by the indivisible predicate ;—they are both declared one and the same with each other, and this as indivisible. They can only, therefore, be properly contradicted by No. 5. No. 8 contradicts the hypothesis of indivisibility on which the proposition proceeds ; it, therefore, annihilates it beforehand, and does not contradict it on the hypothesis admitted.

(c)—SYLLOGISMS.[a]

OBSERVATIONS ON THE MUTUAL RELATION OF SYLLOGISTIC TERMS IN QUANTITY AND QUALITY.

General Canon.—What worst relation of subject and predicate, subsists between either of two terms and a common third term, with which one, at least, is positively related; that relation subsists between the two terms themselves.

[a] In the papers under the headings sign of particularity (some), and the (b) and (c), the two arrows of the logical limitations under which the Author

There are only three possible relations of Terms, (notions, re-presentations, presentations).

1°, The relation of *Toto-total Coinclusion*, (coidentity, absolute convertibility or reciprocation) (AIA).

2°, The relation of *Toto-total Coexclusion*, (non-identity, absolute inconvertibility or non-reciprocation) (AnA).

3°, The relation of *Incomplete Coinclusion*, which involves the counter-relation of *Incomplete Coexclusion*, (partial identity, and non-identity, relative convertibility and non-convertibility, reciprocation, and non-reciprocation). This is of various orders and degrees.

a) Where the whole of one term and the part of another are coinclusive or coidentical (AII). This I call the relation of *toto-partial coinclusion*, as *All men are some animals*. This necessarily involves the counter-relation of *toto-partial coexclusion* (AnI), as *Any man is not some animal*. But the converse of this affirmative and negative affords the relations of

propose to employ them are clearly indicated,—see references in note a, p. 284. As, however, Sir W. Hamilton's application of the definite meaning of *some* has been entirely misrepresented, and as this misrepresentation has been founded mainly on the place which this paper (*c*) occupies in the order of the Appendix, it may be proper to state, that though its date cannot be precisely given, it was written many years before the preceding papers marked (*b*), (*d*), (*2*), (*3*), (*4*). It is, in fact, as stated in the first edition of these Lectures, at p. 474, an early draft of the order of moods printed at the end of this volume. It was in existence in 1845-46, being given in the class teaching of that year, and certainly some years previously. The order of the moods which it contains had evidently been changed before 1849, when the new order was published in the second edition of Archbishop Thomson's *Outline of the Laws of Thought*. The papers under the headings (*b*) and (*d*), (*2*), (*3*), (*4*), though referring to Proposi-tions, and for convenience placed before the paper on Syllogisms, cannot, accord-ingly, be regarded as even introductory to paper (*c*) on Syllogisms, far less as warranting the attribution to the Author of a novel and special meaning of *some* in syllogistic reasoning. See above, p. 284, note e. This latter paper, and the matured Table of Moods given at the end of this volume, are obviously to be viewed in connection with paper (*a*), *Extract from Prospectus of Essay towards a New Analytic of Logical Forms*, p. 251, in which Sir W. Hamilton proposes merely to introduce and apply the principle of a quantified predicate on Aristotelic principles, and thus, among other points, to amplify the number of the ordinary logical moods to thirty-six. The Editors would have thought it wholly unne-cessary to say anything on this point, had it not appeared that these papers which constitute the Appendix—which are fragmentary, and of various dates—have been, notwithstanding what was stated in the Preface to the first edi-tion, most unfairly regarded as the parts of a logical treatise in rigorous consecution.—Ed.

b) *Parti-total Coinclusion* (IfA), and *Parti-total Coexclusion*
(InA), as *Some animal is all man, Some animal is not any
man.*

c) There is still a third double relation under this head, when
two terms partially include and partially exclude each other (IfI
InI), as *Some women are some authors,* and *Some women are not
some authors.* This relation I call that of *Parti-partial Coinclusion,* and *Parti-partial Coexclusion.*

Of those three general relations, the first is [technically styled]
the best ; the second is the worst ; and the third is intermediate.

Former logicians knew only of two worse relations,—a particular, worse than a universal, affirmative, and a negative worse than
an affirmative. As to a better and worse in negatives, they knew
nothing ; for as two negative premises were inadmissible, they had
no occasion to determine which of two negatives was the worse or
better. But in quantifying the predicate, in connecting positive
and negative moods, and in generalising a one supreme canon of
syllogism, we are compelled to look further, to consider the inverse
procedures of affirmation and negation, and to show (*e.g.* in v. a.
and vi. b., ix. a. and x. b) how the latter, by reversing the former,
and turning the best quantity of affirmation into the worst of
negation, annuls all restriction, and thus apparently varies the
quantity of the conclusion. It thus becomes necessary to show
the whole order of best and worst quantification throughout the
two qualities, and how affirmation commences with the whole in
Inclusion, and negation with the parts in Exclusion.[a]

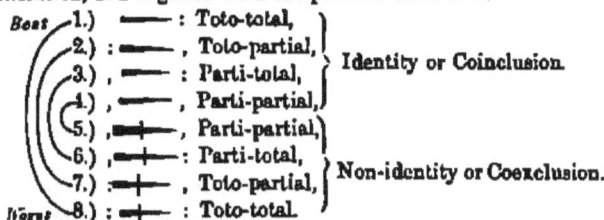

Best 1.) ———— : Toto-total, ⎫
      2.) : ■——— , Toto-partial, ⎪
      3.) , ■——— : Parti-total, ⎬ Identity or Coinclusion.
      4.) , ■——— , Parti-partial,⎪
      5.) ,■—+— , Parti-partial,⎫
      6.) ,—+— : Parti-total,   ⎪
      7.) :—+— , Toto-partial, ⎬ Non-identity or Coexclusion.
Worst 8.) : —+— : Toto-total.   ⎭

As the negation always reduces the best to the worst relation,
in the intermediate relations determining only a commutation from
equal to equal, whilst in both, the symbols of quantity, in their inverse signification, remain externally the same ; it is evident, that

* See Magentinus, (in Brandis, *Scholia,* p. 113, and there the Platonics.)

the quantification of the conclusion will rarely be apparently different in the negative, from what It is in the corresponding positive, mood. There are, indeed, only four differences to be found in the negative from the positive conclusions, and these all proceed on the same principle; viz. in v. a. and vi. b., in ix. a. and x. b. Here the particular quantification of the positive conclusions disappears in the negative moods. But this is in obedience to the general canon of syllogism,—" that the worst relation subsisting between either extreme and the middle, should subsist between the extremes themselves." For what was the best relation in the former, becomes the worst in the latter; and as affirmation comes in from the greatest whole, whilst negation goes out from the least part, so, in point of fact, the *some* of the one may become the *not any* of the other. There is here, therefore, manifestly no exception. On the contrary, this affords a striking example of the universal applicability of the canon under every change of circumstances. The canon would, in fact, have been invalidated, had the apparent anomaly not emerged.

I. Terms each totally coinclusive of a third, are totally coinclusive of each other.

a) A term totally coexclusive, and a term totally coinclusive, of a third, are totally coexclusive of each other.

b) A term totally coinclusive, and a term totally coexclusive, of a third, are totally coexclusive of each other.

II. Terms each parti-totally coinclusive of a third, are partially coinclusive of each other.

a) A term parti-totally coexclusive, and a term parti-totally coinclusive, of a third, are partially coexclusive of each other.

b) A term parti-totally coinclusive, and a term parti-totally coexclusive, of a third, are partially coexclusive of each other.

III. A term totally, and a term parti-totally, coinclusive of a third, are toto-partially coinclusive of each other.

a) A term totally coexclusive and a term parti-totally coinclusive, of a third, are toto-partially coexclusive of each other.

IV. A term parti-totally, and a term totally, coinclusive of a third, or parti-totally coinclusive of each other.

V. A term totally, and a term toto-partially, coinclusive of a third, are parti-totally coinclusive of each other.

VI. A term toto-partially, and a term totally, coinclusive of a third, are toto-partially coinclusive of each other.

VII. A term parti-totally, and a term partially, coinclusive of a third, are partially coinclusive of each other.

VIII. A term partially, and a term parti-totally, coinclusive of a third, are partially coinclusive of each other.

b) A term totally coinclusive, and a term parti-totally coexclusive, of a third, are toto-partially coexclusive of each other.

a) A term parti-totally coexclusive, and a term totally coinclusive, of a third, are parti-totally coexclusive of each other.

b) A term parti-totally coinclusive, and a term totally coexclusive, of a third, are parti-totally coexclusive of each other.

a) A term totally coexclusive, and a term toto-partially coinclusive, of a third, are totally coexclusive of each other.

b) A term totally coinclusive, and a term toto-partially coexclusive, of a third, are parti-totally coexclusive of each other.

a) A term toto-partially coexclusive, and a term totally coinclusive, of a third, are toto-partially coexclusive of each other.

b) A term toto-partially coinclusive, and a term totally coexclusive, of a third, are totally coexclusive of each other.

a) A term parti-totally coexclusive, and a term partially coinclusive, of a third, are partially coexclusive of each other.

b) A term parti-totally coinclusive, and a term partially coexclusive, of a third, are partially coexclusive of each other.

a) A term partially coexclusive, and a term parti-totally coinclusive, of a third, are partially coexclusive of each other.

IX. A term totally, and a term partially, coinclusive of a third, are partially coinclusive of each other.

b) A term partially coinclusive, and a term parti-totally coexclusive, of a third, are partially coexclusive of each other.

a) A term totally coexclusive, and a term partially coinclusive, of a third, are parti-totally coexclusive of each other.

b) A term totally coinclusive, and a term partially coexclusive, of a third, are partially coexclusive of each other.

X. A term partially, and a term totally, coinclusive of a third, are partially coinclusive of each other.

a) A term partially coexclusive, and a term totally coinclusive, of a third, are partially coexclusive of each other.

b) A term partially coinclusive, and a term totally coexclusive, of a third, are toto-partially coexclusive of each other.

XI. A term parti-totally, and a term toto-partially, coinclusive of a third, are parti-totally coinclusive of each other.

a) A term parti-totally coexclusive, and a term toto-partially coinclusive, of a third, are parti-totally coexclusive of each other.

b) A term parti-totally coinclusive, and a term toto-partially coexclusive, of a third, are parti-totally coexclusive of each other.

XII. A term toto-partially, and a term parti-totally, coinclusive of a third, are toto-partially coinclusive of each other.

a) A term toto-partially coexclusive, and a term parti-totally coinclusive, of a third, are toto-partially coexclusive of each other.

b) A term toto-partially coinclusive, and a term parti-totally coexclusive, of a third, are toto-partially coexclusive of each other.

(J)—OBJECTIONS TO THE DOCTRINE OF A QUANTIFIED PREDICATE CONSIDERED.

(1.) GENERAL.

MATERIAL AND FORMAL.—THEIR DISTINCTION.

But it is requisite—seeing that there are such misconceptions prevalent on the point—to determine precisely, what is the *formal* which lies within the jurisdiction of Logic, and which Logic guarantees, and what the *material* which lies without the domain of Logic, and for which Logic is not responsible. This is fortunately easy.

Logic knows,—takes cognisance of, certain general relations; and from these it infers certain others. These and these alone it knows and guarantees; and these are formal. Of all beyond these forms or general relations it takes no cognisance, affords no assurance; and only hypothetically says,—If the several notions applied to these forms stand to each other in the relation of these forms, then so and so is the result. But whether these notions are rightly applied, that is, do or do not bear a certain reciprocal dependence, of this Logic, as Logic, knows nothing. Let A B C represent three notions, A containing B, and B containing C; in that case Logic assures us that C is a part of B, and B a part of A; that A contains C; that C is a part of B and A. Now all is formal, the letters being supposed to be mere abstract symbols. But if we apply to them,—fill them up by,—the three determinate notions,—*Animal—Man—Negro*, we introduce a certain *matter*, of which Logic is not itself cognisant; Logic, therefore, merely says,—If these notions hold to each other the relations represented by A B C, then the same results will follow; but whether they do mutually hold these relations,—that, as *material*, is extra-logical. Logic is, therefore, bound to exhibit a scheme of the forms, that is, of the relations in their immediate and mediate results, which are determined by the mere necessities of thinking,—by the laws of thought as thought; but it is bound to naught beyond this. That, as *material*, is beyond its jurisdiction. However manifest, this has, however, been fre-

quently misunderstood, and the *material* has been currently
passed off in Logic as the *formal*.

But further, Logic is bound to exhibit this scheme full and unexclusive. To lop or limit this in conformity to any circumstance
extrinsic to the bare conditions,—the mere *form*, of thought, is a
material, and, consequently, an illegitimate curtailment. To take,
for instance, the aberrations of common language as a model,
would be at once absurd in itself, and absurd as inconsistent even
with its own practice. And yet this double absurdity the Logic
now realised actually commits. For while in principle it avows
its allegiance to thought alone, and in part has overtly repudiated the elisions of language; in part it has accommodated itself
to the usages of speech, and this also to the extent from which even
Grammar has maintained its freedom. Grammar, the science proper,—the nomology, of language, has not established ellipsis as a
third law beside Concord and Government; nor has it even allowed
Concord or Government to be superseded by ellipsis. And why?
Because the law, though not externally expressed in language, was
still internally operative in thought. Logic, on the contrary, the
science proper,—the nomology, of thought, has established an imperative ellipsis of its abstract forms in conformity to the precarious ellipses of outward speech; and this,'although it professes to
look exclusively to the internal process, and to explicate,—to fill
up, what is implied, but not stated, in the short cuts of ordinary
language. Logic has neglected,—withheld,—in fact openly suppressed, one-half of its forms, (the quantification of the predicate
universally in affirmatives, particularly in negatives), because
these forms, though always operative in thought, were usually
passed over as superfluous in the matter of expression.

Thus has logic, the science of the form, been made hitherto the
slave of the matter, of thought, both in what it has received and in
what it has rejected. And well has it been punished in its servitude. More than half its value has at once been lost, confusion
on the one hand, imperfection on the other, its lot; disgust, contempt, comparative neglect, the consequence. To reform Logic,
we must, therefore, restore it to freedom;—emancipate the form
from the matter;—we must, 1°, Admit nothing material under
the name of formal, and, 2°, Reject nothing formal under the
name of material. When this is done, Logic, stripped of its acci-

dental deformity, walks forth in native beauty, simple and complete; easy at once and useful.

It now remains to show that the quantities of the Predicate denounced by logicians are true logical forms.

  • • • • • •

The logicians have taken a distinction, on which they have defended the Aristotelic prohibition of an overt quantification of the predicate; the distinction, to wit, of the *formal*, in opposition to the *material*,—of what proceeds *vi formæ*, in contrast to what proceeds *vi materiæ*. It will be requisite to determine explicitly the meaning and application of these expressions; for every logical process is *formal*, and if the logicians be correct in what they include under their category of *material*, the whole system which I would propose in supplement and correction of theirs, must be at once surrendered as untenable.

In the *first* place, the distinction is not established, in terms at least, by Aristotle. On the contrary, although the propositional and syllogistic relations which he recognises in his logical precept be all formal, he, as indeed all others, not unfrequently employs some which are only valid, say the logicians, *vi materiæ*, and not *ratione formæ*, that is, in spite of Logic.

But here it is admitted, that a distinction there truly is; it is, consequently, only necessary, in the *second* place, to ascertain its import. What then is meant by these several principles?

The answer is easy, peremptory, and unambiguous. All that is *formal*, is true as consciously necessitated by the laws of thought; all that is *material*, is true, not as necessitated by the laws of thought, but as legitimated by the conditions and probabilities discoverable in the objects about which we chance to think. The one is *a priori*, the other *a posteriori;* the one is necessary, the other contingent; the one is known or thought, the other unknown or unthought.

For example : If I think that the notion Triangle contains the notion Trilateral, and again that the notion Trilateral contains the notion Triangle; in other words, if I think that each of these is inclusively and exclusively applicable to the other; I *formally* say, and, if I speak as I think, must say—"All Triangle is all Trilateral." On the other hand,—if I only think that all Triangles are trilateral, but do not think all Trilaterals to be triangular, and yet say

—" All triangle is all trilateral," the proposition, though materially true, is *formally* false.

Again: If I think, that this, that, and the other iron-attracting stones are *some* magnets, and yet thereon overtly infer,—" All magnets attract iron;" the inference is *formally* false, even though materially not untrue. Whereas, if I think that this, that, and the other iron-attracting stones are *all* magnets, and thence conclude,—"*All* magnets attract iron;" my conclusion is *formally* true, even should it materially prove false.

To give the former example in an abstract notation: If I note [C : ⬛━━ : Γ] I may formally convert the proposition and state [Γ : ⬛━━ : C]. But if I note [C : ━━ Γ], I cannot formally convert it; for the [Γ] may mean either [: Γ] or [, Γ]; and if I do, the product may or may not be true according as it is accidentally applied to this or that particular matter. As to the latter example:

$$C, \text{━━} : (m\ m'\ m'\ \&c.) : \text{━━} : Γ$$

this syllogism is formally legitimate. But, to take the following antecedent: this, if formally drawn, warrants only, (1), a particular conclusion; and if, (2), a universal be drawn, such is logically null:

$$C, \text{━━} : (m\ m'\ m'\ \&c.) : \text{━━}, Γ$$
$$1. \text{━━━━━━}$$
$$2. \text{━━━━━━} :$$

This being the distinction of *formal* and *material*,—that what is formally true, is true by a subjective or logical law,—that what is materially true, is true on an objective or extra-logical condition; the logicians, with Aristotle at their head, are exposed to a double accusation of the gravest character. For they are charged :—1°, That they have excluded, as material, much that is purely formal ; 2°, That they have included, as formal, much that is purely material.—Of these in their order.

1°, I shall treat of this under the heads of *affirmative* and of *negative* propositions.

Of the four *affirmative* relations of concepts, as subject and predicate ; to wit—1, the *toto-total*; 2, the *toto-partial*; 3, the

a For an explanation of the notation   logism, see below, Appendix XII.—
here employed, in reference to Syl-   ED.

parti-total ; 4, the *parti-partial ;* one half (1, 3) are arbitrarily
excluded from logic. These are, however, relations equally necessary, and equally obtrusive in thought, with the other ; and, as
formal realities, equally demand a logical statement and consideration. Nay, in this partial proceeding, logicians are not even self-
consistent. They allow, for example, the *toto-partial* dependency
of notions, and they allow of their conversion. Yet though the
terms, when converted, retain, and must retain, their original relation, that is, their reciprocal quantities ; we find the logicians,
after Aristotle, declaring that the predicate in affirmative propositions is to be regarded as particular ; howbeit, in this instance,
where the *toto-partial* is converted into the *parti-total* relation,
their rule is manifestly false. When I enounce,—" All man is
animal," I mean,—and the logicians do not gainsay me,—" All
man is *some* animal." I then convert this and am allowed to say,
—" Some animal is man." But I am not allowed to say, in words,
though I say,—indeed must say, in thought,—" Some animal is
all man." And why? Simply because there is an old traditionary rule in Logic, which prohibits us in all cases, at least of
affirmative propositions, to quantify the predicate universally ;
and to establish a reason for this exclusion, the principle of
materiality has been called in. But if all is formal which is necessitated by thought, and if all that is formal ought to find an
expression in Logic, in that case, the universal quantification of
the notion, when it stands as predicate, may be, ought, indeed, on
demand, to be, enounced, no less explicitly than when it stood as
subject. This quantification is no more material on the one alternative than on the other ; it is formal in both.

In like manner, the *toto-total* relation is denounced. But a
similar exposition shows that notions, thought as reciprocating or
coequal, are entitled, as predicate, to have a universal quantification, no less than as subject, and this formally, not materially.[a]

In regard to the four *negative* relations of terms ;—1. the *toto-
total*,—2. the *toto-partial*,—3. the *parti-total*,—4. the *parti-
partial ;* in like manner, one half, but these wholly different

---

[a] It is hardly requisite to notice the
blundering doctrine of some authors,
that the predicate is materially quantified, even when predesignated as universal. It is sufficient to observe that
this opinion is explicitly renounced by
the acuter logicians, when they have
chanced to notice the absurdity.—See
Fonseca, *Instit. Dial.*, lib. vi. c. 20.

classes, (3, 4,) are capriciously abolished. I say capriciously; for the relations not recognised in Logic are equally real in thought, as those which are exclusively admitted. Why, for example, may I say, as I think,—" Some animal is not any man;" and yet not say, convertibly, as I still think,—" Any man is not some animal"? For this no reason, beyond the caprice of logicians, and the elisions of common language, can be assigned. Neither can it be shown, as I may legitimately think,—" Some animal is not some animal," (to take an extreme instance,) that I may not formally express the same in the technical language of reasoning.

In these cases, to say nothing of others, the logicians have, therefore, been guilty of extruding from their science much that is purely formal; and this on the untenable plea, that what is formal is material.

(2.) SPECIAL.

Two objections have been taken to the universal quantification of the predicate. It is said to be—1°, *false;* 2°, If not false, *useless.*

I. The first objection may be subdivided into two heads, inasmuch as it may be attempted to establish it, a), on *material;* b), on *formal,* grounds. Of these in their order:—

a). This ground seems to be the only one taken by Aristotle, who, on three (perhaps on four) different occasions, denounces the universal quantification of the predicate (and he but implicitly limits it to affirmative propositions), as "*always untrue.*" * The only proof of this unexclusive denunciation is, however, one special example which he gives of the falsity emerging in the proposition,—" All man is all animal." This must be at once confessed false; but it is only so materially and contingently, and argues, therefore, nothing for the formal and necessary illegitimacy of such a quantification. As extra-logical, this proof is logically incompetent; for it is only because we happen, through an external knowledge, to be aware of the relations of the concepts, *man* and *animal,* that the example is of any import. But, because the universal quantification of the predicate is, in this instance, materially false, is such quantification, therefore, always formally illegal? That this is not the case, let us take other material examples. Is it, then, materially false and formally incompetent, to think and say,—" All human is all rational,"—" All rational is all risible,"—

* See below, p. 305.—ED.

"All risible is all capable of admiration,"—"All trilateral is all triangular,"—"All triangular is all figure with its angles equal to two right angles," &c.? Or, employing Aristotle's material example, is it untrue, as he asserts, to say—"Some animal is all man;" and this either *collectively*,—"A part of the class animal is the whole of the class man,"—or *distributively*,—"Some several animal is every several man."

But the absurdity of such a reasoning is further shown by the fact, that if it were cogent at all, it would equally conclude against the validity of the universal quantification of the *subject*. For this proposition is equally untrue (employing always Aristotle's own material example),—"All animal is man."

After this, it may the less surprise us to find that Aristotle silently abandons his logical canon, and adheres to truth and nature. In fact, he frequently does in practice virtually quantify the predicate, his common reasonings often proceeding on the reciprocation or coextension of subject and predicate. Nay, in his logical system, he expressly recognises this coextension ; unless, indeed, we overtly *supply* the quantification of the predicate, his doctrines of Induction and of Demonstration proper have no logical notation ; and, unless we covertly *suppose* it, they are actually arrested. His definitions of the Universal, as severally given in his *Prior* and *Posterior Analytics*, are, in this respect, conflictive. In the former, his universal, (known in the schools as the *Universale Prioristicum*,) explicitly forbids, whereas the latter, (the *Universale Posterioristicum* of the schoolmen,) implicitly postulates, the quantification of the predicate.

b). The defect in the polemic of their master was felt by his followers. They, accordingly, in addition to, but with no correction of, Aristotle's doctrine, argue the question on broader ground ; and think that they disprove the *formal* validity of such quantification by the following reasoning. Overlooking the case where the subject is particularly, the predicate universally, quantified, as in the instance I have just given, they allege the case of what are called *reciprocating propositions*, where both subject and predicate are taken in their utmost extension,—*vi materiæ*, as subsequent logicians[a] say, but not Aristotle. In this case, then, as in the example,—"All man is all risible," they assert that the overt quantification of the predicate is inept, because, the "all" as applied to the subject being distributively taken, every individual man, as Socrates, Plato, &c., would be all (that is, the whole class) risible. This objection is only respectable by authority, through the great, the all but unexclusive, number of its allegers; in itself it is futile.

Terms and their quantifications are used either in a distributive, or in a collective, sense. It will not be asserted that any quantification is, *per se*, necessarily collective or necessarily distributive; and it remains to ascertain, by rule and relation, in which signification it is, or may be, employed. Now a general rule or postulate of logic is,—That in the same logical unity, (proposition or syllogism), the same term or quantification should not be changed in import.* If, therefore, we insist, as insist we ought, that the quantification here, *all*, should be used *in the same proposition in the same meaning*, that is, as applied to the one term, collectively or distributively, it should be so applied likewise to the other, the objection fails. Thus taken *collectively:*—" All (that is, the whole class) man is all (that is, the whole class) risible," the proposition is valid. Again, taken *distributively:*—" All (that is, every several) man is all (that is, every several) risible," the proposition is, in like manner, legitimate. It is only by violating the postulate, *that in the same logical unity, the same sign or word should be used in the same sense*, that the objection applies; whereas, if the postulate be obeyed, the objection is seen to be absurd.

It is hardly necessary to say anything in confutation of the general doctrine, that in reciprocating propositions the predicate is taken in its full extent, *vi materiæ*. In the *first* place, this doctrine was not promulgated by Aristotle; who frequently allowing—frequently using—such propositions, implicitly abandons the rule which he explicitly lays down in regard to the non-predesignation of the predicate by a universal. In the *second* place, apart from authority, such doctrine is in itself unfounded. For as *form* is merely the necessity of thought, it is as easy to think two notions as toto-totally coinciding, (say, triangle and trilateral), as two notions toto-partially and parti-totally coinciding, (say, triangle

* See above, p. 255.—En.

and figure). Accordingly, we can equally abstractly represent their relations both by geometric quantities, (lines or figures), and by purely logical symbols. Taking lines:—the former ⊏⊐ ; the latter ⌐⎯. Taking symbols: the former C :▬ : Γ; the latter A, ⎯ : B.—But if the reciprocation were determined by the mere matter, by the object contingently thought about, all abstract representation would be impossible. So much for the first objection,—that the universal quantification of the predicate would, at least in affirmative propositions, be *false*.

II. As to the second objection, that such quantification would be useless and superfluous, disorderly, nay confusive, this only manifests the limited and one-sided view of the objectors, even though Aristotle be at their head.

Is it useless in any case, theoretical or practical, that error be refuted, truth established? And in this case :—

1°, Is it disorderly and confusive, that the doctrine of *Exponibles*, as they are called, should be brought back from anomaly and pain to ease and order,—that propositions Exclusive and Exceptive, now passed over for their difficulty, and heretofore confessedly studied as " opprobria and excruciations," should be shown to be, not merely reducible by a twofold and threefold tortuosity, through eight genera and eight rules, but simple, though misunderstood, manifestations of the universal quantification of the predicate ? [a]

2°, Is it useless to demonstrate that every kind of proposition may be converted, and not some only, as maintained by Aristotle and the logicians? And is it disorderly and confusive, in all cases, to abolish the triple (or quadruple) confusion in the triple (or quadruple) processes of Conversion, and to show, that of these processes there is only one legitimate, and that, the one simple of the whole ?

3°, Is it disorderly and confusive to abolish the complex confusion of Mood and Figure, with all their array of rules and exceptions, general and special ; and thus to recall the science of reasoning to its real unity ?

4°, Is it useless and superfluous to restore to the science the many forms of reasoning which had erroneously, ineffectually, and even inconsistently, been proscribed ?

[a] See above, p. 283.—Ed.

5°, Is it useless or superfluous to prove, that all judgment, and, consequently, all reasoning, is simply an equation of its terms, and that the difference of subject and predicate is merely arbitrary?

6°, In fine, and in sum, is it useless or superfluous to vindicate Logic against the one-sided views and errors of logicians, to reconcile the science with truth and nature, and to re-establish it, at once, in its amplitude and simplicity?

(g) HISTORICAL NOTICES OF DOCTRINE OF QUANTIFIED PREDICATE.

I.—ARISTOTLE.

It will be sufficient to make one extract from Aristotle in illustration of his doctrine upon this point, and I select the following passage from his *Categories*, c. v. § 7.

" Further, the primary substances, [πρῶται οὐσίαι,—individual existences], because they are subjects to all the others, and as all the others are predicated of, or exist in, them,—are, for this reason, called *substances* by pre-eminence. And as the primary substances stand to all the others, so stands the Species to the Genus. *For genera are predicated of species, but not, conversely, species of genera;* so that of these two, the species is more a substance than the genus."

Ammonius, who has nothing in his Commentary on the *Categories* relative to the above passage of Aristotle, states, however, the common doctrine, with its reasons, in the following extract from his Commentary on Porphyry's *Introduction*, (f. 29, ed. Ald. 1546).

" But confining ourselves to a logical consideration, it behoves us to inquire,—of these, which are subjected to, which predicated of, the others; and to be aware, that Genera are predicated of Differences and Species, but not conversely. These, as we have said, stand in a certain mutual order,—the genus, the difference, and the species; the genus first, the species last, the difference in the middle. And the superior must be predicated of the inferior; for to predicate the inferior of the superior is not allowable. If, for example, we say,—' All man is animal,' the proposition is true ; but if we convert it, and say,—'All animal is man,' the enouncement is false.[a] Again, if we say,—'All horse is irrational,' we are right ; but if conversely we say,—'All irrational is horse,' we are wrong. For it is not allowed us to make a subject of the accidental. Hence is it incompetent to say that 'Animal is man,' as previously stated."

[*Categ.*, c. ii. § 1.

" When one thing is predicated of another as of its subject, all that is said [truly] of the predicate will be said [truly] also of the subject. Thus *man* is predicated of this and that man,β and *animal* of *man ; animal* will therefore be predicated of this and that individual, for this and that individual is both man and animal."

De Interpret, c. vii. § 2-4 ; see also c. x.

"To enounce something of a universal universally, I mean as, 'All or every (πᾶς)[γ] man is white,' ' No man is white.' . . . . To enounce something of universals not universally, I mean as 'Man is white,' ' Man is not white ;' for whilst the term *man* is universal, it is not used in these enouncements *as* universal. For ' *All* ' or ' *Every* ' does not indicate the universal [itself], but that [it is applied to a subject] universally. Thus, in reference to an universal predicate, to predicate the universal, is not true. For no affirmation is true, in which the universal is predicated [of an universal predicate], as, 'All man is all animal.'" (See Ammonius, Boethius, Psellus, Magentinus, &c.)

Prior. Analyt., L. I. c. 27, § 9. "The consequent [i.e. the predicate] is not to be taken as if it wholly followed [from the antecedent, or subject, exclusively]. I mean, for example, as if *all*

a The converse of a true proposition is always true ; but the false propositions which are here given, as conversions of the true, are not conversions at all. The true propositions, if explicitly stated, are,—"All man is some animal," and,—"All horse is some irrational." Convert these,—"Some animal is all man," and,—"Some irrational is all horse ;" the truth remains, but the one-sided doctrine of the logicians is exploded.

β [For the τὶς here, as elsewhere, denotes the *individuum signatum*, not the *individuum vagum*.]

γ The Greek πᾶς (as the Latin *omnis*) indifferently denotes *all* collective, and *every* distributive. The English *all* may be, in like manner, used ambiguously, both in a collective and a distributive sense ; and in this ambiguous sense, the reader will observe that it is always used in the following translations. To have repeated the *all* or *every* on all occasions would have been nauseous

[or *every*] *animal* [were consequent] *on man*, or *all* [or *every*]
science on music. The consequence *simply* [is to be assumed], as
in our propositions has been done ; to do otherwise, (as to say
that *all* [or *every*] *man is all* [or *every*] *animal*, or that *justice is
all* [or *every*] *good*,) is useless and impossible ; but to the antecedent [or subject] the *all* [or *every*] is prefixed."

Post. Analyt. I., I. c. xii. § 10. "The predicate is not called
all [or *every*] ;" [that is, the mark of universality is not annexed
except to the subject of a proposition].

In refutation of Aristotle's reasoning against the universal predesignation of the predicate—it will equally disprove the universal predesignation of the subject. For it is absurd and impossible to say, *All animal is man ; All (every) immortal is the soul ;
All pleasure is health ; All science is music ; All motion is pleasure.*[a] But in point of fact such examples disprove nothing ; for
all universal predesignations are applicable neither to subject
nor predicate, nor to both *subject* and *predicate*—are *thoughts* not
things ; and so are all *predesignations ;* therefore, &c. It is only
marvellous that such examples and such reasoning could satisfy
the acutest of intellects ; that his authority should have imposed
on subsequent logicians is less wonderful.[b]

---

[a] Examples from Wegelin, *In Gregorii Aneponymi Comp. Phil. Synt.*, L.
iv. c. 1, p. 473; L. vi. c. 1, p. 673.

[b] And here I may correct an error,
as I conceive it to be, which has descended from the oldest to the most
recent interpreters of the *Organon*, and
been adopted implicitly by logicians in
general. It is found in Alexander and
Ammonius, as in Trendelenburg, Saint-
Hilaire, and Waitz; nor, indeed, as
far as I know, has it ever been called
in question during the interval. It
regards the meaning of the definition
elevated into a twofold axiom, the
cur in toto, &c., and *dici de omni*, &c.,
toward the conclusion of the first
chapter of the first book of the *Prior
Analytics.* Τὸ δὲ ἐν ὅλῳ εἶναι ἕτερον
ἑτέρῳ καὶ τὸ κατὰ παντὸς κατηγορεῖσθαι
θατέρου θάτερον ταὐτόν ἐστιν. This,
with its ambiguity, may be thus literally, however awkwardly, translated:

—"But (to say) that one thing is in a
whole other, and (to say) that one thing
is *predicated of all another*, are identical." — Now, the question arises, —
What does Aristotle here mean by "a
whole other" ? — for it may signify,
either the class or higher notion under
which an inferior concept comes, or
the inferior concept itself, of which, as
of a subject, the higher is predicated.
The former is the sense given by all
the commentators ; the latter, the
sense which, I am confident, was intended by Aristotle.

There are only two grounds of interpretation. The rule must be expound-
ed in consistency—1°, with itself ; 2°,
with the analogy of Aristotelic usage.

1°. On the former ground, the common doctrine seems untenable ; for
what Aristotle declares to be identical,
by that doctrine become different, nay
opposed. An inferior concept may be

Quantification of Predicate—Aristotle.

1. Admits that syllogism mental not oral (*An. Post.* I. 10). This to be borne in mind.
2. That individual is never predicated, (*Cat.* c. 2), refuted by reciprocation of singular, (*An. Pr.* ii. 23, § 4).
3. That affirmative universal not [to] be added to predicate, incompatible with what he says of reciprocation, (*An. Pr.* ii. cc.

in a higher whole or class, either partially or totally; and the definition on the prevalent interpretation virtually runs—"To say that one thing is, *all or part*, in the whole of another, and to say that this other is predicated of it unexclusively, are convertible." Had Aristotle, therefore, used the expression in the signification attributed to him, he must, to avoid the contradiction, have said—Τὸ δὲ τὰν ἕτερον ἐν ὅλῳ εἶναι ἑτέρῳ, κ.τ.λ. ("But to say that one thing is *all* in a whole other," &c.)

2°. On the second ground, it may, however, be answered, that the ambiguity of the word, as it stands, is superseded, its signification being determined by other passages. I join issue; and on this ground am well content to let the question be decided.

In the *first* place, the meaning I attribute to the expression, "*whole other*," that is, whole subject or inferior notion, is in strict conformity with Aristotle's ordinary language. There are, I admit, sundry passages in his logical writings, where the term *whole* is clearly used as synonymous with *class or higher notion*; as, to limit ourselves to the *Prior Analytics*, in Book I. iv. § 2; and II. i. § 4. But *every single text*, in which the term *whole* appears in this relation, is overruled by *more than five others*, in which it is no less clearly applied to denote *the totality of a lower notion*, of which a higher is predicated—passages in which the word *whole* (ὅλος) is used convertibly with *all* (πᾶς). See, for example, *An. Pr.* II. ii. § 5, § 16—iii. § 5, § 7 (*bis*), § 13 (*bis*), § 14, § 15—iv.

§ 6 (*bis*), § 8, § 10, § 12 (*bis*)—xxii. § 7, § 8—xxiii. § 4.

But in the *second* place, (and this is directly subversive of the counter-opinion, even in the principal of the few passages where the term *whole* is used for *class*,) the lower notion may be in or under the higher *only particularly*; and this manifestly shows that Aristotle could not possibly mean, by merely saying, that one thing is in another, as in a class, that it is so, *unexclusively* or *universally*. Compare *An. Pr.* I. iv. §§ 2, 3, 10. On this interpretation, *Darii* and *Ferio* would there be annulled; a special result which ought to have startled the logicians into a doubt of the accuracy of the received doctrine in general. (See, *inter omnium*, Pacius, in his relative Notes and Commentary.)

That doctrine must, therefore, be abandoned, and the rule reduced to a definition, read in the following signification:—"But to say that one thing *is in the whole of another, as in a subject*, and *to predicate one thing universally of another*, are merely various expressions of the same meaning." This, in fact, is just the preliminary explanation of the two ordinary modes of stating a proposition, subsequently used by Aristotle. Here, in both convertibles, be descends from extension to comprehension, from the predicate to the subject; and the ingenious exposition by the commentators, old and new, of the inverse intention of the philosopher in the two clauses, must be regarded as erroneous.

22, 23, *et alibi*). That his custom to draw universal conclusions in Third Figure and affirmative in Second[a] with allowance of simple conversion in certain universal affirmatives.

4. That particular not in negative predicate, absurd in ού πας, *non omnia.*

Aristotle's doctrine of Predesignation.

1°, How can Aristotle, on his doctrine, make universal terms taken indifferently, or without predesignation, be tantamount to particulars? (*An. Prior.*, I. c. 4, § 13, *Org. Pacii*, p. 135, *alibi.*)

2°, *An. Prior.*, I. c. 27, § 7. He says, as elsewhere, "a proposition being indefinite, [preindesignate], it is not clear whether it be universal; when, however, it is definite, [predesignate], that is manifest." Contrast this statement with his doctrine of the *all.*

3°, There are syllogisms in Aristotle which are only valid through the quantity of the predicate.[b]

4°, Aristotle requires, though he does not admit, the universal predesignation of the predicate in his syllogism of Induction. Vide *An. Prior.*, L. ii., c. 23, § 4, *Organon Pacii*, p. 399. (Compare also his doctrine, p. 396.)

II.—ALEXANDER APHRODISIENSIS.

Alexander Aphrodisiensis, in his commentary on the first book of the *Prior Analytics*, and in reference to the second passage of Aristotle, states as follows :—

"And in the book of *Enouncement* Aristotle explains, why he there says :—that to predicate the universal of a universal predicate is not true; for there will be no proposition, if in it we predicate the universal of the universal, as, 'All man is all animal.' He repeats the same also here; showing how it is useless to attempt thus to express the consecution [of higher from lower notions]; and adds, that it is not only useless, but impossible. For it is *impossible,* that 'all man' should be 'all animal,' as [*useless* to say, (άχρηστον είπεῖν must have dropt out),] that 'all man is all risible.' We must not, therefore, apply the 'all' to the consequent, [or predicate,] but to that from which it fol-

---

lows, [or subject.] For 'man' is to be taken universally, as that from which 'animal' follows, supposing this to be the consequent of 'all man.' Thus shall we obtain a stock of universal propositions. The process is the same, in making 'man' the consequent on its proper 'all;' but 'man' is not consequent on 'all biped,' but on 'all rational.'

"The words, 'as we express ourselves,' mean—as we express ourselves in common usage. For we say, that 'all man is,' simply, 'animal,' and not 'all animal,' and that 'all pleasure is natural,' not 'all natural;' prefixing the 'all,' not to the consequent, but to the subject from which the predicate follows." (Edd. Ald. f. 100 a; Junt., f. 122 a; Comparo Ald, f. 86 a; Junt., f. 105 a.)

III.—AMMONIUS HERMIÆ.

Ammonius Hermiæ, *In de Interp.* c. vii. § 4. (Aldine editions, of 1503, sig. C. vii. *et seq.*, of 1546, ff. 70, 74.)

" In these words Aristotle inquires :—Whether, as the annexation of the Affirmative Predesignation (προσδιορισμός) to the Subject constitutes one distinct class of propositions, the same annexation to the Predicate, may not, likewise, constitute another; and he answers, that the supposition is absolutely groundless. Thus the enouncement—'all [or every] man is all [or every] animal' (πᾶς ἄνθρωπος πᾶν ζῶον ἐστι)—asserts that 'each man is all animal '—as horse, ox, &c. But this proposition is impossible; as is shown by Aristotle in his here omitting the word 'true.' For no affirmation can be true, in which the universal is predicated of a universal predicate ; that is, in which the universal predesignation is added to a universal prodicate ; as when we say that 'man' (of whom 'all,' or, as he says, *universally,* 'animal' is predicated), is not simply 'animal,' but 'all animal.' He, therefore, teaches, that such an affirmation, as utterly untrue, is utterly incompetent . . . . . .

" Neither does Aristotle allow the predesignation 'some' to be annexed to the predicate, that propositions may, thereby, become true, always or occasionally. For logicians, (as they do not propose to themselves every superfluous variety of enunciation,) are prohibited from considering propositions, (not only those always

true or always false, but those) which express no difference in
reference to necessary or impossible matter, and afford us absolutely no discrimination of truth from falsehood. Thus, particular propositions, which may be alternatively true and false,
ought not to have a predesignated predicate. For in a proposition, which has all their power, without any predesignation of
its predicate,—why should we prefer to the simpler expression,
that which drags about with it a superfluous additament? Why,
for example, instead of—' All man is some animal,' [I read, τι
ζῶον], or, ' All man is not all animal,' ª should we not say—
' All man is animal,' and in place of—' All man is no stone,' not
say,—' All man is not stone ;' or, what is a simpler and more
natural enouncement still,—' No man is stone.'

" And when we find some of the *ancients* teaching that the
particular affirmative predesignation is to be connected with the
predicate, as when Aristotle himself styles the soul a certain
(some) entelechy (ἐντελέχειάν τινα), and Plato, rhetoric a certain (some) experience (ἐμπειρίαν τινὰ), it is to be observed
that the ' some ' is there added for the sake of showing, that the
predicate is not convertible with the subject, but is its genus,
and requires the adding on of certain differences in order to
render it the subject's definition.

" But, add they, is not the reasoning of Aristotle refuted by
fact itself, seeing that we say, ' All man is capable of all science ;'
thus truly connecting the universal predesignation with the universal predicate? The answer is this :—that, in truth it is not
the predicate to which we here annex the ' all.' For what is
predicated, is what is said of the subject. But what is here said
of man is, not that he is ' science,' but that he is ' capable of
science.' If, therefore, the ' all ' were conjoined with the ' capable ' and the proposition then to remain true, as when we say—
' all man is all capable of science ;' in that case, the reasoning of
Aristotle would be refuted. But this proposition is necessarily
false. It, in fact, asserts nothing less, than that of men, each
individual is all the kind ;—that Socrates is not Socrates only, but

---

ª It will be observed, that Ammonius does not attempt an equivalent
for this proposition. In fact it is impossible on the common or Aristotelic
doctrine ; and this impossibility itself
ought to have opened his eyes upon
the insufficiency of the view he maintained.

also Plato, Alcibiades, and, in short, every other man. For, if 'all man is all capable of science,' Socrates [in either case] being one of the 'all,' is, therefore, himself 'all capable of science ;' so that Socrates will be Plato, Alcibiades, &c., since they also are capable of science. For if Socrates be not, at once, Plato, Alcibiades, &c., neither will he be ' all capable of science.' . . . . . .

" Now, that we ought not to prefix the universal affirmative predesignation to the predicate, (whether the predicate be more general than the subject, as 'All man is all animal,' or, whether they be coadequate, as 'All man is all risible,')—this is manifest from what has been said. Even when the terms are coadequate or reciprocating, the proposition runs into the absurd. For, declaring that ' all man is all risible,' it virtually declares, that each individual man is identical with all men ; that Socrates, in that he is a man, is 'all risible,' consequently, 'all man.' . . . . . .

" But why is it, that the predicate is intolerant of the predesignation ' all,' though this be akin to the counter-predesignation ' no ' or 'none' ? Is it because the affirmative predicate, if predicated universally, tends always to contain under it the subject, and this not only when itself coadequate with the subject, but when transcending the subject in extension ; while, moreover, through a participation in its proper nature, it is suited to bind up and reduce to unity the multitude of individuals of which the subject is the complement ? For, as Aristotle previously observed,—'The *all* does not indicate the universal, but that [the universal predicate inheres in, or is attributed to, the subject] *universally.*' If, therefore, the affirmative predicate thus tend to collect into one what are by nature distracted, in virtue of having been itself previously recognised as simple ; in this case, the ' all,' [superadded to this universal predicate, in fact,] enounces not a unity, but a multitude of several things,—things which it is manifestly unable to complicate into reciprocity.— But, on the other hand, since what is negatively predicated of, is absolutely separated from, the subject ; we are, consequently, enabled to deny of the subject all under the predicate, as in saying, ' All man is no stone.' We may indeed condense this proposition, and say more simply, ' All man is not stone ;' or more simply still, ' No man is stone ;' thus dispensing with the affirmative predesignation in a negative proposition.

IV.—Boethius.

Boethius, *In Librum de Interpretatione*, editio secunda, et in textum laudatum. *Opera*, p. 348.

"What he says is to this purport:—Every simple proposition consists of two terms. To these there is frequently added a determination either of universality or of particularity; and to which of the two parts these determinations are to be added, he expounds. It appears to Aristotle that the determination ought not to be conjoined to the predicate term; for in this proposition,—'Man is animal' (*Homo est animal*,) it is inquired whether the determination ought to be coupled with the subject, so that it shall be—'All man is animal' (*Omnis homo animal est*); or with the predicate, so that it shall be,—'Man is all animal' (*Homo omne animal est*); or with both the one and the other, so that it shall be, 'All man is all animal' (*Omnis homo omne animal est*). But neither of these latter alternatives is competent. For the determination is never joined to the predicate, but exclusively to the subject; seeing that all predication is either greater than the subject, or equal. Thus in this proposition—'All man is animal' (*omnis homo animal est*), 'animal' [the predicate] is greater than 'man' [the subject]; and, again, in the proposition—'Man is risible' (*homo risibilis est*), 'risible' [the predicate] is equated to 'man' [the subject]; but that the predicate should be less and narrower than the subject is impossible. Therefore, in those predicates which are greater than the subject, as, for example, where the predication is 'animal,' the proposition is manifestly false, if the determination of universality be added to the predicate term. For if we say, 'Man is animal,' (*homo est animal*), we contract 'animal,' which is greater than 'man,' by this determination to [an identity of extension with] 'man,' the subject, although the predicate, 'animal,' may be applied not only to 'man,' but to many other objects. Moreover, in those [subjects and predicates] which are equal, the same occurs; for if I say, 'All man is all risible,' (*omnis homo omne risibile est*),—in the first place, in reference to the nature of man itself, it is *superfluous* to adject the deter-

mination ; and, again, if this be added to all several men, the proposition becomes *false*, for when I say, ' All man is all risible,' by
this I seem to signify that the several men are [each of them] ' all
risible,' which is absurd. The determination is, therefore, to be
placed not to the predicate but to the subject. But the words of
Aristotle are thus reduced to the following import:— *In those predicates which are universal, to add to them aught universal, so that
the universal predicate may be predicated universally, is not true.*
For this is what he says—' In the case of a universal predicate,'
(that is, in a proposition which has an universal predicate), ' to
predicate the universal, itself universally, is not true.' For in an
universal predicate, that is, which is universal and is itself predicated, in this case universally to predicate the predicate, which is
universal, that is, to adject to it a determination of universality, is
not true: for it cannot be that any affirmation should be true in
which a universal determination is predicated of a predicate universally distributed; and he illustrates the conception of the matter by the example, ' All man is all animal ' (*omnis homo omne animal est*), of the incompetency of which we have already spoken."

Boethius, *In Librum de Interpretatione*, editio prima. *Opera*
p. 296. (Text so wretchedly printed that the sense must be constituted by the reader.)

[*Aristotle*, c. vii. § 4.] " ' In what is predicated as an universal, to predicate the universal universally is not true.'

"In this sentence he instructs us what is the place to which the
determination of universality should be rightly added. For he
teaches that the universality, which we call the universal determination, is to be connected with the subject term, never with the
predicate. For were we to say—' All man is animal,' (*omnis
homo animal est*), we should say rightly, annexing the ' all '
to the subject, that is, to the term ' man.' But if we thus
speak—' All man is all animal,' (*omnis homo omne animal est*),
we should speak falsely. He, therefore, does not say this, [in the
words]—' in what is predicated as an universal,' as ' animal ' of
' man ;' for animal is universal, being predicated of ' all man.'
[But he says]—To predicate this universal itself, ' animal,' to wit,
universally, so that we enounce—' All animal is man, (*omne animal esse hominem*), is not true ; for he allows this to be rightly

doue neither in these nor in any other affirmation.ᵃ He adds, therefore:—' For no affirmation will be true in which a universal predicate shall be universally predicated, as " All man is all animal" (*omnis homo est omne animal*).'

" Why this happens, I will explain in a few words. The Predicate is always greater than the Subject, or equal to it. Greater, as when I say, ' Man is animal,' (*homo animal est*): here ' animal ' is predicated, ' man' is subjected; for animal is predicated of more objects than man. Again, it is equal when we thus speak—' Man is risible' (*homo risibilis est*); here ' man' is the subject, 'risible' the predicate. But 'man' and 'risible' are equal; for it is proper to man to be a risible animal. But that the Predicate should be found less than the Subject is impossible. Is the Predicate the greater ? Then, to adject the universal to the Predicate, is *false*, as in the example he himself has given—' All man is all animal,' (*omnis homo omne animal est*). Is it equal ? Then, the adjection is superfluous, as if one should say, 'All man is all risible,' (*omnis homo omne risibile est*). Wherefore, to predicate a universal predicate universally is incompetent."

V.—AVERROES.

Averroes, *Perihermenias*, L. I. c. v.

" Propositions are not divided from the conjunction of the predesignation (*clausurae*) with the predicate; because the predesignation, when added to the predicate, constitutes a *false* or a *superfluous* proposition :—*False*, as ' All man is all animal' (*omnis homo est omne animal*); *superfluous*, as ' All man is some ' or ' a certain animal,' (*omnis homo est quoddam animal*)." Vide Conimbricenses, *In Arist. Dial.*, ii. 158.

ᵃ The Coimbra Jesuits (Sebastianus Contus, 1606), erroneously make Iluethius and Averroes oppose Aristotle, "thinking that the sign of universality may be annexed to the predicate of a universal proposition, when it is coextensive with the subject," (*ad locum*, ii. p. 158). This, a mistake, has been copied by their brother Jesuit, P. Val-lius of Rome, in his mighty Logic (*ad locum*). With Boethius he joins Levi Gersonides;—he means the Rabbi Levi Ben Gerson, of Catalonia, who died at Perpignan in 1370, who wrote on Theology, Philosophy, Mathematics, and Logic. See Jöcher v. *Levi*, from Bartolocci and Wolf.

VI.—ALBERTUS MAGNUS.

Albertus Magnus, *Perihermenias*, L. I., *Tractatus* v. c. 1, (*Op.*
d. Lugd. 1651, t. i. p. 261).

[" Ly '*omnis*' non est universale, sed signum universalitatia.
Quare ly '*omnis*' et hujusmodi signa distributiva non sunt universalia, secundum Avicennam.] Hoc enim signum distributivum, quod est *omnis*, non est universale, proprie loquendo :
sed est signum per quod stat pro particularibus universaliter universale, cui tale signum est adjunctum. Causa autem, quare non
sit universale, est :—quia, quamvis secundum grammaticum sit
nomen appellativum, hoc est, multis secundum naturæ suæ apti-
tudinem conveniens ; tamen est, secundum formam, infinitum,
nullam enim naturam unam dicit. Propter quod *omnis* naturæ
communis est distributivum. Universale autem est, quod est in
multis et de multis, suæ naturæ suppositis. Ideo *omnis*, et
nullus, et hujusmodi signa universalia esse non possunt ; sed
sunt signa designantia utrum universale sit acceptum universaliter vel particulariter, secundum sua supposita. Et hæc sunt
verba Avicennæ.

["Quare signum universale non sit ponendum a parte
prædicati.] In subjecto universali signum distributivum ordi-
nandum : quia per divisionem subjecti prædicatum partibus
attribuitur subjecti, ut divisim participent id per prædica-
tionem, et non in prædicato ponendum : quia quum prædicatum
formaliter sit acceptum, non proprie dividitur, nisi alterius, hoc
est, subjecti divisione : sed inæqualiter redditur subjecto et
partibus ejus. Unde id quod est universale, prædicari potest, ut
Omnis homo est animal; sed universale universaliter acceptum
non potest prædicari : nulla enim vera affirmatio esse potest, in
qua de universali aliquo prædicato prædicetur sive prædicatio
fiat ; quoniam universaliter sic patet, quod falsum est, *Omnis
homo est omne animal*, et si ponatur, quod *Nullum animal sit
nisi homo*. Cum enim *homo* subjiciatur gratia partium suarum,
et prædicata formaliter accipiantur, oportet quod *Quilibet homo
esset omne animal, quod falsum est.*"

VII.—LEVI BEN GERSON.

Levi Ben Gerson (or Levi Gersonides), a Jewish philosopher, who died in 1370 at Perpignan, wrote commentaries on Averroes' Commentary upon the logical books of Aristotle. The following is what he says on Averroes' doctrine touching the quantification of the predicate, as it is found (f. 39) in the Venice (folio) edition, of 1552,[a] of the works of Aristotle and Averroes :—"Although it be not necessary that when the quantitative note is attached to the predicate, this should be false or superfluous, seeing that it may be neither, as when we say, 'All man is all rational ;' and the same holds good in all other reciprocating propositions ;—nevertheless, as in certain matters it may so happen, Aristotle has declared that the quantitative note is not to be joined to the predicate in any language. But it may be here objected, that if this be the case, the quantitative note should not be annexed even to the subject, since there too it may be either false or superfluous. Superfluous :—As when we say, 'Some animal is rational.' For the very same follows here, as if we simply say, 'Animal is rational ;' the ' some,' therefore, is superfluous. False :—As when we say, 'All animal is rational.' The reason, therefore, assigned by Aristotle why the quantitative note should not be annexed to the predicate, is futile, seeing that for the same reason it should not be connected with the subject. To this we may answer : That the cause why the quantitative note is not usually conjoined with the predicate, is, that there would thus be two quæsita at once,—to wit, whether the predicate were affirmed of the subject, and, moreover, whether it were denied of everything beside. For when we say, 'All man is all rational,' we judge that all man is rational, and judge, likewise, that rational is denied of all but man. But these are in reality two different quæsita ; and therefore it has become usual to state them, not in one, but in two several propositions. And this is self-evident ; seeing that a quæsitum, in itself, asks only—Does, or does not, this inhere in that ? and not—Does this inhere in that, and, at the same time, inhere in nothing else ? "

a Not in the 8vo edition of these works. Venice, 1560.

VIII.—THE MASTERS OF LOUVAIN.

Facultatis Artium in Academia Lovaniensi Commentaria in Aristotelis Libros de Dialectica (1535), Tr. iii. c. 1, p. 162, ed. 1547.

Speaking of the text in the *De Interpretatione*, the Masters, *inter alia*, allege : " But if it be even elegantly said by a poet— ' Nemo est omnis homo,'—' Non omnes omnibus artes '—[proverb, ' Unus homo nullus homo '], why may we not contradict this aptly, howbeit falsely,—' Aliquis est omnis homo'? Why, (they say), do you determine the predicate by the note of universality, seeing that the quantity of the proposition is not to be sought from the predicate, but from the subject ? We answer,— Because we wish to express a certain meaning in words, which by no others can be done. But if the mark of universality could only be employed in changing the quantity of propositions, it would not be lawful to annex it to the part of the predicate. We have, therefore, thought these few cautions requisite to evince that what is condemned by these critics for its folly, is not incontinently sophistical or foolish babbling. But as to the universal rule which Aristotle enounces,—' No affirmation will be true,' &c.—it is sufficient if it hold good in the majority of cases ; whether the predicate exceed the subject, as ' All man is all animal,' be its equal, as, ' All man is all risible,'—or its inferior, as [' Some] animal is all man.' In a few cases, however, the exception is valid ; as,—' This sun is every sun,' ' One phœnix is all phœnix,' and some others. Nor are these futile subtleties, since reason herself approves."

IX.—TITIUS AND RIDIGER.

The only notice of these speculations of Titius,[a] which I have met with in any subsequent philosopher, (and I speak from an

a [Titius, *Ars Cogitandi*, c. vi. § 36 et sq., has the following relative to the quantification of the predicate :— " Licet autem Propositionem quantitas ex Subjecto æstimetur, attamen Præ-

dicatum non penitus negligendam videbatur, cen vulgo in hoc tractatione fieri solet, nam et hujus quantitatem observamus utile est, et crediderim et disqui- sitionis hujus neglecta varios errores

inspection of several hundred logical systems, principally by
Germans), is his friend Ridiger's; who in his elaborate work *De
Sensu Veri et Falsi*, first published some eight years subsequently,
(in 1709, but I have only the second edition of 1722), attempts a
formal refutation of the heresy of a quantified predicate. It was
only, however, after "the most manifest demonstrations of the
falsehood of this novel prejudice had been once and again pri-
vately communicated to his very learned friend," (Titius?), that
Ridiger became at length tired, as he expresses it, "of washing
a brick," and laid the polemic before the public. It was not certainly the cogency of this refutation which ought to have thrown
the counter opinion into oblivion; but this refutation, such as it
is, though with nothing new, is deserving of attention, as presenting the most elaborate discussion of the question to be met
with, after Ammonius, and in modern times. But the whole
argument supposes certain foundations; and it will be sufficient

tam in doctrina Conversionis, quam
Syllogistica cum exortos, quos suis locis
videbimus. Breviter itaque observan-
dam, in propositionibus affirmativis,
licet universalibus, praedicatum plerumque sms *particulare*, tribuique subjecto secundum totam quidem suam
comprehensionem, non vero *extensionem*.
. . . E contrario in propositionibus
negativis, licet particularibus, plerumque praedicatum est *universale*, so
tam secundum comprehensionem quam
extensionem suam totam, a subjecto
removetur. . . . Interim non pe-
tarem affirmationem vel negationem
ipsam diversam illam praedicati quantitatem necessario postulare, sed crediderim potius, id omne a diverso rerum
et idearum habitu oriri, affirmationi
vero et negationi praedicati quantitatem esse velut indifferentem. Nam
plerumque praedicata subjectis sunt
latiora; quodsi igitur illa cum his com-
ponas, non poterit non praedicatem particulare inde emergere, dum unico ad
subjectum restringi nequit, sed ad alia
quoque extendi aptum manet. Ast si
praedicatum a subjecto removeas; universale illud erit, cum quicquid in ejus
vel comprehensione vel extensione est

ab hoc sejungatur, nec imminuit uni-
versalitatem, quod idem ab aliis subjectis quoque removeatur, nam si praedicatum aliis etiam conveniat, tum
quidem uni subjecto non potest dici
universaliter tributum, verum si de
multis negetur, potest nihilominus de
certo aliquo subjecto universaliter
quoque negari. Quodsi habitus attri-
buti permittat, poterit aliquando propositio affirmativa praedicatum universale, et negativa particulare habere;
nihil enim obstat, quo minus aliquando
totum alteri jungere, vel partem ab
eodem removere queas. Hæc itaque
propositio :—*Omnis homo est visibilis*,
habet praedicatum universale, si visibi-
litatem pro hominis proprio habeas;
sicut hæ,—*Nullus Turca est homo*,
(scil. *Christianus*), vel *Quidam medicus non est homo quidam*,—praedicatum
particulare continent, dum pars solum
comprehensionis et extensionis removetur." For the application by Titius,
of the principle of a quantified predicate to the doctrine of Conversion, see
above, pp. 276, 277; and to the theory
of Syllogism, see below, pp. 382, 383,
and Appendix XI.—Ed.]

to show that these are false, to dispose of the whole edifice erected upon them. I ought to mention, that it was Ridiger's criticism which first directed my attention to the original of Titius.

" Origo autem hujus erroris neglectus notissimæ acquivocationis signorum *omnis* et *quidam* esse videtur, qua hæc signa, vel *collective* sumi possunt, vel *distributive.* Priori modo, quantitas in prædicato concepta sensum quidem infert non penitus absurdum, cæterum propositionem constituit *identicam et frustraneam.*" Ridiger then goes on to a more detailed statement of what he supposes to be the grounds on which the erroneous opinion proceeds.[a]

First Case.—" Verbi gratia, *Quoddam animal est omnis homo ;* hoc est, *Species quædam animalis, homo nempe, omne id, quod homo est :* quod alium sensum habere nullum potest, quam *quod omnis homo sit homo :* sic autem *collective* sumitur et signum subjecti et signum prædicati.*" This objection is absurd, for it is suicidal ; applying equally to the proposition which the objector holds for good, and to that which he assails as bad. *All man is (some) animal.* Here, is not *animal* or *some animal,* just a certain species of animal, and is not this species, *man,* to wit, *all that is man,* and nothing else? There is, consequently, the same tautology in the one case as in the other ; and if we are blamed for only virtually saying, by the former, " All man is man," does the objector say a whit more than this, by the latter? Ridiger goes on : " Quodsi vel alterum signum, vel utrumque, *distributive* sumatur, semper absurdus erit propositionis sensus."

Second Case.—" Verbi gratia, sumatur *utrumque* signum *distributive,* sensus erit, *Quoddam individuum animalis,* (v. g. *Petrus,) est omne individuum hominis,* (v. g. *Davus, Oedipus*)." This is a still higher flight of absurdity ; for, to refute the proposition, it is first falsely translated into nonsense. Its true meaning, *both* quantified terms being taken *distributively,* is :—" All several men are some several animals," or, " Every several man is some several animal."

In these two cases, therefore, all is correct, and the objection from the identity or absurdity of a quantified predicate, null.

Third Case.—" Sumatur signum subjecti *distributive,* signum prædicati *collective,* sensus erit: *Quoddam individuum animalis est universa species hominis.*"

a Second Edition, pp. 232, 302.

Fourth Case.—"Sumatur, denique, signum subjecti *collective*, signum prædicati *distributive*, sensus erit: *Quædam species animalis, ut universale et prædicabile, est omne individuum hominis.*"

In regard to these last two cases, it is sufficient to refer to what has been already said in answer to Ammonius (p. 302); or simply to recall the postulate, that in the same logical unity (proposition or syllogism) the terms should be supposed in the same sense. If this postulate be obeyed, these two cases are inept, and, consequently, the objections superfluous.

Ridiger then proceeds to treat us with four long "demonstrations *a priori*," and to one elaborate "demonstration *a posteriori*;" but as these are all founded on the blunders now exposed, it would be idle to refute them in detail.

Ridiger, it may well surprise us, howbeit the professed champion of "the old and correct doctrine," is virtually, perhaps unconsciously, a confessor of the truth of "the new and false prejudice;" for I find him propounding four several syllogistic forms, three of which are only valid through the universal quantification of the predicate in affirmatives, and two, (including the other one), proceed on a correct, though partial, view, opposed to that of the logicians, touching the conclusion of the Second Figure, (L. II. c. vi.) I shall insert the quantities, operative but not expressed.

In the First Figure.—"At, aut ego nihil video, aut *longe naturalior* est hic processus:—*Quoddam fluidum est* [*quoddam*] *lere; quoddam corpus est* [*omne*] *fluidum; ergo quoddam corpus est quoddam lere;* quam si dicas, &c.*" (§ 34.)—Here the middle term is, and must be, affirmatively distributed as predicate.

C,————,M:————,Γ.

In the Second Figure.—"Verbi gratiâ:—*Quoddam ens est* [*omne*] *animal; omnis homo est* [*quoddam*] *animal; ergo, omnis homo est* [*quoddam*] *ens.* Hæc conclusio verissima, &c.*" (§ 39.) In like manner the middle is here universally quantified in an affirmative.

C,————: M,————: Γ.

The following, Ridiger (p. 330) gives, as "two new moods, which cannot be dispensed with."—"*Quoddam animal est* [*omnis*] *homo ; nullum brutum est* [*ullus*] *homo ; ergo, quoddam animal non est* [*ullum*] *brutum.*" Item :—"*Quoddam animal non est* [*ullus*] *homo ; omnis civis est* [*quidam*] *homo ; ergo, quoddam animal non est* [*ullus*] *civis.*"—In the first of these, the middle, as predicate, is affirmatively distributed ; and in both syllogisms, one conclusion, denied by the logicians, is asserted by Ridiger, although the other, which involves a predicate, particular and negative, is recognised by neither.

C,———— : M :———: P   C,———: M, ———— : Γ

X.—GODFREY PLOUCQUET.

Godfrey Ploucquet, a philosopher of some account, Professor of Logic and Metaphysic in the University of Tübingen, by various writings, from the year 1759, endeavoured to advance the science of reasoning ; and his failure was perhaps owing more to the inadequacy and limitation of his doctrine, than to its positive error. To say nothing about his attempt to reduce Logic to a species of computation, in which his one-sided views came into confliction with the one-sided views of Lambert,—he undoubtedly commenced auspiciously, on the principle of a quantified predicate. This, like a few preceding logicians, he certainly saw afforded a mean of simplifying the conversion of propositions ;* but he did not see that it could accomplish much more, if properly applied, in the theory of syllogism. On the contrary, in syllogistic he professedly returns, on mature consideration, to the ordinary point of view, and thinks himself successful in recalling the common doctrine of inference to a single canon. That canon is this :— The terms in the conclusion are to be taken absolutely in the same extension which they hold in the antecedent."—"In conclusione sint termini plane iidem, qui in præmissis, intuitu quantitatia." (*Methodus tam demonstrandi directe omnes syllogismorum species, quam vitia formæ detegendi, ope unius regulæ ;—Methodus calculandi in Logicis ;* passim. Both in 1763). This rule, as applied to his logical calculus, he thus enounces : " Arrange the terms in syllogistic order ; strike out the middle ; and the

* An extract from his *Fundamenta Philosophia Speculativa,* 1759, containing Ploucquet's doctrine touching the quantification of the predicate, will be found in Mr Baynes' *Essay,* p. 128.

extremes then afford the conclusion."—" Deleatur in præmissis medius ; id quoad restat indicat conclusionem." (*Methodus calculandi*, passim ; *Elementa Philosophiæ Contemplativæ, Logica*, § 122, 1778.) This rule is simple enough, but, unfortunately, it is both *inadequate* and *false*. Inadequate (and this was always sufficiently apparent); for it does not enable us to ascertain, (and these the principal questions), how many terms,—of what identity—of what quantity—and of what quality, can be legitimately placed in the antecedent. But it is not true, (though this was never signalised); for its peculiar principle is falsified by eight of the thirty-six moods, to wit, in affirmatives, by ix., x., xL, xii., and in negatives, by ix. b, x. a, xi. b, xii. a.ª In all these the quantity of an extreme in the conclusion is less than its quantity in the antecedent.—We can hardly, therefore, wonder that Ploucquet's logical speculations have been neglected or contemned ; although their author be an independent and learned thinker, and his works all well worthy of perusal. But, though dismissed by Hegel and other German logicians, not for its falsity, with supreme contempt, Ploucquet's canon has, however, found its admirers in this country, where I have lately seen it promulgated as original.

XI.—ULRICH.

Institutiones Logicæ et Metaphysicæ, § 171, 1785.—" Non tantum subjecto sed et *prædicato*, ad subjectum relato, sua constat quantitas, suumque igitur signum quantitatis præfigere licet. Sed hæc prædicati quantitas ex veterum præceptis sæpe justo minor invenitur. In loco de conversione distinctius de eo exponetur." In that place, however, nothing of the kind appears.ᵝ

ª See Table of Moods below, Appendix XI.—ED.

ᵝ [That the Extension of Predicate is always reduced to Extension of Subject, *i.e.*, is equivalent to it, see Purchot, *Instit. Phil. Logica*, L pp. 123, 125. Tracy, *Elémens d'Idéologie*, t. III., Disc. Prel. pp. 99, 100. Crousaz, *Logique*, t. iii. p. 190. Derodon, *Logica Restituta*, P. II. c. v. art. 4, p. 224. Boethius, *Opera*, p. 348, (see above, p. 513). Sargeant, *Method to Science*, B. II. Lem. i., p. 127. Beneke, *Lehrbuch*

der *Logik*, § 130, p. 100. Stattler, *Logica*, § 196.

That the Predicate has quantity, and potential designation of it, as well as the Subject, see Hoffbauer, *Analytik der Urtheile und Schlüsse*, § 31 et seq. Lambert, *Deutscher Gelehrter Briefwechsel*, Brief vi. vol. L p. 395. Platner, *Philosophische Aphorismen*, i. § 546. Carvinus, *Instit. Phil. Rat.*, § 413. Conimbricenses, *In Arist. Dial.*, t. ii. pp. 158, 283. Scotus, *In An. Prior.*, L. L qu. 4, l. 240 ; qu. 13, ff.

VII.

CANONS OF SYLLOGISM; GENERAL HISTORICAL NOTICES AND CRITICISM.

A. HISTORICAL NOTICES.

(a) QUOTATIONS FROM VARIOUS LOGICIANS.

(Collected and Translated Autumn 1844.   See above, Vol. I. p. 303.— Ed.)

I.—DAVID DERODON.

David Derodon (who died at Geneva in 1664, and had previously been Professor of Philosophy at Die, Orange and Nismes), was a logician of no little fame among the French Huguenots; the study of his works was (if I recollect aright) even formally recommended to the brethren of their communion, by one of the Gallican Synods. "Either the Devil or Doctor Derodon," was long a proverbial expression in France for the authorship of an acute argument; and the "*Sepulchre of the Mass*" has been translated into the vernacular of every Calvinist country.—Derodon has left two systems of Logic; a larger (*Logica Restituta*, 1659), and a smaller (*Logica Contracta*, 1664), both published in 4to soon after his decease.* I shall quote only from the former.

It is impossible to deny Derodon's subtlety, but his blunders unfortunately outweigh his originality. Leaving Conversion as he found it, after repeating, with approbation, the old rules,—that the predicate is not to be overtly quantified universally, (p. 573), but to be taken, in affirmative propositions, particularly, as in

254 b, 255 a; qu. 14, f. 256 b; qu. 23, f. 273 a.

For instances of Aristotle virtually using distributed predicate, see *An. Post.*, l. 6, § 1. Cf. Zabarella, *ad loc. Opera Logica*, p. 735. The same, *In An. Post.*, l. 2. *Opera*, p. 827, and *De Quarta Figura Syllog.*, *Op.*, p. 123. The adding mark of universality to predicate is, Aristotle says, "useless and impossible" (*An. Prior.*, i. c. 27, § 9); yet see ii. c. 22, §§ 7, 8; c. 23, §§ 4, 5. On this question, see Bolzano, *Logik*, § 131, p. 27, (and above, pp. 301, 308, 309.)

That the predesignation of the predicate by *all* collectively, in fact, reduces the universal to a singular proposition, see Purchot, *Instit. Phil.*, l. p. 124. Cf. *Logica Contracta Trajectina*, P. ii. c. 5. (1707.)]

a Derodon seems wholly unknown to the German logicians, and, I need hardly add, to those of other countries. In Scotland his works are not of the rarest; a considerable number in the same binding must have been imported at once, probably in consequence of the Synodical recommendation.

negative propositions, universally, (p. 623 ; we are surprised to
find him controverting, in detail, the special rules of syllogism.
This polemic, as might be expected, is signally unsuccessful ; for
it is frequently at variance with all principle, and uniformly in
contradiction of his own. It is, indeed, only interesting as a
manifestation, that the old logical doctrine was obscurely felt by
so original a thinker to be erroneous; for the corrections attempted
by Derodon are, themselves, especially on the ground which he
adopts, only so many errors. He unhappily starts with a blunder;
for he gives as "*rectus*," an example of syllogism, in which the
middle term is, even of necessity, undistributed ; and he goes on
(pp. 627, 628, 636, 637, 638, 639, 649) either to stumble in the
same fashion, or to adduce reasonings, which can only be vindi-
cated as inferential, by supplying a universal quantity to the
predicate in affirmative propositions, or by reducing it to particularity in negatives ;—both in the teeth of Derodon's own laws.
—I have, however, recorded, in my Table of Syllogisms, some of
his examples, both the *two* forms which he has named, and *four*
others which he only enounces ; according, by liberal construction, what was requisite to give them sense, and which, without
doubt, the author would himself have recognised.

II.—RAPIN.

Rapin, *Réflexions sur la Logique*, § 4, 1684.

" Before Aristotle there had appeared nothing on logic systematic and established. His genius, so full of reason and intelligence, penetrated to the recesses of the mind of man, and laid
open all its secret workings in the accurate analysis which he
made of its operations. The depths of human thought had not
as yet been fathomed. Aristotle was the first who discovered
the new way of attaining to science, by the evidence of demonstration, and of proceeding geometrically to demonstration, by
the infallibility of the syllogism, the most accomplished work
and mightiest effort of the human mind," &c.

Rapin errs in making Aristotle lay the rule of proportion along
with the *Dictum de Omni* as a principle of syllogism.

III.—LEIBNITZ.

Leibnitz, *De la conformité de la Foi avec la Raison*, § 22. *Op.*

t. i., p. 81. " Hence the facility of some writers is too great, in
conceding that the doctrine of the Holy Trinity is repugnant
with that great principle which enounces—*What are the same
with the same third, are the same with each other;* that is, if A
be the same with B, and C be the same with B, it is necessary
that A and C should also be the same with one another. For
this principle flows immediately from the principle of Contradiction, and is the ground and basis of all Logic; if that fail,
there is no longer any way of reasoning with certainty."

IV.—REUSCH.

Reusch, *Systema Logicum,* 1734.

§ 506. " That dictum of the Aristotelians *de Omni et Nullo*
(§ 503) evinces, indeed, a legitimate consequence, but it only
regulates one species of syllogisms, at least immediately. By
this reason, therefore, logicians have been induced to prove the
consequence of the other species by means of the first, to which
they are reduced. But, that we may be able to supersede this
labour, I have endeavoured to give a broader basis to the Dictum
de Omni et Nullo, or by whatever name that rule is called, to
which, in the construction of syllogisms, the order of thought is
conformed."

§ 507. " For the whole business of ordinary reasoning is accomplished by the substitution of ideas in place of the subject
or predicate of the fundamental proposition. This some call the
equation of thoughts. Now, the fundamental proposition may be
either affirmative or negative, and in each the ideas of the terms
may be considered either agreeing or diverse, and according to
this various relation there obtains a various substitution, which
we shall clearly illustrate before engaging with our doctrine of
the Dictum de Omni et Nullo." [Having done this at great
length, he proceeds.]

§ 510. " From what has been now fully declared, the following
Dictum de Omni et Nullo may be formed, which the definition
itself of reasoning and syllogism (§ 502) supports, and to which
all syllogisms in every figure and mood may be accommodated.

" *If two ideas* (two terms) *have, through a judgment,* (proposition), *received a relation to each other, either affirmative or negative, in that case it is allowable, in place of either of these,* (that

ie, the subject or predicate of that judgment or proposition), *to substitute another idea*, (term), *according to the rules given of Equipollence or Reciprocation* (§ 508, a. 9), *of Subordination, of Co-ordination*," (see Waldin, below, p. 332).

IV.—CRUSIUS.

Crusius, *Weg zur Gewissheit.* Ed. i. 1747; Ed. ii. 1762.

§ 256. "The supreme law of all syllogism is, *What we cannot otherwise think than as true, is true, and what we absolutely cannot think at all, or cannot think but as false, is false.*" [a]

§ 259. Of necessary judgments, of judgments which we cannot but think, "which are not identical, and which constitute, in the last result, the positive or the kernel in our knowledge; to which we apply the principle of Contradiction, and thereby enrich the understanding with a knowledge of real judgments,"—such judgments are principally the following: *Every power or force is inherent in a subject; All that arises*, (begins to be), *arises in virtue of a sufficient cause; All whose non-existence cannot be thought, has its cause, and has at some time arisen*, (begun to be); *Every substance exists somewhere; All that exists, exists at some time; Two material things cannot exist at the same time and in precisely the same place.* There are also many other propositions, which treat of the determinate qualification of things as present; for example—*The same point of a body cannot be at once red and green; A man cannot be in two places at once*, and so forth.

§ 261. "All the judgments previously alleged, (§ 259), may be comprehended under these two general propositions,—*What cannot in thought be separated from each other, cannot be separated*

a Kant, (*Über die Evidenz in metaphysischen Wissenschaften*, 1763, *Verm. Schrift.*, ii. 43), has hereon the following observation:—"In regard to the supreme rule of all certainty which this celebrated man thought of placing as the principle of all knowledge, and, consequently also of the metaphysical, —*What I cannot otherwise think than as true is true*, &c. ; it is manifest that this proposition can never be a principle of truth for any knowledge whatever. For if it be agreed that no other principle of truth is possible than inasmuch as we are incapable of holding a thing not for true, in this case it is acknowledged that no other principle of truth is competent, and that knowledge is indemonstrable. It is indeed true that there are many indemonstrable knowledges, but the feeling of conviction in regard to them is a confession, but not a ground of proof, that they are true."—See also Reid, *Intellectual Powers*, Essay iv. ch. 4.

from each other in reality; and, *What cannot in thought be connected into a notion, cannot in reality be connected;* to wit, although no contradiction shows itself between the notions, but we are only conscious of a physical necessity to think the thing so and so, clearly and after a comparison of all the circumstances with each other. For we now speak of propositions which are not identical with the Principle of Contradiction, but of such as primarily afford the matters on which it may be applied. Hence we see that the supreme principle of our knowledge given above, (§ 256), has two determinations; inasmuch as the impossibility to think a something arises, either because a contradiction would ensue, or because we are positively so compelled by the physical constitution of our thinking faculties."

§ 262. "The highest principle of all syllogism thus resolves itself into the three capital propositions;

"1. *Nothing can at once be and not be in the same point of view.*

"2. *Things which cannot be thought without each other, without each other cannot exist.*

"3. *What cannot be thought as with and beside each other, cannot exist with and beside each other, on the supposition even that between the notions there is no contradiction.*

"The second of these capital propositions I call the *Principle of Inseparables, (principium inseparabilium)*; and the third the *Principle of Inconjoinables (principium inconjungibilium).* They may be also termed the three *Principles of Reason.*"

Ch. VIII. *Of the different species of syllogisms,* he says, (§ 272):—"Among the higher principles of syllogisms it is needful only to enumerate the *Principle of Contradiction* and the *Principle of Sufficient Reason,* which is subsumed from the principle of Inseparables (§ 262). We shall state the laws of syllogism in this order,—Consider those which flow, 1°, From the *Principle of Contradiction;* 2°, From the *Principle of Sufficient Reason;* and, 3°, From both together."

V.—FRANCIS HUTCHESON.

[Francisci Hutcheson.] *Logicæ Compendium. Glasguæ,* in ædibus academicis, excudebant *Robertus et Andreas Foulis, Academiæ Typographi.* 1764.

Part III., Ch. ii. p. 58.

" The whole force of syllogism may be explicated from the following axioms.

" First Axiom.—*Things which agree in the same third, agree among themselves.*

" Second Axiom.—*Things whereof the one agrees, the other does not agree, in one and the same third, these things do not agree among themselves.*

" Third Axiom.—*Things which agree in no third, do not agree among themselves.*

" Fourth Axiom.—*Things which disagree in no third, do not disagree among themselves.*"

" Hence are deduced the general rules of syllogisms.

" Of these the three first regard the *Quality* [not alone] *of Propositions.*

" Rule 1.—*If one of the premises be negative, the conclusion will be negative.* [By Ax. 2].

" Rule 2.—*If both premises be affirmative, the conclusion will be affirmative.* [By Ax. 1].

" Rule 3.—*If both premises be negative, nothing follows* : because of things mutually agreeing and mutually disagreeing, both may be different from a third thing. [By Ax. 3, 4].

" Two Rules regard the *Quantity of Terms.*

" Rule 4.—*Let the middle be once at least distributed, or taken universally;* for the common term frequently contains two or more species mutually opposed, of which it may be predicated according to various parts of its extension ; these [specific] terms do not, therefore, truly agree in one third, unless one at least of them agrees with the whole middle. [By Ax. 3, 4].

" Rule 5.—*No term ought to be taken more universally in the conclusion than in the premises* : because no consequence is valid from the particular to the universal. [Because we should, in that case, transcend the agreement or disagreement of the two terms in a third, on which, *ex hypothesi*, we found].

" [In like manner there are two rules] concerning the *Quantity of Propositions.*

" Rule 6.—*If one of the premises be particular, the conclusion will also be particular.*

" For, Case 1.—If the conclusion be affirmative, therefore both premises will be affirmative (by Rule 1). But, in a particular proposition, there is no term distributed; the middle is, therefore, to be distributed in one or other of the premises (by Rule 4). It will, therefore, be the subject of a universal affirmative proposition; but the other extreme is also taken particularly, when it is the predicate of an affirmative proposition, the conclusion will, therefore, be particular (by Rule 5).

"Case 2.—Let the conclusion be negative; its predicate is, therefore, distributed: hence, in the premises, the major and the middle terms are to be distributed (by Rules 5 and 4).

"But when one of the premises is negative, the other is affirmative (by Rule 3). If one premise be particular, these two terms only can be distributed; since one premise affirms, whilst the other is particular. The minor extreme, the subject of the conclusion, is not, therefore, distributed in the premises; it cannot, therefore, (by Rule 5), be distributed in the conclusion.

"Rule 7.—*From two particular premises nothing follows;* at least according to the accustomed mode of speaking, where the predicate of a negative proposition is understood to be distributed. For, 1°, If the conclusion affirm, both premises will affirm, and, consequently, no term is distributed in the premises; contrary to Rule 4. 2°, Let the conclusion be negative, its predicate is therefore distributed; but in particular premises there is only distributed the predicate of a negative proposition; there is, therefore, necessarily a vice, (either against Rule 4 or Rule 5)." *

a Rules 1 and 7 are thus contracted into one: *The conclusion follows the weaker part;* that is, the negative or the particular. All these Rules are included in the following verses:—

Distribuas medium, nec quartus terminus adsit,
Utraque nec præmissa negans, nec particularis.
Sectetur partem conclusio deteriorem ;
Et non distribuat nisi cum præmissa, negetur.

In an unusual mode of speaking, a certain negative conclusion may be effected with a non-distributive predicate. As in this example :—

A   B
Some Frenchmen are [some] learned ;
C   B
Some Englishmen are not [any] learned,
Therefore, some Englishmen are not some Frenchmen.

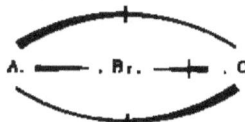

A. —— . B r. —— . C

(What are within [ ] are by me).
[Written Autumn 1844. In the latest notation (,) is substituted for (.), and (:) for (:.). See below, Appendix XII. — Ed.]

VI.—SAVONAROLA.

Savonarola, *Compendium Logices*, L iv. p. 115, ed. Venetiis, 1542.—" In whatever syllogism any proposition can be concluded, there may also be concluded every other proposition which follows out from it." On this he remarks :—" When any syllogism infers a conclusion flowing from its immediate conclusion, it is not to be called *one* syllogism but *two*. For that other conclusion does not follow simply in virtue of the premises, but in virtue of them there first follows the proper conclusion, and from this conclusion there follows, by another syllogism, the conclusion consequent on it. Hence there are tacitly two syllogisms ; otherwise the moods of syllogisms would be almost infinite."

VII.—BAUMGARTEN.

Baumgarten, *Acroasis Logica*. Ed. Töllner. Ed. I. 1765.

§ 297. "Every reasoning depends on this proposition :—*A and B connected with a third C, are connected with each other* : in affirmation immediately, in negation mediately. This proposition is, therefore, the foundation and principle of all reasoning ; which, however, is subordinate to the principle of Contradiction.

§ 324. "Every ordinary syllogism concluding according to the *Dictum*, either *de Omni*, or *de Nullo*. This *Dictum* is thus the foundation of all ordinary syllogisms. (It had been previously announced, §§ 319, 321.)

"Whatever is truly affirmed of a notion universally, is also truly affirmed of all that is contained under it. Whatever is truly denied of a notion universally, is also truly denied of all that is contained under it."

VIII.—REIMARUS.

Reimarus, *Vernunftlehre*. 1766.

§ 176. "The fundamental rules of syllogism are, consequently, no other than the rules of *Agreement* [Identity] and of *Contradiction*. For what the geometer in regard to magnitudes takes as the rule of equality or inequality, that the reasoner here adopts as the universal rule of all mediate insight :—*If two things be identical with a third, they are also in so far identical with each other. But if the one be, and the other be not, identical with the third, then they are not mutually identical, but rather mutually repugnant.*"

§ 177. Here he notices that the *Dictum de Omni et Nullo* is not properly a rule for all figures, but for the first alone.

IX.—WALDIN.

Waldin, *Novum Logicæ Systema.* 1766.

§ 335. "Since the syllogism requires essentially nothing but a distinct cognition of the sufficient reason of some proposition, the most universal rule of all syllogisms is,—*The sufficient reason of a given proposition is to be distinctly cognised.*"

§ 364. "The most general rule of all reasonings, (§ 335), remains also the rule of all reasonings as well in synthesis as in analysis. But in the synthesis of the ordinary syllogism, the middle term in the major proposition is referred to the major term, in the minor proposition to the minor term. (§ 360). Wherefore, from this relation we must judge whether the middle term be or be not the sufficient reason of the conclusion. Wherefore, the synthesis of the ordinary syllogism is to be cognised from the relation of its ideas. This you may thus express:

"1.) *After the true proposition, the relation of whose extremes you distinctly apprehend;*

"2.) *Add to its subject or predicate another idea different from both, whether agreeing or disagreeing;*

"3.) *Inquire into the relation of the added idea, to the end that you may know whether the middle term in the given relation infer the conclusion; and this is known by the application of the rules of Reciprocation, Subordination, Co-ordination, and Opposition.* If any one wish to call this the *Dictum de Omni et Nullo,* I have no objections."

"*Observation.* This they call the *Dictum de Omni et Nullo* of the celebrated Reusch. It stands true indeed ; but is beset with difficulties, inasmuch as it is rather a complexus of all rules than one only, which as yet is to be referred to the class of *pia desideria.* Logicians have, indeed, taken pains to discover one supreme rule of all ordinary reasonings; but no one has as yet been so happy as to find it out." Then follows a criticism of the attempts by the Port Royal and Syrbius.

X.—STATTLER.

Stattler, *Philosophia,* P. I. *Logica,* 1769.

§ 237. "In this comparison of two ideas with a third, six different cases may in all occur: for either,

1.) "*One of the two ideas contains that same third, which again contains the other;* or,

2.) "*Both of the two are contained in the third;* or,

3.) "*Each of the two contains the third;* or,

4.) "*One of the two contains the third, the other being repugnant with it;* or,

5.) "*One of the two is contained in the third, with which the other is repugnant;* or,

6.) "*Both of the two are repugnant to the third.*

"The former three cases generate an affirmative conclusion, the latter three a negative." In a note Stattler eliminates a seventh case, in which neither may contain, and neither be repugnant to the third.

§ 244. General Law of all Reasonings. "*In all reasonings, as often as a consequent is, by legitimate form, inferred from an antecedent, so often is there included in the antecedent what the consequent enounces; either the congruity and reciprocal containment, or the repugnance of A and O; and if such be not included in one or other of the antecedents, whatever is inferred in the consequent is void of legitimate form.*"

XI.—SAUTER.

Sauter, *Institutiones Logicæ,* 1798.

§ 123. "*Foundations of Syllogism.*—In every syllogism there are two notions compared with a third, to the end that it may appear whether they are to be conjoined or sejoined. There are, therefore, here, three possible cases. For there *agree* with the assumed third, either *both* notions, or *one*, or *neither*. In reasoning, our mind, therefore, reposes on these axioms, as on fundamental principles :—

1.) " *Where two notions agree with the same third, they agree with one another.* . . . .

2.) " *Where one is contained by the third, with which the other is repugnant, they are mutually repugnant.*

3.) " *When neither notion agrees with the third, there is between them neither agreement nor repugnance.*"

XII.—SUTER.

Suter, *Logica.*

§ 61. " Quæ eidem tertio conveniunt vel disconveniunt, etiam conveniunt vel disconveniunt inter se."

XIII.—SEGUY.

Seguy, *Philosophia ad Usum Scholarum Accommodata*, Tom. I. *Logica.* Paris, 1771.

P. 175, ed. 1785. " Concerning the rule of recent philosophers."

Having recited the general rule of the *Port Royal Logic*, he thus comments on it :—

" 1°. This is nothing else than the principle of reasoning ; therefore, it is improperly adduced as a new discovery, or a rule strictly so called.

" 2°. It may be useful to the rude and inexperienced, to recognise whether a syllogism be legitimate or illicit.

" But the principal fault of this rule is, that it contains no certain method whereby we may know when, and when not, one of the premises contains a conclusion ; for the discovery of which we must frequently recur to the general rules." a

P. 176. Seguy exposes Father Buffier's error in saying " that,

a Followed by Larroque, *Élémens de Philosophie*, p. 231 ; Galluppi, *Lezioni de Logica e di Metafisica*, L. 348. E contra, *Philosophia Lugdunensis*, L. 159, where the rule is called " optima " on various accounts. Troxler, *Logik*, ii. 41.

according to Aristotle and the common rules of Logic, the middle
term ought absolutely to be the predicate in the first or major
proposition;" seeing that the middle term is not the predicate
in the first and third Figures. This must be a mistake; for I
cannot find such a doctrine in Buffier, who in this respect, in
many places, teaches the correct.

XIV.—HOFFBAUER.

Hoffbauer, *Anfangsgründe der Logik*, 1794, 1810.

§ 317. *Fundamental Principles.*

"I. 1.) An attribute which belongs to all and every of the
objects contained under a notion, may also be affirmed of these
objects so contained. (Dictum de Omni.)

"2.) An attribute which belongs to none of the objects contained under a notion, must also be denied of these objects so
contained. (Dictum de Nullo.)

"II. When, of the objects X and Z, the one contains an attribute which the other does not contain, and they are thus different from each other, then X is not Z, and Z is not X.

"III. 1.) When objects which are contained under a notion
a are also contained under another notion *b*, then this last notion
contains under it some at least of the objects which are contained
under the first.

"2.) If certain objects which are not contained under a notion
a are contained under *b*, then *b* contains under it some at least
of the objects which are not contained under *a*.

"IV. 1.) If objects which are contained under a notion *a*
belong to those which are contained under another notion *b*,
then this second notion *b* contains under it some at least of the
objects which are contained under *a*.

"2.) If all objects which are contained under a notion *a*
belong to those which are not contained under a certain other
notion *b*, then this notion *b* contains under it no object which is
contained under the notion *a*.

" 3.) If all the objects contained under a certain notion *a* are different from certain other objects contained under *b*, then *b* contains under it at least some objects which are not contained under *a*.

XV.—Kant.

Kant, *Logik.* 1800-6. II. Syllogisms.

" § 56. *Syllogism in general.*—A syllogism is the cognition that a certain proposition is necessary, through the subsumption of its condition under a given general rule.

" § 57. *General principle of all Syllogisms.*—The general principle whereon the validity of all inference, through the reason, rests, may be determinately enounced in the following formula :—

" *What stands under the condition of a rule, that stands also under the rule itself.*

" *Observation.*—The syllogism premises a *General Rule,* and a *Subsumption* under its *Condition.* Hereby we understand the conclusion *a priori,* not as manifested in things individual, but as universally maintained, and as necessary under a certain condition. And this, that all stands under the universal, and is determinable in universal laws, is the Principle itself of *Rationality* or of *Necessity,* (*principium rationalitatis seu necessitatis.*)

" § 58. *Essential constituents of the Syllogism.*—To every syllogism there belong the three following parts :—

" 1.) A general rule, styled the *Major Proposition,* (*propositio major, Obersatz.*)

" 2.) The proposition which subsumes a cognition under the condition of the general rule, called the *Minor Proposition,* (*propositio minor, Untersatz*) ; and, finally,

" 3.) The proposition which affirms or denies the predicate in the rule of the subsumed cognition,—the *Concluding proposition,* or *Conclusion (Conclusio, Schlussatz*).

" The two first propositions, taken in connection with each other, are called the *Antecedents,* or *Premises* (*Vordersätze*).

" *Observation.*—A rule is the assertion of a general condition. The relation of the condition to the assertion, how, to wit, this stands under that, is the *Exponent* of the rule. The cognition, that the condition, (somewhere or other), takes place, is the *Subsumption.*

. "The nexus of what is subsumed under the condition, with the assertion of the rule, is the *Conclusion.*"

Having shown the distribution of syllogisms into *Categorical, Hypothetical,* and *Disjunctive,* he proceeds to speak of the first class.

"§ 63. *Principle of Categorical Syllogisms.*—The principle whereon the possibility and validity of Categorical Syllogisms rest, is this,—What pertains to the attribute of a thing, that pertains to the thing itself; and what is repugnant to the attribute of a thing, that is repugnant to the thing itself, (*Nota notœ est nota rei ipsius ; Repugnans notœ, repugnat rei ipsi*).

" *Observation.*—From this principle, the so-called Dictum de Omni et Nullo is easily deduced, and cannot, therefore, be regarded as the highest principle either of the Syllogism in general, or of the Categorical Syllogism in particular. *Generic and Specific Notions* are in fact the general notes or attributes of all the things which stand under these notions. Consequently the rule is here valid—*What pertains or is repugnant to the genus or species, that also pertains or is repugnant to all the objects which are contained under that genus or species.* And this very rule it is which is called the Dictum de Omni et Nullo."

XVI.—CHRISTIAN WEISS.

Christian Weiss, *Logik,* 1801.

"§ 216. *Principle for all Syllogisms.*—The principle of every perfect Syllogism consists in *the relation of one of the notions contained in the conclusion to a third notion (terminus medius), to which the other notion of the conclusion belongs. Now the relation which the first of these holds to the middle notion, the same must hold to the second, just because the second coincides with the middle notion to the same extent as the first.*

" *Remark.*—'*Relation to*' means only any determinately thought relation, expressed in a judgment . . . . .

"The older logicians adopt, some of them, the principle *Nota notœ est nota rei ipsius,—quod repugnat notœ, repugnat ipsi rei ;*

this, however, is only properly applicable to the first figure. The expression of others is preferable, *Quæcumque conveniunt (vel dissentiunt) in uno tertio, eadem conveniunt (vel dissentiunt) inter se.* Others, in fine, among whom is Wolf, give the Dictum de Omni et Nullo (cf. § 233) as the principle of syllogisms in general; compare *Philosophical Aphorisms* [of Platner], P. i. § 546. All inference takes place according to a universal rule of reason, here only expressed in reference to syllogism, to which, however, some have chosen to give a more mathematical expression;—*If two notions be equal to a third, they are also equal to each other.*

[*N.B.*—Weiss's mistake (§ 231) in supposing that Aristotle "designated the syllogistic moods with words, like his learned followers."]

"§ 231. Categorical Syllogisms, Figure 1.—The first figure concludes by means of a subordination of the minor term in the conclusion under the subject of another judgment.

"§ 233. This takes place under the general principle:—

"1.) *What pertains to all objects contained under a notion, that pertains also to some, and to each individual, of their number.*

"2.) *What belongs to none of the objects contained under a notion, that also does not pertain to some, or to any individual, of their number.*

"These are the celebrated Dicta de Omni and de Nullo,—*Quidquid prædicatur de omni, idem etiam de aliquo,* and, *Quidquid prædicatur de nullo, id nec de aliquo prædicatur.*"

XVII.—FRIES.

Fries, *System der Logik.*

"§ 52. Hitherto we have maintained two views of the Syllogism in connection. The end in view of reasoning is this,—that cases should be subordinated to general rules, and through them become determined. For example, the general law of the mutual attraction of all heavenly bodies has its whole significance, for my knowledge, in this, that there are given individual heavenly bodies, as Sun and Earth, to which I apply it. To enounce these relations, it is, in the first place, necessary that I have a general rule, as Major Proposition (*Obersatz*); in the second, a Minor Proposition (*Untersatz*), which subordinates cases to the

rule, and, finally, a Concluding Proposition, which determines
the cases through the rule. On the other hand, we see that
every Conclusion is an analytico-hypothetic judgment, and this
always flows from the Dictum de Omni et Nullo, inasmuch as
the relation of subordination of particular under universal notions, is the only relation of Reason and Consequent given in the
form of thought itself. Now, if the conclusion, as syllogism,
combines a plurality of judgments in its premises, in this case
the principle of the inference must lie in a connection of the
thoughts,—a connection which is determined by the matter of
these judgments. In the simplest case, when taking into account
only a single syllogism, I thus would recognise in the premises
the relation of subordination between two notions by reference
to the same third notion, and therethrough perceive in the conclusion the relation of these two notions to each other. I know,
for example, that *all men are mortal*, and that *Caius is a man*.
Consequently, through the relation of the notion of *mortality*,
and of my imagination of *Caius*, to the notion *man*, the relation
of *Caius to mortality* is likewise determined:—*Caius is mortal*.
The first of these views is a mere postulate; but in conformity
to the second we are enabled immediately to evolve the general
form of syllogisms, and from this evolution does it then become
manifest that all possible syllogisms satisfy the postulate. We,
therefore, in the first instance, attach ourselves to the second
view. Through this there is determined as follows:—

"1.) Here the determination of one notion is carried over to
another, superordinate or subordinate to itself. To every syllogism there belong three notions, called its *terms* (*termini*). (We
say *notions* (*Begriffe*), because they are, in general, such, and
when individual representations [or images] appear as terms, in
that case there is no inter-commutation possible). A *major term*,
or *superior notion* (*Oberbegriff*), P, is given as the logical determination of a *middle term* or *notion* (*Mittelbegriff*), M, and,
through this, it is positively or negatively stated as the determination of a *minor term* or *notion* (*Unterbegriff*), S.

"2.) If, then, we regard the propositions in which these relations are enounced; there is, firstly, in the *conclusion* (*Schlussatz*),
the minor term, or inferior notion, subordinated to the major
term, or superior notion, (S *is* P). Farther, in one of the pre-

mises, the middle must be connected with the major term or
notion, (M *is* P). This is called the *major proposition (Obersatz).*
In the other, again, the minor is connected with the major term
or notion, (S *is* M); this is called the *minor proposition (Untersatz).*

"The form of every syllogism is therefore :—

Major Proposition	M *is* P.
Minor Proposition,	S *is* M.
Conclusion,	S *is* P.

"In the example given above, *man* is the middle term; *mortality* the major term; and *Caius* the minor term. The syllogism is :—

Major Proposition,	*All men are mortal;*
Minor Proposition,	*Caius is a man;*
Conclusion,	*Caius is mortal.*

"The fundamental relation in all syllogisms is that of the
middle term to the major and minor terms, in other words, that of
the carrying over of a logical determination from one notion to
another, through certain given subordinations. For howbeit the
Dictum de Omni et Nullo, as a common principle of all syllogisms
in the formula,— *What holds good of the universal, holds also good
of the particulars subordinate thereto,* and still more in that other,
—*The attribute of the attribute is also the attribute of the thing
itself,*—is proximately only applicable to the categorical subordination of a representation [or notion] under a notion; still,
however, the law of mental connection is altogether the same in
syllogisms determined by the subordination of consequence under
a reason, [Hypothetic Syllogisms], or of the complement of parts
under a logical whole, [Disjunctive Syllogisms]. The displayed
form is the form of every possible syllogism. In fact, it also coincides with the first requirement that, in the syllogism, a case
should always be determined by a rule, inasmuch as every syllogism proposes a universal premise, in order rigorously to infer its
conclusion. This will be more definitely shown, when we treat of
syllogisms in detail. Only, the declaration, *that the rule is always
the major proposition,* is sometimes at variance with the declara-

tion, *that the major proposition contains the relation of the middle term to the major term.* We must, however, in the first place, always follow the determination of the latter. For every syllogism properly contains the three processes:—1), The subordination of a particular under a universal ; this is the function of the minor proposition, and the relation between the minor and major terms ; 2), Postulate of a logical determination for one of these two ; this is the function of the major proposition, and the relation of the middle to the major term ; 3), The carrying over this determination to that other ; this is the function of the conclusion and the relation of the minor to the major term.

§ 53.—" The subordination of a particular to a universal must, therefore, in every syllogism, be understood wholly in general. Here either a particular may be determined through its superordinated universal, and such an inference from universal to particular we shall call a *syllogism in the first figure;* or there is a universal known through its subordinated particular, and this inference from the particular to the universal is called *a syllogism in the second [third] figure.* If, for example, the subordination is given me,—*All gold is metal;* I can either transfer an attribute of metal, for instance *fusibility,* to the gold, or enounce an attribute of gold, *ductility,* for instance, of some metal. In the first case, I draw a conclusion in the first figure, from the universal to the particular :—

> *All metal is fusible;*
> *All gold is metal;*
> _____
> *All gold is fusible.*

" In the other case, I conclude in the second [third] figure from the particular to the general :—

> *All gold is ductile;*
> *All gold is metal;*
> _____
> *Some metal is ductile.*"

Then, after distribution of the Syllogism into Categorical, Hypothetical, and Divisive, (Disjunctive), he proceeds with the first class.

XVIII.—KIESEWETTER.

Kiesewetter, *Allgemeine Logik*, 1801, 1824. I. Theil.

§ 228.—" All pure Categorical Syllogisms, whose conclusion is an affirmative judgment, rest on the following principle:— *What pertains to the attribute of an object, pertains to the object itself.* All syllogisms, whose conclusion is a negative judgment, are based upon the principle:—*What is repugnant to the attributes of an object, is repugnant to the object itself.* Two principles which can be easily deduced,—the first from the principle of Identity, the second from the principle of Contradiction.

§ 229.—" If we take into consideration that the major proposition of every categorical syllogism must be a universal rule,—from this there flow the following rules:—

" 1. Whatever is universally affirmed of a notion, that is also affirmed of everything contained under it. The *Dictum de Omni.*

" 2. What is universally denied of a notion is denied also of everything contained under it. The *Dictum de Nullo.*

" These rules are also thus expressed :

" What pertains to the genus or species, pertains also to whatever is contained under them. What is repugnant to the genus or species, is repugnant also to whatever is contained under them."

See also the *Writers Auseinandersetzung* on these paragraphs.

XIX.—LARROQUE.

Larroque, *Élémens de Philosophie*, Paris, 1830. *Logique*, Ch. i, p. 202. " The attribute of an affirmative proposition is taken sometimes particularly, sometimes universally. It is taken particularly, when it has a greater extension than the subject; universally, when it has not a greater extension, which occurs in every proposition where the two terms are identical. The reason of this difference is palpable. If the attribute be a term more general than the subject, we affirm that the subject is a species or individual contained in the extension of the attribute:—*Man is mortal; Paul is learned:*—that is, *man* is one, and not the only, species contained in the extension of the term *mortal; Paul* is an individual, and not every individual, contained in the extension of the term *learned.* If, on the contrary, the attribute be not more general than the subject, the attribute is the same thing with the subject, and, consequently, we affirm that the subject is all that is contained in the extension of the attribute:—*A circle is a plane surface, which has all the points, in [a line called] its circumference, at an equal distance from a point called its centre* —that is, a circle is *all* or *every* plane surface, &c.

"The attribute of a negative proposition is always taken universally. When we deny an attribute of a subject, we deny of this subject everything that has the nature of that attribute, that is to say, all the species, as all the individuals, contained in its extension: *The soul is not extended;* to wit the *soul is not any* of the species, *not any* of the individuals, contained in the extension of the term *extended*."

Ch. ii., p. 230. "We have supposed, in the demonstration of these rules [the general rules of the Categorical Syllogism], that the attribute of an affirmative premise is always taken particularly. It would, therefore, seem that the calculations on which this demonstration rests, are erroneous, whensoever the attribute is not a term more general than the subject, for we have seen that, in these cases, the attribute can be taken universally. But it is to be observed, that when the two terms of a proposition are identical, if the one or the other may be taken universally, they cannot both be so taken at once; and that, if it be the attribute which is taken universally, it ought to be substituted for the subject, which then affords a particular attribute. *A triangle is a figure which has three sides and three angles.* We cannot say, '*All* triangle is *all* figure, which,' &c.; but we can say, '*All* triangle is *some* figure, which,' &c.; or, '*All* figure which has three sides and three angles, is *some* triangle.' Now, in adopting either of these last expressions of the proposition, the attribute is particular."

Ch. ii., p. 231. "We have seen that the Syllogism inferred from its premises a proposition to be proved; now this conclusion cannot be inferred from, unless it be contained in, the premises. From this incontestable observation, the author of the Port Royal Logic has endeavoured to draw the following pretended rule, by aid of which we may detect the vice of any fallacious reasoning whatsoever: *Thus, should one of the premises contain the conclu-*

sion, and the other show that it is so contained. A great many treatises on Logic call this *the single rule of the moderns.* This pompous denomination seems to point to some marvellous discovery, of which the ancients had no conception,—at some consummative result of the efforts of the human intellect. It is true, indeed, that a syllogism is invalid, if the conclusion be not contained in the premises; but a fine discovery forsooth! This all the world already knew,—Aristotle among the rest; but he justly noted that it is not always easy to see whether the conclusion be contained in the premises, and it is to assure ourselves of this that he laid down his rules. The pretended rule of Port Royal is, therefore, not one at all; it enounces only an observation, true but barren."

XX.—GALLUPPI.

Galluppi, *Lezioni di Logica e di Metafisica*, 1832. Lez. xlvii., p. 353, ed. 1841.

"In a reasoning there must be an idea, common to the two premises; and a judgment which affirms the identity, either partial or perfect, of the other two ideas."

In the same Lecture, (p. 348), he shows that he is ignorant of the law quoted from the *Philosophia Lugdunensis* being by the authors of the *L'Art de Penser.*

XXI.—BUFFIER.

Buffier, *Première Logique*, about 1725. The following is from the Recapitulation, § 109 :—

The Syllogism is defined, a tissue of three propositions so constituted, that if the two former be true, it is impossible but that the third should be true also. (§ 62.)

The first Proposition is called the *Major;* the second the *Minor;* the third the *Conclusion,* which last is the essential end in view of the syllogism. (§ 65.)

Its art consists in causing a consciousness, that in the conclusion the idea of the *subject* comprises the idea of the *predicate;* and this is done by means of a third idea, called the *Middle Term,* (because it is intermediate between the subject and predicate), in such sort that it is comprised in the subject, and comprises the predicate. (§ 67.)

If the first thing comprise a second, in which a third is comprised, the first comprises the third. If a *fluid* comprise *chocolate*, in which *cocoa* is comprised, the *fluid* itself comprises *cocoa.* (§ 68.)

To reach distant conclusions, there is required a plurality of syllogisms. (§ 71.)

Our rule of itself suffices for all syllogisms,—even for the negative; for every negative syllogism is equivalent to an affirmative. (§ 77.)

Hypothetical syllogisms consist in the enouncement by the major premise, that a proposition is true, in case there be found a certain condition; and the minor premise shows that this condition is actually found. (§ 79.)

Disjunctive syllogisms, to admit of an easy verification, ought to be reduced to hypotheticals. (§ 81.)

Although the single rule, which is proposed for all syllogisms, be subject to certain changes of expression, it is nevertheless always the most easy; in fact, all logical laws necessarily suppose this condition. (§ 87.)

The employment of Grammar is essential for the practice of Logic. (§ 50.)

By means of such practice, which enables us to estimate accurately the value of the terms in every proposition, we shall likewise obtain the rule for the discovery of all sophisms, which consist only of the mere equivocation of words, and of the ambiguity of propositions. (§ 92 *et seq.*)

XXII.—VICTORIN.

Victorin, *Neue natürlichere Darstellung der Logik,* Vienna, 1835.

II. Simple Categorical Syllogisms. § 94. The fundamental rule of all such syllogisms :—

" *In what relation a concept stands to one of two reciprocally subordinate concepts, in the same relation does it stand to the other.*"

§ 94. First Figure; fundamental rule :—" *As a notion determines the higher notion, so does it determine the lower of the same;* " or, " *In what relation a notion stands to one notion, in the same relation it stands to the lower of the same.*"

§ 96. Second Figure; fundamental rule:—" *When two notions are oppositely determined by a third notion, they are also themselves opposed;* " or, " *If two notions stand to a third in opposed relations, they also themselves stand in a relation of opposition.*"

§ 98. Third Figure; fundamental rule:—" *As a notion determines the one of two [to it] subordinate notions, so does it determine the other;* " or, " *In what relation a notion stands to the one of two [to it] subordinate notions, in the same relation stands it also to the other.*"

§ 100. Fourth Figure; fundamental rule:—" *As a notion is determined by the one of two subordinate notions, [two notions in the relation to each other of subordination,] so does it determine the other;* " or, " *In what relation one of two subordinated notions, [notions reciprocally subordinate or superordinate], stands to a third, in the same relation stands it also to the other.*"

(b) Fundamental Laws of Syllogism.—References.

(See Galluppi, *Lezioni di Logica e di Metafisica*, Lez. xlvii., vol. i. p. 345 *et seq.;* Troxler, *Logik*, i. p. 33; Bolzano, *Wissenschaftslehre, Logik*, vol. ii. § 263, p. 543.)

I. Logicians who confound the Nota notæ and the Dictum de Omni, being ignorant of their several significances; making them—

a) Co-ordinate laws without distinction.

Jäger, *Handb. d. Logik*, § 68, (1839). Prochazka, *Gesetzb. f. d. Denken*, § 217, (1842). Calker, *Denklehre*, § 143, (1822). Troxler, *Logik*, ii. p. 40.

b) Derivative; the Dictum de Omni, to wit, from the Nota notæ. This supreme or categorical.

Wenzel, *Elem. Philos. Log.*, §§ 253, 256; *Canonik*, § 64. Kant, *Die falsche Spitzf.*, § 3; *Logik*, § 63. Krug, *Logik*, § 70. Bachmann, *Logik*, § 123. Jakob, *Logik*, § 262, 4th ed. 1800; 1st ed. 1788.

II. Logicians who enounce the law of Identity, (Proportion,) in the same third, by the mathematical expression *Equality.*

Reimarus, *Vernunftlehre*, § 176. Mayer, *Vernunftschlusse*, i. p. 290. Arriaga, *In Sum.*, D. iii. § 3, p. 23.

III. Logicians who make the Dictum de Omni the fundamental rule of syllogisms in general.

Aristot., *An. Prior.*, L. i. c. 1, § 11. Wolf, *Phil. Rat.*, § 353. Scheibler, *Op. P.* iv., *De Syll.* c. ii. § 12. Jac. Thomasius, *Erot. Log.*, c. 395. Buttner, *Cursus Philos., Log.*, § 146. Conimbricenses, *In Arist. Dial.*, L. ii. pp. 240, 243.

IV. Logicians who confound or make co-ordinate the law of Proportion or Analogy, and the Dictum de Omni.

Wyttenbach, *Præc. Philos. Log.*, P. iii. c. 6, § 4. Whately, *Logic*, Intr., Ch. ii. P. iii. § 2. Leechman, *Logic*, P. iii. ch. 2. Kockermann, *Systema Logicæ Minus*, L. iii. c. 2; *Syst. Log. Majus*, L. iii. c. 5.

V. Logicians who make the Law of Identity the one supreme.

Suter, *Logica*, § 61, calls this the principle of Identity and Contradiction. Aldrich, *Comp.*, L. i. c. 3, § 2, p. 2. Hutcheson, *Log. Comp.*, P. iii. c. 2. Arriaga, *Curs. Phil., In Sum.*, D. iii. §§ 16-22, pp. 23, 24. Larroque, *Logique*, p. 224. Mayer, *Vernunftschlusse*, i. p. 203. Troxler, *Logik*, ii. p. 33-40. Reimarus, *Vernunftlehre*, § 176. Mendoza, *Disp. Log. et Met.*, I. p. 470. Derodon, *Log. Rest., De Log.*, p. 639-644. Darjes, *Via, &c.*, § 271, p. 97. Smiglecius, *Logica*, D. xiii. qu. 14, p. 517. Frau. Bonæ Spei, *Com. Prim. in Log. Arist.*, D. vii. d. 2, p. 25. *Cursus Complet., De Arg.*, L. iii. c. 4, p. 57. Alstedius, *Enc., Logica*, § ii. c. 10, p. 435. Havichorst, *Inst. Log.*, § 323. Poncius, *Cursus Philos., In An. Prior.*, D. xx. qu. 5, p. 282.

VI. Logicians who restrict the Dictum de Omni to the first Figure (immediately).

Aldrich, *Comp.* L. i. c. 3, § 7. Noldius, *Log. Rec.*, c. xii. p. 290. Grosser, *Pharus Intellectus*, Sect. I. Pars iii. memb. iii., p. 160.

VII. Logicians who make the Dicta de Omni et de Nullo the supreme canons for Universal Syllogisms; the law of Proportion for Singular Syllogisms.

Burgersdicius, *Inst. Log.*, L. ii. c. 8, p. 171. Melanchthon, *Erot. Dial., De Syll. Expos.*, L. iii. p. 172, ed. 1586. Fonseca, *Instit. Dial.*, L. vi. cc. 21, 24, pp. 363, 373.

VIII. What name given by what logicians to the Law of Proportion, &c.

Law of Proportion, or of Analogy, Keckermann, *Syst. Log.*, L. iii. c. 5, *Op.*, p. 746. Alstedius, *Encycl.*, p. 435, τὸ ἀναλογίας.

Dictum de Omni et Nullo Majus, Noldius, *Log.*, p. 288.
Law of Identity, Zedler's *Lexicon.*
Principium convenientiæ, Darjes, *Via ad Verit.*, § 270, p. 96.
Law of proportional Identity and Non-Identity, Self.

IX. Logicians erroneously supposing Aristotle to employ, besides the Dictum de Omni, the rule of Proportion as a fundamental law of syllogism.

Rapin, *Réflexions sur la Logique*, § 4.

X. Terms under which the law of Proportion has been enounced. *Agree with. Coincide with. The same with. Cohere*, (Syrbius). *Co-exist* (bad). *Co-identical with. Equal to*, (No. ii.) *In combination with*, Darjes, *Via ad Ver.*, p. 97, (includes negative.) *Convertible.*

(c) ENUNCIATIONS OF THE HIGHER LAWS OF SYLLOGISM.

Law of Proportion.

Aristotle, *Elench.*, c. vi. § 8. " Things the same with one and the same, are the same with one another." Compare *Topica*, L. vii. c. I, § 6. Thus Scotus, *In An. Prior.*, L. i. qu. 9, f. 248.

Some say, "Uni tertio *indivisibili* "—some others, " Uni tertio indivisibili, *indivisibiliter sumpto.*" Others, in fine, say, " Uni tertio, *adæquate sumpto.*" See Irenæus, *Intrg. Philos. Log.*, §§ 3, 5. Some express it, " Things that are equal to the same third are equal to each other. See Irenæus, *ib.* So Reimarus, Mayer.

Some express it, " Quæcunque conveniunt (vel dissentiunt) in uno tertio, eadem conveniunt (vel dissentiunt) inter se."

" Quæ duo conveniunt cum uno quodam tertio, eatenus conveniunt inter se; quando autem duorum unum convenit cum tertio, et alterum huic repugnat, repugnant quoque eatenus sibi invicem," Wynpersse, *Inst. Logicæ*, § 272, Lug. Bat. 3d ed. 1806.

Noldius (*Logica*, p. 288), calls these the Dicta de Omni et de Nullo. The former is, " Quæcunque affirmantur in aliquo tertio, (*singulari* identice, *universali* et identice et complete distributive) affirmantur inter se." The latter, "Quorum unum [totaliter] affirmatur in aliquo tertio, alterum negatur, ea inter se neguntur."

Noldius—" Whatever is affirmed essentially of a subject, is affirmed of all that is inferior or reciprocal to that subject. Whatever is denied of a subject, is denied of all inferior or reciprocal."

(See Noldius against the universal application of these Dicta, *Log. Rec.* p. 290).

Reusch, (*Syst. Logicum*, ed. i. 1734, § 503) makes the Dicta de Omni et de Nullo the rule of ordinary syllogisms, and thus enunciates them :—" Si quid prædicatur de omni, illud etiam prædicatur de aliquo: et, Si quid predicatur de nullo, illud etiam non prædicatur de aliquo. Sensus prioris est, Quidquid de genere, vel specie omni prædicari potest, illud etiam prædicatur de quovis sub illo genere, vel sub illa specie, contento ; Item,—Cuicunque competit definitio, illi quoque competit definitum :" (and so *vice versâ* of the other).

Syrbius gives these two rules :—

1) " If certain ideas cohere with a third, they also cohere in the same manner with each other ; "

2) " Ideas which do not cohere with the same third, these do not cohere with each other." (Given in the original by Waldin, *Systema*, p. 162. See also *Acta Eruditorum*, 1718, p. 333.) Syrbius thinks that the law of Proportion, unless limited, is false.

Darjes, *Via ad Veritatem*, (1755), § 270, p. 96, 2d. ed. 1764, "Two [things or notions] in combination with the same third, may be combined together in the same respect (ea ratione), wherein they stood in combination with that third." (See further ; shows that other rules are derived from this.)

Dictum de Omni, &c.

Aristotle, *Anal. Pr.*, L. i., c. i. § 11.

" To be predicated, de Omni, universally is, when we can find nothing under the subject of which the other [that is, the predicate] may not be said ; and to be predicated de Nullo, in like manner."

Jac. Thomasius, *Erotemata Logica*, 1670.

" 40. What do you call the foundation of syllogism ?—The Dictum de Omni et Nullo.

" 41. What is the Dictum de Omni ?—When nothing can be subsumed under the subject of the major proposition of which its predicate may not be affirmed.

" 42. What is the Dictum de Nullo ?—When nothing can be subsumed under the subject of the major proposition of which its predicate is not denied."

Thomasius notices that the first rule applies only to the affirmative moods of the first figure, Barbara and Darii ; the second only to the negative moods of the same figure, Celarent and Ferio.

(d) Objections to the Dictum de Omni et Nullo.

I. As a principle of syllogism in general.

II. As a principle of the First Figure, as enounced by Aristotle.

1°, Only applies to syllogisms in extension.

2°, Does not apply to individual syllogisms; as, *Peter is running; but some man is Peter; therefore, some man is running.* Arriaga, *In Summ.*, p. 24.

3°, Does not apply in co-extensive reasonings; as, *All trilateral is (all) triangular; but all triangular has three angles equal to two right angles; ergo*, &c. Arriaga, *ib.*

Dictum de Omni et Nullo does not apply,

1°, To the other Figures than the First.

2°, Not to all the moods of First Figure, for in many of these the higher class is subjected to the lower.

3°, The form of the First Figure does not depend upon the principle of the Dictum de Omni et Nullo. This imperfect; not upon the thoroughgoing principle, that in this figure one motion is compared to a second, and this second with a third.

(e) General Laws of Syllogism in verse.

(1) Partibus ex puris sequitur nil, (2) sive negatis.
(3) Si qua præit partis, sequitur conclusio partis.
(4) Si qua negata præit, conclusio sitque negata.
(5) Lex generalis erit, medius concludere nescit.[α]
(6) Univocusque; (7) triplex; (8) ac idem terminus esto.[β]

(1) Distribuas medium; (2) nec quartus terminus adsit.
(3) Utraque nec præmissa negans; (4) nec particularis.
(5) Sectetur partem conclusio deteriorem;
(6) Et non distribuat nisi cum præmissa; (7) negetve.[γ]

α Petrus Hispanus, *Summulæ*. [Tr. iv. c. 3, f. 45 b.—Ed.

β Campanella, *Dialect.*, p. 394.

γ Hutcheson, *Log. Comp.* [P. iii. c. 2, p. 53.—Ed.]

(1) Terminus esto triplex : medius, majorque, minorque :
(2) Latius hunc quam præmissæ, conclusio non vult.
(3) Nequaquam medium capiat conclusio oportet.
(4) Aut semel aut iterum medium generaliter esto.
(5) Nil sequitur geminis ex particularibus unquam.
(6) Utraque si præmissa neget, nihil inde sequetur.
(7) Ambæ affirmantes nequeunt generare negantem.
(8) Est parti similis conclusio deteriori.
Pejorem sequitur semper conclusio partem. } a

(1) Terminus est geminus, mediumque accedit utrique.
(2) Præmissis dicat ne finis plura, caveto.
(3) Aut semel, aut iterum medium genus omne capessat ;
(4) Officiique tenax rationem claudere nolit. β

1) Terminus est triplex. (2) Medium conclusio vitet.
(3) Hoc ex præmissis altera distribuat.
(4) Si præmissa simul fuit utraque particularis,
(5) Aut utrinque negans, nulla sequela venit.
(6) Particulare præit ? sequitur conclusio partis.
(7) Ponitur ante negans ? Clausula talis erit.
(8) Quod non præcessit, conclusio nulla requirit. γ
Tum re, tum sensu, triplex modo terminus esto.

{ Argumentari non est ex particulari.
{ Neque negativis recte concludere si vis.
{ Nunquam complecti medium conclusio debet.
{ Quantum præmissæ, referat conclusio solum.
{ Ex falsis falsum verumque aliquando sequetur ;
{ Ex veris possunt nil nisi vera sequi. δ

(f) Special Laws of Syllogism in Verse.

1. Fig. Sit minor affirmans, nec major particularis.
2. Fig. Una negans esto, major vero generalis.
3. Fig. Sit minor affirmans, conclusio particularis.

a Purchot, with variations of Saguy, PA. Legd., Gallupp. [Purchot, Inst. Phil., vol. i., Logica, P. iii. c. 3, p. 171.—Ed.]
β Facciolati, Rudimenta.
γ Isendonra, Logica, L. iii. c. 6, p. 27, 8°. (1652). Chauvin and Walch,

Lex. v. Syllog.
δ Crakanthorpe, Logica, L. iii. c. 16, p. 210. Ubaghs, Logicæ Elementa, § 225. Bancruvius, Dialectica ad Mentem Dort. Subtilis, L. i. c. 3, p. 103. Lond. 1673.

4. Fig. a) Major ubi affirmat, generalem sume minorem.
   b) Si minor affirmat, conclusio sit specialis.
   c) Quando negans modus est, major generalis habetur.

B. CRITICISM.

(a) CRITICISM OF THE SPECIAL LAWS OF SYLLOGISM.

The Special Laws of Syllogism, that is, the rules which govern the several Figures of Categorical Reasonings, all emerge on the suspension of the logical postulate—To be able to state in language what is operative in thought. They all emerge on the refusal or neglect to give to the predicate that quantity in overt expression, which it possesses in the internal operations of mind. The logicians assert, 1°, That in affirmative propositions the predicate must be always presumed particular or indefinite, though in this or that proposition it be known and thought as universal or definite; and, 2°, That in negative propositions this same predicate must be always presumed absolutely, (i.e. universally or definitely), excluded from the sphere of the subject, even though in this or that proposition it be known and thought as partially, (i.e. particularly or indefinitely), included therein. The moment, however, that the said postulate of Logic is obeyed, and we are allowed to quantify the predicate in language, as the predicate is quantified in thought, the special rules of syllogism disappear, the figures are all equalised and reduced to unessential modifications ; and while their moods are multiplied, the doctrine of syllogism itself is carried up to the simplicity of one short canon. Having already shown that the general laws of syllogism are all comprised and expressed in this single canon,[a] it now only remains to point out how, on the exclusive doctrine of the logicians, the special rules became necessary, and how, on the unexclusive doctrine which is now propounded, they become at once superfluous and even erroneous. It is perhaps needless to observe, that the following rules have reference only to the whole of Extension.

The double rule of the First Figure, that is, the figure in which the middle term is subject in the sumption, and predicate in the subsumption, is,—

    " Sit minor affirmans ; nec major particularis."

   [a] See above, p. 290, and below, p. 357.—ED.

Here, in the first place, it is prescribed that the minor premise must be affirmative. The reason is manifest. Because if the minor premise were negative, the major premise behoved to be affirmative. But in this figure, the predicate of the conclusion is the predicate of the major premise; but, if affirmative, the predicate of that premise, on the doctrine of the logicians, is presumed particular, and as the conclusion following the minor premise is necessarily negative, a negative proposition thus, contrary to logical law, has a particular predicate. But if we allow a negative proposition to have in language, as it may have in thought, a particular or indefinite predicate, the rule is superseded.

The second rule, or second part of the rule, of this First Figure, is, that the major premise should be universal. The reason of this is equally apparent. For we have seen, that, by the previous rule, the minor premise could not be negative, in which case certainly, had it been allowable, the middle term would, as predicate, have been distributed. But whilst it behoved that the middle term should be once at least distributed, (or taken universally), and, as being the subject of the major premise, it could only be distributed in a universal proposition, the rule, on the hypothesis of the logicians, was compulsory. But as we have seen that the former rule is, on our broader ground, inept, and that the middle term may be universally quantified, as the predicate either of an affirmative or negative subsumption, it is equally manifest that this rule is, in like manner, redundant, and even false.

In the Second Figure, that is, the figure in which the middle term is predicate both in sumption and subsumption, the special rule is,—

"*Una negans esto ; major vero generalis.*"

In regard to the first rule, or first half of the rule,—That one or other of the premises should be negative,—the reason is manifest. For, on the doctrine of the logicians, the predicate of an affirmative proposition is always presumed to be particular; consequently, in this figure the middle term can, on their doctrine, only be distributed, (as distributed at least once it must be), in a negative judgment. But, on our doctrine, on which the predicate is quantified in language as in thought, this rule is abolished.[a]

a [For examples from Aristotle of Figure, see *De Cœlo*, L. ii. c. 4, § 4, affirmative conclusions in the Second text 23, *ibi* Averrœs. *Phys.*, L. ii. c.

The second rule, or second moiety of the rule,—That the sumption should be always universal,—the reason of this is equally clear. For the logicians, not considering that both extremes were in equilibrio in the same whole of extension, and, consequently, that neither could claim [in either quantity] the place of major or minor term, and thereby constitute a true major or a true minor premise;—the logicians, I say, arbitrarily drew one instead of two direct conclusions, and gave the name of major term to that extreme which formed the predicate in that one conclusion, and the name of major premise to that antecedent proposition which they chose to enounce first. On their doctrine, therefore, the conclusion and one of the premises being always negative, it behoved the sumption to be always general, otherwise, contrary to their doctrine, a negative proposition might have a particular predicate. On our doctrine, however, this difficulty does not exist, and the rule is, consequently, superseded.

In the Third Figure, that is, the figure in which the middle term is subject of both the extremes, the special rule is,—

"*Sit minor affirmans; conclusio particularis.*"

Here the first half of the rule,—that the minor must not be negative,—is manifestly determined by the common doctrine. For, (major and minor terms, major and minor propositions, being in this figure equally arbitrary as in the second), here the sumption behoving to be affirmative, its predicate, constituting the major term or predicate of the conclusion, behoved to be particular also. But the conclusion, following the minor premise, would necessarily be negative; and it would have,—what a negative proposition is not allowed on the common doctrine,—an undistributed predicate.

The second half of the rule,—That the conclusion must be particular,—is determined by the doctrine of the logicians, that the particular antecedent, which they choose to call the minor term, should be affirmative. For, in this case, the middle term being the subject of both premises, the predicate of the subsump-

2, § 12, text 23, *ibi* Averroes; c. 4, § 8, text 33, *ibi* Averroes; c. 7, § 1, text 42, *ibi* Averroes. *An. Post.*, L. I. c. 12, § 12, text 92, *ibi* Averroes et Pacius. Argoes himself, like Censeus, from two affirmative propositions in Second Figure, and does not give the reason why the inference is good or bad in such syllogisms. Cf. Ammonius and Philoponus *ad loc; An. Prior.*, L. ii. c. 22, §§ 7, 8; *An. Post.*, L. I. c. 6, § 1, *et ibi* Themistius, Pacius, Zabarella. Cf. also Zabarella, *De Quarto Fig. Syll.*, c. x.]

tion is the minor extreme; and that, on their doctrine, not being distributed in an affirmative proposition, it, consequently, forms the undistributed subject of the conclusion. The conclusion, therefore, having a particular subject, is, on the common doctrine, a particular proposition. But as, on our doctrine, the predicate of an affirmative proposition may have an universal quantification, the reason fails.

(b) LAWS OF SECOND FIGURE—ADDITIONAL.[a]

By designating the quantity of the predicate, we can have the middle term, (which in this figure is always a predicate), distributed in an affirmative proposition. Thus :—

All P is all M;
All S is some M;
Therefore, all S is some P.

All the things that are organised are all the things that are endowed with life ;
But all plants are some things endowed with life ;
Therefore, all plants are some things organised.

This first rule (see above, Vol. I. p. 408) must, therefore, be thus amplified :—The middle term must be of definite quantity, in one premise at least, that is, it must either, 1°, Be a singular,—individual,—concept, and, therefore, identical in both premises; or, 2°, A universal notion presumptively distributed by negation in a single premise; or, 3°, A universal notion expressly distributed by designation in one or both premises.

But the second rule, which has come down from Aristotle, and is adopted into every system of Logic, with only one exception, an ancient scholiast, is altogether erroneous. For, 1°, There is properly no sumption and subsumption in this figure; for the premises contain quantities which do not stand to each other in any reciprocal relation of greater or less. Each premise may, therefore, stand first. The rule ought to be, "One premise must be definite ;" but such a rule would be idle ; for what is here given

a What follows to page 357 was an early written interpolation by the Author in *Lectures* (vol. i. p. 408), being an application of the principle of a quantified predicate to syllogism. The interpolation appears in students' notes of the Lectures of session 1841-42 ; and was probably given still earlier.—ED.

as a special canon of this figure, was already given as one of the laws of syllogism in general. 2°, The error in the principle is supported by an error in the illustration. In both the syllogisms given,[a] the conclusion drawn is not that which the premises warrant. Take the first or affirmative example. The conclusion here ought to have been, "No S is some P," or "Some P is no S;" for there are always two equivalent conclusions in this figure. In the concrete example, the legitimate conclusions, as necessitated by the premises, are,—" No horse is some animal," and, "Some animal is no horse." This is shown by my mode of explicating the quantity of the predicate,—combined with my symbolical notation. In like manner, in the second or negative syllogism, the conclusion ought to have been either of the two following : In the abstract formula,—" All S are not some P ;" or " Some P are not all S ;"—in the concrete example, " All topazes are not some minerals," i. e., " No topazes are some minerals ;" or, "Some minerals are not all topazes," i. e., "Some minerals are no topazes."

The moods Cesare and Camestres may be viewed as really one, for they are only the same syllogism, with premises placed first or second, as is always allowable in this [Figure], and one of the two conclusions, which are always legitimately consequential, assigned to each.

A syllogism in the mood Festino, admits of either premise being placed first ; it ought, therefore, to have had another mood for its pendant, with the affirmative premise first, the negative premise second, if we are to distinguish moods in this figure by the accidental arrangement of the premises. But this was prohibited by the second Law of this Figure,—that the Sumption must always be universal. Let us try this rule in the formula of Festino now stated, reversing the premises.

Some S are M ; (i. e., some M.)
No P is M ;

{ No P is some S. }
{ Some S are no P. }
Some actions are praiseworthy ;
No vice is praiseworthy ;

{ No vice is some action. }
{ Some action is no vice. }

a See above, vol. i. p. 409.—Ed.

From what I have now said, it will be seen that the Dictum de Omni et de Nullo cannot afford the principle of the Second Figure.

The same errors of the logicians, on which I have already commented, in supposing that the sumption or major premise in this figure must always be universal,—an error founded on another error, that there is, (properly speaking), either sumption or subsumption in this figure at all,—this error, I say, has prevented them recognising a mood corresponding to Baroco, the first premise being a particular negative, the second a universal affirmative, i. e., Baroco with its premises reversed. That this is competent is seen from the example of Baroco now given. Reversing it we have:

| [Some á are not D ; | Some animals are not (any) oviparous ; |
| All a are B. | All Birds are (some) oviparous. |

| No a is some á ; | No bird is some animal ; |
| Some á are no a.] | Some animal is no bird. |

(c) AUTHOR'S SUPREME CANONS OF CATEGORICAL SYLLOGISMS.

[The supreme Canon or Canons of the Categorical Syllogism, finally adopted by Sir W. Hamilton, are as follows:—]

I. " For the Unfigured Syllogism, or that in which the terms compared do not stand to each other in the reciprocal relation of subject and predicate, being, in the same proposition, either both subjects or (possibly) both predicates,—the canon is :—*In so far as two notions, (notions proper, or individuals), either both agree, or one agreeing, the other does not, with a common third notion ; in so far, these notions do or do not agree with each other.*"

II. " For the Figured Syllogism, in which the terms compared are severally subject and predicate, consequently, in reference to each other, containing and contained in the counter wholes of Intension and Extension ;—the canon is :—*What worse relation of subject and predicate subsists between either of two terms and a common third term, with which one, at least, is positively related ; that relation subsists between the two terms themselves.*

" Each Figure has its own Canon.

"First Figure ;—*What worse relation of determining, (predicate), and of determined, (subject), is held by either of two notions to a third, with which one at least is positively related ; that rela-*

*tion do they immediately, (directly), hold to each other, and indi-
rectly, (mediately), its converse.*

"Second Figure ;— *What worse relation of determined, (subject),
is held by either of two notions to a third, with which one at least
is positively related ; that relation do they hold indifferently to
each other.*

"Third Figure ;— *What worse relation of determining (predicate), is held by either of two notions to a third, with which one at
least is positively related ; that relation do they hold indifferently
to each other."* [a]

(d) ULTRA-TOTAL QUANTIFICATION OF MIDDLE TERM.
(1.) LAMBERT'S DOCTRINE.

Lambert, *Neues Organon.*

Dianoiologie, § 193.　"If it be indetermined how far A does,
or does not, coincide with B, but on the other hand we know that
A and B, severally, make up *more than half,*[b] the individuals
under C, in that case it is manifest, that a [linear] notation is
possible, and that of the two following kinds :—

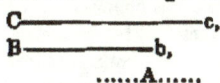

$$
\begin{array}{l}
\text{C}\!\!-\!\!-\!\!-\!\!-\!\!-\!\!-\!\!-\text{c,} \\
\text{B}\!\!-\!\!-\!\!-\!\!-\!\!-\text{b,} \\
\quad\ldots\ldots\text{A}\ldots\ldots
\end{array}
$$

For since B and A are each *greater than the half* of C, A is
consequently greater than C less by B ; and in this case it is of
necessity that some A are B, and some B are A.[γ] We may
accordingly so delineate :—

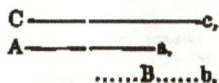

$$
\begin{array}{l}
\text{C}\!\!-\!\!-\!\!-\!\!-\!\!-\!\!-\!\!-\text{c,} \\
\text{A}\!\!-\!\!-\!\!-\!\!-\!\!-\!\!-\text{a,} \\
\quad\ldots\ldots\text{B}\ldots\ldots\text{b,}
\end{array}
$$

seeing that it is indifferent, whether we commence with A or with
B. I may add, that the case which we have here considered does
not frequently occur, inasmuch as *the comparative extension of*

---

[a] *Discussions,* pp. 654, 655.—ED.

[b] It is enough if either A or B exceed the half ; the other need be only
half. This, which Lambert here and
hereafter overlooks, I have elsewhere
had occasion to show.

[γ] In the original for A there is, by
a typographical erratum, C. See *Ph.,*
§ 203.

our several notions is a relation which remains wholly unknown.[a]
I, consequently, adduce this only as an example, that a legitimate
employment may certainly be made of these relations."

Phänomenologie, § v. *Of the Probable—*

" § 188. In so far as such propositions are particular, they may,
like all other particular propositions, be syllogistically employed ;
but no farther, unless we look to their *degree of particularity,* or
other *proximate determination,* some examples of which we have
adduced in the *Dianoiologie,* (§ 235 *et seq.*) Thus the degree of
particularity may render a syllogism valid, which, without this,
would be incompetent. For example—

> *Three-fourths of A are B ;*
> *Two-thirds of A are C ;*
> *Therefore, some C are B.*

The inference here follows, because three-fourths added to two-
thirds are greater than unity ; and, consequently, there must be,
at least, five-twelfths of A, which are at once B and C.

" § 204. In the Third Figure we have the middle term, subject
in both premises, and the conclusion, particular. If now, the
subjects of the two premises be furnished with fractions [*i.e.* the
middle term on both sides], both premises remain, indeed, particular, and the conclusion, consequently, indetermined. But, inasmuch as, in both premises, the degree of particularity is determined,
there are cases where the conclusion may be drawn not only with
probability, but with certainty. Such a case we have already
adduced, (§ 188). For, if both premises be affirmative, and the

---

[a] In reference to this statement, see
above, *Dian.*, § 179, and below, *Ph.*, §
157, where it is repeated and confirmed.
Lambert might have added, that as
we rarely can employ this relation of
the comparative extension of our notions, it is still more rarely of any import that we *should.* For in the two
abstract, or notional, wholes, the
two wholes correlative and counter to
each other, with which Logic is always
conversant, (the Universal and Formal),
if the extension be not complete, it
is of no consequence to note its comparative amount. For Logic and Philosophy tend always to an unexclusive
generality ; and a general conclusion
is invalidated equally by a single ad-
verse instance, as by a thousand. It is
only in the concrete or real whole,—
the whole quantitative or integrate,
and, whether continuous or discrete,
the whole in which mathematics are
exclusively conversant, but Logic and
Philosophy little interested,—that this
relation is of any avail or significance.

sum of the fractions with which their subjects are furnished greater
than unity, in that case a conclusion may be drawn.   In this sort
we infer with certainty :—

> *Three-fourths of* A *are* B ;
> *Two-thirds of* A *are* C ;
> *Therefore, some* C *are* B.

"§ 205. If, however, the sum of the two fractions be less than
unity, as—

> *One-fourth of* A *are* B ;
> *One-third of* A *are* C,

in that case there is no certainty in any affirmative conclusion,
[indeed in any conclusion at all].   But if we state the premises
thus determinately,—

> *Three-fourths of* A *are not* B ;
> *Two-thirds of* A *are not* C ;

in that case, a negative conclusion may be drawn.   For, from the
propositions,—

> *Three-fourths of* A *are not* B ;
> *One-third of* A *are* C ;

there follows—*Some* C *are not* B.   And this, again, because the
sum of the two fractions, (three-fourths added to one-third,) is
greater than unity."   And so on ; see the remainder of this section and those following, till § 211.

(2.) AUTHOR'S DOCTRINE.

Aristotle, followed by the logicians, did not introduce into his
doctrine of syllogism any quantification between the absolutely
universal and the merely particular predesignations, for valid reasons.—1°, Such quantifications were of no value or application in
the one whole (the universal, potential, logical), or, as I would
amplify it, in the two correlative and counter wholes, (the logical,—
and the formal, actual, metaphysical,) with which Logic is conversant.   For all that is out of classification, all that has no reference to genus and species, is out of Logic, indeed out of Philosophy ; for Philosophy tends always to the universal and necessary.
Thus the highest canons of deductive reasoning, the *dicta de*

Omni et de Nullo, were founded on, and for, the procedure from
the universal whole to the subject parts ; whilst, conversely, the
principle of inductive reasoning was established on, and for, the
(real or presumed)collection of all the subject parts as constituting
the universal whole.—2°, The integrate or mathematical whole, on
the contrary, (whether continuous or discrete), the philosophers
contemned.   For whilst, as Aristotle observes, in mathematics
genus and species are of no account, it is, almost exclusively, in the
mathematical whole, that quantities are compared together, through
a middle term, in neither premise equal to the whole.   But this
reasoning, in which the middle term is never universal, and the
conclusion always particular, is—as vague, partial, and contingent
—of little or no value in philosophy.   It was accordingly ignored
in Logic ; and the predesignations *more, most,* &c., as I have said,
referred to universal, or (as was most common) to particular, or to
neither, quantity.*   This discrepancy among logicians long ago
attracted my attention ; and I saw, at once, that the possibility of
inference, considered absolutely, depended exclusively on the quantifications of the middle term, in both premises, being, together,
more than its possible totality—its distribution, in any one.   At
the same time I was impressed—1°, With the almost utter inutility
of such reasoning, in a philosophical relation; and, 2°, Alarmed with
the load of valid moods which its recognition in Logic would introduce.   The mere quantification of the predicate, under the two
pure quantities of *definite* and *indefinite*, and the two qualities of
affirmative and *negative*, gives (abstractly) in each figure, *thirty-
six* valid moods ; which, (if my present calculation be correct),
would be multiplied, by the introduction of the two hybrid or ambiguous quantifications of *a majority* and *a half*, to the fearful
amount of *four hundred and eighty* valid moods for each figure.
Though not, at the time, fully aware of the strength of these objections, they however prevented me from breaking down the old
limitation ; but as my supreme canon of Syllogism proceeds on the
mere formal possibility of reasoning, it of course comprehends all
the legitimate forms of quantification.   It is ;—*What worst relation of subject and predicate, subsists between either of two terms*

a [Cf. Corvinus, *Instit. Phil.*, c. v. §   Wallis, *Instit. Log.*, L. II. c. 4, p. 100.
370, p. 121.   Jena, 1742.   Reusch,   5th ed.—Ed.]
Wallis.]   [Reusch, *Syst. Log.*, § 360.

and a common third term, with which one, at least, is positively related;—that relation subsists between the two terms themselves: in other words;—In as far as two notions both agree, or, one agreeing, the other disagrees, with a common third notion;—in so far, those notions agree or disagree with each other. This canon applies, and proximately, to all categorical syllogisms,—in extension and comprehension,—affirmative and negative,—and of any figure. It determines all the varieties of such syllogisms; is developed into all their general, and supersedes all their special, laws. In short, without violating this canon, no categorical reasoning can, formally, be wrong. Now this canon supposes, that the two extremes are compared together through the *same common middle;* and this cannot but be, if the middle, whether subject or predicate, in both its quantifications together, exceed its totality, though not taken in that totality in either premise.

But, as I have stated, I was moved to the reconsideration of this whole matter; and it may have been Mr De Morgan's syllogism in our correspondence, (p. 19), which gave the suggestion. The result was the opinion, that these two quantifications should be taken into account by Logic, as authentic forms, but then relegated as of little use in practice, and cumbering the science with a superfluous mass of moods.[a]

Author's Doctrine—*continued.*

No syllogism can be formally wrong in which, (1°), Both premises are not negative; and, (2°), The quantifications of the middle term, whether as subject or predicate, taken together, exceed the quantity of that term taken in its whole extent. In the former case, the extremes are not compared together; in the latter, they are not necessarily compared through the same third. These two simple rules, (and they both flow from the one supreme law), being obeyed, no syllogism can be bad; let its extremes stand in any relation to each other as major and minor, or in any relation to the middle term. In other words, its premises may hold any mutual subordination, and may be of any Figure.

On my doctrine, Figure being only an unessential circumstance,

---

[a] Extract from *A Letter to A. de   ton, p. 41.*—Ed.
Morgan, Esq., from Sir W. Hamil-

and every proposition being only an equation of its terms, we may discount Figure, &c., altogether; and instead of the symbol (———) marking subject and predicate, we might use the algebraical sign of equality (=).

The rule of the logicians, that the middle term should be once at least distributed, [or indistributable], (i. e., taken universally or singularly = definitely), is untrue. For it is sufficient if, in both the premises together, its quantification be more than its quantity as a whole, (Ultratotal). Therefore, a *major part*, (a *more* or *most*), in one premise, and a *half* in the other, are sufficient to make it effective. It is enough for a valid syllogism, that the two extreme notions should, (or should not), of necessity, partially coincide in the third or middle notion; and this is necessarily shown to be the case, if the one extreme coincide with the middle, to the extent of a half, (Dimidiate Quantification); and the other, to the extent of aught more than a half, (Ultradimidiate Quantification). The first and highest quantification of the middle term (:) is sufficient, not only in combination with itself, but with any of all the three inferior. The second (.,) suffices in combination with the highest, with itself, and with the third, but not with the lowest. The third (.) suffices in combination with either of the higher, but not with itself, far less with the lowest. The fourth and lowest (,) suffices only in combination with the highest. [1. Definite; 2. Indefinito-definite; 3. Semi-definite; 4. Indefinite.]

(1st *March* 1847.—Very carefully authenticated.)

There are four quantities (, | . | ., | :), affording (4 x 4) 16 possible double quantifications of the middle term of a syllogism.

Of these 10 are legitimate equivalents, (: $\overset{\frown}{M}$: | : $\overset{\frown}{M ., | ., M}$.

: $\overset{\frown}{M . | . M}$: | : $\overset{\frown}{M, | , M}$: | ., $\overset{\frown}{M ., | ., M . | . M .,}$); and 6 illegitimate, as not, together, necessarily exceeding the quantity of that term, taken once in its full extent (., M, | , M., | . M . | . M, | , M . | , M,).

Each of these 16 quantified middle terms affords 64 possible moods; to wit, 16 affirmative, 48 negative; legitimate and illegitimate.

Altogether, these 16 middle terms thus give 256 affirmative and 768 negative moods; which, added together, make up 1024 moods, legitimate and illegitimate, for each figure. For all three figures = 3072.

The 10 legitimate quantifications of the middle term afford, of legitimate moods, 160 affirmative and 320 negative (=480) i. e. each 16 affirmative and 32 negative moods, (=48); besides of illegitimate moods, from double negation, 160, i. e., each 16. The 6 illegitimate quantifications afford, of affirmative moods, 96; of simple negative moods, 192; of double negative moods, 96 (=384). Adding all the illegitimates = 544.

The 1024 moods, in each figure, thus afford, of legitimate, 480 moods, (1440 for all 3 Figs.); being of affirmative 160 (480 for 3 Figs.), of negative 320 (960 for 3 Figs.), of illegitimate 544 moods; there being excluded in each, from inadequate distribution alone, (§), 288 moods, (viz. 96 affirmative, 192 negative); from double negation alone, (‡), 160 moods; from inadequate distribution and double negation together, (§ ‡), 96 moods.

(3.) MNEMONIC VERSES.

A it affirms of *this, these, all*—
　Whilst E denies of *any*:
I, it affirms, whilst O denies,
　Of *some* (or few or many).

Thus A affirms, as E denies,
　And definitely either:
Thus I affirms, as O denies,
　And definitely neither.

A *half*, left semi-definite,
　Is worthy of its score;
U, then, affirms, as Y denies,
　This, neither less nor more.

Indefinito-definites,
　To UI and YO we come;
And that affirms, and this denies,
　Of *more, most*, (half plus some.)

UI and YO may be called Indefinito-definite, either, (1°), Because they approximate to the whole or definite, [forming] more than its moiety, or, (2°), Because they include a half, which, in a certain sense, may be regarded as definite, and something, indefinite, over and above.

VIII.

INDUCTION AND EXAMPLE.

(See above, Vol. I. p. 318.)

(a) QUOTATIONS FROM AUTHORS.

I.—ARISTOTLE.

Aristotle, *Anal. Pri*, L. ii. c. 23. After stating that "we believe all things either through [Deductive] Syllogism or from Induction," he goes on to expound the nature of this latter process.

"Now, Induction, and the Syllogism from Induction, is the inferring one extreme [the major] of the middle through the other; if, for instance, B is the middle of A C, and, through C, we show that A inheres in B. Thus do we institute Inductions. In illustration:—Let A be *long-lived*, B, *wanting-bile*, and C, *individual long-lived animals, as man, horse, mule, &c.* A, then, inheres in the whole of C, (for all animal *without bile* is [at least some] *long-lived*); but B, *wanting bile*, also [partially, at least,] inheres in all C.[a] If now C reciprocate with B, and do not go

---

[a] I have, however, doubts whether the example which now stands in the Organon, be that which Aristotle himself proposed. It appears, at least, to have been considerably modified, probably to bring it nearer to what was subsequently supposed to be the truth. This I infer, as likely, from the Commentary of Ammonius on the *Prior Analytics*, occasionally interpolated by, and thus erroneously quoted under the name of, a posterior critic.—Joannes, surnamed Philoponus, &c. His words are, in reference to Aristotle, as follows:—"He wishes, through an example, to illustrate the Inductive process; it is of this intent. Let A be *long-lived*; B, *wanting bile*; C, as *crow, and the like*. Now he says:—that the *crow and the stag* being animals without bile and long-lived; therefore, animal wanting bile is long-lived. Thus, through the last [or minor], do we connect the middle term with the [major] extreme. For I argue thus:—The individual animals wanting bile are [all] long-lived; consequently, [all] animals wanting bile are long-lived." F. 107 a,

beyond that middle, [if C and B, subject and predicate, are each all the other], it is of necessity that A [some, at least,] should inhere in [all] D. For it has been previously shown,[a] that if any two [notions] inhere in the same [middle notion], and if the middle[b] reciprocate with either [or with both] ; then will the

ed. Ald. Compare also the greatly later Leo Magentinus, on the *Prior Analytics*, f. 41 a, ed. Ald. On the age of Magentinus, historians, (as Saxius and Fabricius,) vary, from the seventh century to the fourteenth. He was certainly subsequent to Michael Psellus, junior, whom he quotes, and, therefore, not before the end of the eleventh century ; whilst his ignorance of the doctrine of Conversion, introduced by Boethius, may show that he could hardly have been so recent as the fourteenth.

Aristotle, *De Part. Animal.* (L. iv. c. 2), says :—"In some animals the gall [bladder] is absolutely wanting, as in the horse, mule, ass, stag, and roe. . . . It is, therefore, evident that the gall serves no useful purpose, but is a mere excretion. Wherefore those of the ancients say well, who declare that the cause of longevity is the absence of the gall ; and this from their observation of the solidungula and deer, for animals of these classes want the gall, and are long-lived." *Hist. An.*, L. ii. c. 11 Sehn., 18 Scal., 15 vul. Notices that some animals have, others want, the gall-bladder (χολή, v. Schn. iii. p. 106) at the liver. Of the latter, among viviparous quadrupeds, he notices stag, roe, horse, mule, ass, &c. Of birds who have the gall-bladder apart from the liver and attached to the intestines, he notices the pigeon, crow, &c.

a Aristotle refers to the chapter immediately preceding, which treats of the Reciprocation of Terms, and in that to the fifth rule which he gives, and of the following purport. "Again, when A and B inhere in all C [i. e. all C is A and is B], and when C reciprocates [i. e. is of the same extension and com-

prehension] with B, it is necessary that A should inhere in all B [i.e. that all B should be A]."

b For ἄκρον, I read μέσον ; but perhaps the true lection is—πρὸς τοῦτο θάτερον αὐτῶν ἀντιστρέφῃ τῶν ἄκρων. The necessity of an emendation becomes manifest from the slightest consideration of the context. In fact, the common reading yields only nonsense ; and this on sundry grounds.—1°, There are *three* things to which θάτερον is here applicable, and yet it can only apply to *two*. But if limited, as limited it must be, to the two inherents, two absurdities emerge.—2° For the middle, or common notion, in which both the others inhere, that, in fact, *here exclusively wanted, is alone excluded.*—3°, One, too, of the inherents is made to reciprocate with either ; that is, *with itself*, or other.—4°, Of the two inherents, the *minor* extreme is that which, on Aristotle's doctrine of Induction, is alone considered as reciprocating with the middle or common term. But, in Aristotle's language, τὸ ἄκρον, "The Extreme," is (like ἡ πρότασις, *The Proposition* in the common language of the logicians) a synonym for the *major* term, in opposition to, and in exclusion of, the minor, term. In the two short correlative chapters, the present and that which immediately follows, on Induction and on Example, the expression, besides the instance in question, occurs at least seven times ; and in all as the *major* term.—5°, The emendation is required by the demonstration itself, to which Aristotle refers. It is found in the chapter immediately preceding, (§ 5) ; and is as follows :—"Again, when A and B inhere in all C ; *and when C reciprocates with* other of the predicates, [the syllogism being in the third figure,] inhere in the co-reciprocating extreme. But it behoves us to conceive C as a complement of the *whole* individuals; for Induction has its inference *through* [as it is of] all.[a]

" This kind of syllogism is of the primary and immediate proposition. For the reasoning of things mediate is, through their medium; of things immediate, through Induction. And in a certain sort, Induction is opposed to the [Deductive] Syllogism. For the latter, through the middle term, proves the [major] extreme of the third [or minor]; whereas the former, through the third [or minor term, proves] the [major] extreme of the middle. Thus, [absolutely] in nature, the syllogism through a medium is the prior and more notorious; but [relatively] to us, that through Induction is the clearer."

An. Pr., L. ii. c. 24. Of example.—§ 1. " Example emerges, when it is shown that the [major] extreme inheres in the middle, by something similar to the third [or minor term]" . . . § 4. " Thus it is manifest that the Example does not hold the relation either of a whole to part [Deduction], or of a part to whole [Induction], but of part to part; when both are contained under the same, and one is more manifest than the other." § 5. " And [Example] differs from Induction, in that this, from all the individuals, shows that the [major] extreme inheres in the middle, and does not [like Deduction] hang the syllogism on the major extreme; whereas that both hangs the syllogism [on the major extreme], and does not show from all the individuals [that the major extreme is inherent in the minor.]"

An. Post., L. i. c. 1, § 3.—" The same holds true in the case of reasonings, whether through [Deductive] Syllogisms or through Induction; for both accomplish the instruction they afford from information foreknown, the former receiving it as it ware from the tradition of the intelligent, the latter manifesting the uni-

---

B; It necessarily follows that A should [partially, at least,] inhere in all B. For whilst A [some, at least,] inheres in all C; and [all] C, by reason of their reciprocity, inheres in [all] B; A will also [some, at least,] inhere in all B." The mood here given is viii. of our Table. (See below, Appendix XII.)

[a] This requisite of Logical Induction, —that it should be thought as the result of an agreement of all the individuals or parts,—is further shown by Aristotle in the chapter immediately following, in which he treats the reasoning from Example. [See passage quoted on this page (§ 5).—ED.]

versal through the light of the individual." (Pacii, p. 413. See the rest of the chapter).

An. Post., L. i. c. 18, § 1.—" But it is manifest that, if any sense be wanting, some relative science should be wanting likewise, this it being now impossible for us to apprehend. For we learn everything either by induction or by demonstration. Now, demonstration is from universals, and induction from particulars; but it is impossible to speculate the universal unless through induction, seeing that even the products of abstraction will become known to us by induction."

A. Aristotle's Errors regarding Induction.

Not making Syllogism and its theory superior and common to both Deductive and Inductive reasonings.

A corollary of the preceding is the reduction of the genus Syllogism to its species Deductive Syllogism, and the consequent contortion of Induction to Deduction.

B. Omissions.

Omission of negatives.

Of both terms reciprocating.

C. Ambiguities.

Confusion of Individuals and Particular. See Scheibler, [*Opera Logica*, P. iii. *De Prop.*, c. vi. tit. 3, 5.—Ed.]

Confusion or non-distinction of Major or Minor extremes.

———

The subsequent observations are intended only to show out Aristotle's authentic opinion, which I hold to be substantially the true doctrine of Induction; to expose the multiform errors of his expositors, and their tenth and ten times tenth repeaters, would be at once a tedious, superfluous, and invidious labour.— I shall, first of all, give articulately the correlative syllogisms of Induction and Deduction which Aristotle had in his eye; and shall employ the example which now stands in the *Organon*, for, though physiologically false, it is, nevertheless, (as a supposition,) valid, in illustration of the logical process.

ARISTOTLE'S CORRELATIVE SYLLOGISMS.

(a) OF INDUCTION.

All C (man, horse, mule, &c.) is
some A (long-lived);

All C (man, horse, mule, &c.) is
all B (wanting-bile);

All B (wanting-bile) is some A
(long-lived).

A , —— : C (p, q, r, &c.) : ——: B

(b) OF DEDUCTION.

All B (wanting-bile) is some A
(long-lived);

All C (man, horse, mule, &c.) is
all B (wanting-bile);

All C (man, horse, mule, &c.) is
some A (long-lived).

A , —— : B : ——: C (p, q, r, &c.)

These syllogisms, though of different figures, fall in the same
mood; in our table they are of the eighth mood of the third and
first Figures. Both unallowed. (See Ramus, quoted below, p. 370.)

The Inductive syllogism in the first figure given by Schegkius,
Pacius, the Jesuits of Coimbra, and a host of subsequent repeaters, is altogether incompetent, in so far as meant for Aristotle's
correlative to his Inductive syllogism in the third. Neither
directly nor indirectly does the philosopher refer to any Inductive reasoning in any other figure than the third. And he is
right; for the third is the figure in which all the inferences of
Induction naturally run. To reduce such reasonings to the first
figure, far more to the second, is felt as a contortion; as will be
found from the two following instances, the one of which is
Aristotle's example of Induction, reduced by Pacius to the first
figure, and the other the same example reduced by me to the
second. I have taken care also to state articulately what are
distinctly thought,—the quantifications of the predicate in this
reasoning; ignored by Pacius and logicians in general, and admitted only on compulsion, among others, by Derodon, (below,
p. 371), and the Coimbra commentator.[a]

ARISTOTLE'S INDUCTIVE SYLLOGISM IN FIGURES.

(c) Fig. I.

All C (man, horse, mule, &c.) is
some A (long-lived);

All B (wanting-bile) is all C
(man, horse, mule, &c.);

All B (wanting-bile) is some A
(long-lived).

(d) Fig. II.

Some A (long-lived) is all C
(man, horse, mule, &c.);

All B (wanting-bile) is all C
(man, horse, mule, &c.);

All B (wanting-bile) is some A
(long-lived).

a [In An. Prior., L. II. p. 403. Cf.
Pacius, Dialectica, L. iii. p. 256
(1544). Timpl. Comp. Phil., Logica,
T. I. L. iii. c. 1, p. 118.]

IL.—PACHYMERES.

Pachymeres, *Epitome of Aristotle's Logic*, (Title viii. ch. 3, c. 1280).—" Induction, too, is celebrated as another instrument of philosophy. It is more persuasive than Deductive reasoning; for it proposes to infer the universal from singulars, and, if possible, from all. But as this is frequently impossible, individuals being often in number infinite, there has been found a method through which we may accomplish an Induction, from the observation even of a few. For, after enumerating as many as we can, we are entitled to call on our adversary to state on his part, and to prove, any opposing instances. Should he do this, then [for ' data instantia, cadit inductio '] he prevails; but should he not, then do we succeed in our Induction.—But Induction is brought to bear in the third figure; for in this figure is it originally cast. Should, then, the minor premise be converted, so that the middle be now predicated of all the minor extreme, as that extreme was predicated of all the middle; in that case, the conclusion will be, not of *some*, but of *all*. [In Induction] the first figure, therefore, arises from conversion,—from conversion of the minor premise, —and this, too, converted into *all*, and not into *some*. But [an Inductive syllogism] is drawn in the third figure as follows:— Let it be supposed that we wish to prove—*Every animal moves the lower jaw*. With that intent, we place as terms:—the major *moves the under jaw;* the minor, [*all*] *animal;* and, lastly, the middle, *all contained under animal*, so that *these contents* reciprocate with *all animal*. And it is thus perfected [?] in the first figure—*To more the lower jaw* is predicated of *all individual animals; these all* are predicated of *all animal ;* therefore, *moving the lower jaw* is predicated of *all animal*. In such sort induction is accomplished."

III.—RAMUS.

Ramus, *Scholæ Dialecticæ*, I. viii. c. 11. " Quid vero sit inductio perobscure [Aristoteli] declaratur: nec ab interpretibus intelligitur, quo modo syllogismus per medium concludat majus extremum de minore: inductio majus de medio per minus." Ramus has confirmed his doctrine by his example. For, in his expositions, he himself is not correct.

IV.—DERODON.

Derodon, *Logica Restituta*, 1659, p. 602. *Philosophia Contracta*, 1664, *Logica*, p. 91. "Induction is the argumentation in which, from *all* the particulars, their universal is inferred ; as—*Fire, air, water, earth, are bodies ; therefore, every element is body.* It is recalled, however, to syllogism, by assuming all the particulars [including singulars] for the middle term, in this manner : —*Fire, air, water, and earth are bodies ; but fire, air, water, and earth are every element ; therefore, every element is body.* Again : —*The head, chest, feet, &c., are diseased ; but the head, chest, feet, &c., are the whole animal ; therefore, the whole animal is diseased.* Thus Induction is accomplished, when, by the enumeration of *all* the individuals, we conclude of the species what holds of all its individuals ; as—*Peter, Paul, James, &c., are rational ; therefore, all man is rational ;* or when, by the enumeration of *all* the species, we conclude of the genus what holds of all its species ; as—*Man, ass, horse, &c., are sensitive ; therefore, all animal is sensitive ;* or when, by the enumeration of *all* the parts, we conclude the same of the whole ; as—*Head, chest, feet, &c., are diseased ; therefore, the whole animal is diseased.*"

V.—THE COLLEGE OF ALCALA.

A curious error in regard to the contrast of the Inductive and the Deductive syllogism stands in the celebrated *Cursus Complutensis,*—in the *Disputations on Aristotle's Dialectic,* by the Carmelite College of Alcala, 1624, (L. iii. c. 2). We there find surrendered Aristotle's distinctions as accidental. Induction and Deduction are recognised, each as both ascending and descending, as both from, and to, the whole ; the essential difference between the processes being taken, in the existence of a middle term for Deduction, in its non-existence for Induction. The following is given as an example of the descending syllogism of Induction :— *All men are animals ; therefore, this, and this, and this, &c., man is an animal.* An ascending Inductive syllogism is obtained from the preceding if reversed. Now all this is a mistake. The Syllogism here stated is Deductive ; the middle, minor, and major terms, the minor premise and the conclusion being confounded

together. Expressed as it ought to be, the syllogism is as follows :
—*All men are (some) animals ; this, and this, and this, &c., are
(constitute) all men ; therefore, this, and this, and this, &c., are
(some) animal.* Here the middle term and three propositions reappear ; whilst the Deductive syllogism in the first figure yields,
of course on its reversal, an Inductive syllogism in the third.

The vulgar errors, those till latterly, at least, prevalent in this
country,—that Induction is a syllogism in the Mood Barbara of
the first figure, (with the minor or the major premise usually
suppressed) ; and still more that from a *some* in the antecedent,
we can logically induce an *all* in the conclusion ; these, on their
own account, are errors now hardly deserving of notice, and
have been already sufficiently exposed by me upon another
occasion. (*Edinburgh Review*, LVII. p. 224 *et seq.*) [*Discussions*, p. 158 *et seq.*—ED.]

VI.—FACCIOLATI.

Facciolati, *Rudimenta Logica*, p. iii. c. 3, defines Induction as
" a reasoning *without a middle*, and concluding the universal by
an enumeration of the singulars of which it is made up." His
examples show that he took it for an Enthymeme.—" *Prudence,
Temperance, Fortitude, &c., are good habits [these constitute all
virtue] ; therefore, [all] virtue is a good habit.*"

VII.—LAMBERT.

Lambert, *Neues Organon*, i. § 287. " When, in consequence of
finding a certain attribute in all things or cases which pertain to
a class or species [genus (?)], we are led to affirm this attribute of
the notion of the class or genus ; we are said to find the attribute
of a class or genus through Induction. There is no doubt that
this succeeds, so soon as the induction is complete, or so soon as
we have ascertained that the class or species A contains under it
no other cases than C, D, E, F,......M, and that the attribute B
occurs in each of the cases C, D, E, F,......M. This process now
presents a formal syllogism in *Caspida*. For we thus reason—

> C, as well as D, E, F,..........M are all B ;
> But A is either C, or D, or E, or F..........or M ;
> Consequently, all A are B.

"The example previously given of the syllogistic mood *Caspida* may here serve for illustration. For, to find whether every syllogism of the Second Figure be negative, we go through its several moods. These are *Cesare, Camestres, Festino, Baroco.* Now, both the first conclude in E, both the last in O. But E and O are negative, consequently all the four, and herewith the Second Figure in general, conclude negatively.* As in most cases, it is very difficult to render the minor proposition, which has the disjunctive predicate for its middle term, complete, there are, therefore, competent very few perfect inductions. The imperfect are [logically] worthless, since it is not in every case allowable to argue from *some* to *all*. And even the perfect we eschew whensoever the conclusion can be deduced immediately from the notion of the genus, for this inference is a shorter and more beautiful."

Strictures on Lambert's doctrine of Induction.

1°, In making the minor proposition disjunctive.

2°, In making it particular.

3°, In making it a minor of the First Figure instead of the Third.

Better a categorical syllogism of the Third Figure, like Aristotle, whom he does not seem to have been aware of. Refuted by his own doctrine in § 230, (below, p. 437).

The recent German Logicians,β following Lambert, (*N. Org.* i. § 287), make the inductive syllogism a hybrid disjunctive. Lambert's example is :—" C, as well as D, E, F,......M, all are B; but A is either C, or D, or E, or F,...... or M; therefore, all A is B." Or, to adapt it to Aristotle's example :—" Man, as well as horse, mule, &c., all are *long-lived* animals; but animal *void of gall* is either man, or horse, or mule, &c.; therefore, all animal *void of gall* is *long-lived.*"

This, I find, was an old opinion; and is well invalidated by the commentators of Louvain.γ

α It is given in § 285, as follows :—
" *The syllogisms, as well as Cesare as in Camestres, Festino, and Baroco, are all negative;*
" *Now every syllogism of the Second Figure is either in Cesare, or Camestres,*

or *Festino,* or *Baroco;*
" *Consequently every syllogism in the Second Figure is negative.*"
β As Herbart, *Lehrbuch der Logik*, § 68. Twesten, Drobisch, H. Ritter.
γ "I am aware of the opinion of many

The only inducement to the disjunctive form is, that the predicate is exhausted without the predesignation of universality, and the First Figure attained. But as these crotchets have been here refuted, therefore, the more natural, &c.

Some logicians, as "Oxford Crakanthorpe," (*Logica*, L. iii., c. 20, published 1622, but written long before), hold that Induction can only be recalled to a *hypothetical* syllogism. As,—*If Sophocles be risible, likewise Plato and all other men, then all man is risible ; but Socrates is risible, likewise Plato and all other men ; therefore, all man is risible.* Against the Categorical syllogism in one or other figure he argues :—" This is not a universal categorical, because both the premises are singular ; nor a singular categorical, because the conclusion is universal." It is sufficient to say, that, though the *subjects* of the premises be singular, (Crakanthorpe does not contemplate their being particular), as supposed to be *all* the constituents of a species or relatively universal whole, they are equivalent to that species ; their universality, though contrary to

that the singulars in the Inductive syllogism should be enumerated by a disjunctive conjunction, in so much that the premises of such a syllogism are commonly wont to be thus cast : —' Whatsoever is John, or Peter, or Paul, &c., is capable of instruction.' But they err, not observing that the previous proposition is manifestly equivalent to the following,—John and Peter, and Paul, &c., are capable of instruction.' " (Lovaniensss, *Comm. in An. Pr.*, L. ii. tr. 3, c. 2, p. 286, ed. 1547.) This here said of the major is true of Lambert's minor.

This doctrine—that the Inductive syllogism should be drawn in a disjunctive form—was commonly held, especially by the scholastic commentators on Petrus Hispanus. Thus Versor, (to take the books at hand,) whose *Expositio* first appeared in 1487, says : —" In the fourth place, Induction is thus reduced to syllogism, seeing that in the conclusion of the Induction there are two terms, of which the subject forms the minor, and the predicate the major, extreme in the syllo-

gism ; whilst the singulars, which have no place in the conclusion, constitute the middle term. Thus the Induction,—'Socrates runs, Plato runs, (and so of other men) ; therefore, all man runs,' is thus reduced : ' All that is Socrates, or Plato, (and so of others), runs ; but all man is Socrates, or Plato, (and so of others) ; therefore, all man runs.' And these singulars ought to be taken disjunctively, and disjunctively, not copulatively, verified of their universal." (*In Hisp. Summul.* Tr. v.)

The same doctrine is held in the *Reparationes* of Arnoldus de Tungeri and the Masters Regent in the Burse (or College) of St Lawrence, in Cologne, 1496. (Tr. iii. c. ii., Sec. Pri.)

It is also maintained in the *Copulata* of Lambertus de Monte, and the other Regents in the Burse Montis of Cologne, 1490. They give their reasons, which are, however, not worth stating and refuting.

But Tataretus, neither in his Commentaries on Hispanus nor on Aristotle, mentions this doctrine.

Aristotle's canon), is, indeed, overtly declared, in one of the premises, by the universal predesignation of the *predicate*. Our author further adds, " that Induction cannot be a categorical syllogism, because it contains *four* terms;" this quaternity being made by the " *all men* " (in his example) of the premises being considered as different from the " *all man* " of the conclusion. This is the veriest trifling. The difference is wholly factitious: *all man*, *all men*, &c., are virtually the same; and we may indifferently use either or both, in premises and conclusion.

(b) MATERIAL INDUCTION.

Material or Philosophical Induction is not so simple as commonly stated, but consists of two syllogisms, and two deductive syllogisms, and one an Epicheirema. Thus:—

I.— *What is found true of some constituents of a natural class, is to be presumed true of the whole class, (for nature is always uniform)*; *a a′ a″ are some constituents of the class* A ; *therefore, what is true of a a′ a″ is to be presumed true of* A.

II.— *What is true of a a′ a″, is to be presumed true of* A ; *but s is true of a a′ a″* ; *therefore, s is true of* A.

It will be observed, that all that is here inferred is only a presumption, founded, 1°, On the supposed uniformity of nature; 2°, That A is a natural class ; 3°, On the truth of the observation that *a a′ a″* are really constituents of that class A ; and, 4°, That s is an essential quality, and not an accidental. If any be false, the reasoning is naught, and, in regard to the second, *a a′ a″*, (*some*), cannot represent A, (*all*), if in any instance it is found untrue. " *Data instantia cadit inductio.*" In that case the syllogism has an undistributed middle.

IX.

HYPOTHETICAL AND DISJUNCTIVE REASONING— IMMEDIATE INFERENCE.

A.—AUTHOR'S DOCTRINE—FRAGMENTS.

(See above, Vol. I. p. 321.)

All Mediate inference is one ; that incorrectly called *Categorical ;* for the *Conjunctive* and *Disjunctive* forms of *Hypothetical* reasoning are reducible to immediate inferences.

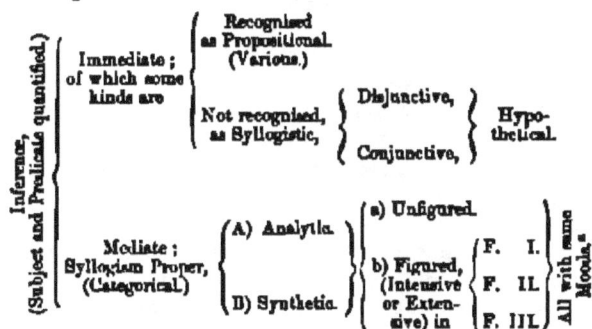

§ 1. Reasoning is the showing out explicitly that a proposition, not granted or supposed, is implicitly contained in something different, which is granted or supposed.

§ 2. What is granted or supposed is either a single proposition, or more than a single proposition. The Reasoning, in the former case, is Immediate, in the latter, Mediate.

§ 3. The proposition implicitly contained, may be stated first or last. The Reasoning, in the former case, is Analytic, in the latter, Synthetic.

Observations.—§ 1. " A proposition," not a truth ; for the proposition may not, absolutely considered, be true, but relatively to what is supposed its evolution, is and must be necessary.

a Reprinted from *Discussions,* p. 651.—ED.

All Reasoning is thus hypothetical; hypothetically true,
though absolutely what contains, and, consequently, what
is contained, may be false.[a]

Observations.—§ 2. Examples: Immediate—*If* A *is* B, *then*
B *is* A; Mediate—*If* A *is* B, *and* B *is* C, *then* A *is* C.

Observations.—§ 3. Examples: Analytic—B *is* A, *for* A *is* B;
A *is* C, *for* A *is* B, *and* B *is* C. Synthetic—A *is* B; *therefore* B *is* A; A *is* B, *and* B *is* C; *therefore,* A *is* C.

ON THE NATURE AND DIVISIONS OF INFERENCE OR SYLLOGISM IN GENERAL.

(November 1848.)

I. Inference, what . . . . . .

II. Inference is of three kinds; what I would call—1°, the
Commutative; 2°, the *Explicative*; and, 3°, the *Comparative.*

1°, In the first, one proposition is given; and required, what
are its formal commutations?

2°, In the second, two or more connected propositions are
given, under certain conditions, (therefore, all its species are conditionals); and required, what are the formal results into which
they may be explicated? Of this genus there are two species,—
the one the Disjunctive Conditional, the other the Conjunctive
Conditional. In the Disjunctive, (the Disjunctive also of the Logicians), two or more propositions, with identical subjects or predicates, are given, *under the disjunctive condition of a counter
quality, i. e.* that one only shall be affirmative; and it is required
what is the result in case of one or other being affirmed, or one
or more denied. (Excluded Middle.) In the Conjunctive, (the

a [That all logical reasoning is hypothetical, and that categorical Syllogism is really, and in a higher signification, hypothetical, see Maimon, *Versuch einer neuen Logik*, § vi. 1, pp. 82, 83. E. Reinhold, *Logik*, § 103, p. 283 *et seq.* Smiglecius, *Logica*, Disp. xiii., q. 5, p. 495, (1st ed. 1618).

On the nature of the Necessity in Syllogistic Inference; distinction of Formal and Material Necessity, or of *necessitas consequentiæ* and *necessitas consequentis*, see Boetius, *Quæstiones, Super Elenchos*, qu. iv., p. 227, ed. 1639, and that all logical inference hypothetical, *In An. Prior.*, L. ii. qu. i. p. 231. Apuleius, *De Hab. Doct. Plat.*, p. 34. Aristotle, *An. Prior.*, L. 22, § 5. Smiglecius, *Logica*, loc. cit. Balforeus, *In Arist. Org., An. Prior.*, L. t. 8, p. 454.] [See also *Discussions*, p. 149, note.—ED.]

Hypotheticals of the Logicians), two or more propositions, convertible or contradictory, with undetermined quality, are given, *under the conjunctive condition of a correlative quality,* i. e. that the affirmation or negation of one being determined, determines the corresponding affirmation or negation of the other or others; and it is required what is the result in the various possible cases. (Identity and Contradiction, not Sufficient Reason, which in Logic is null as a separate law.)

3°, In the third, three terms are given, two or one of which are positively related to the third; and required, what are the relations of these two terms to each other?[a]

III. All inference is hypothetical.

IV. It has been a matter of dispute among logicians, whether the class which I call *Explicative* (viz. the Hypothetical and Disjunctive Syllogisms) be of Mediate or Immediate inference. The immense majority hold them to be mediate; a small minority, of which I recollect only the names of Kant, [Fischer, Weiss, Bouterwek, Herbart],[b] hold them to be immediate.

The dispute is solved by a distinction. Categorical Inference is mediate, the medium of conclusion being a *term;* the Hypothetical and Disjunctive syllogisms are mediate, the medium of conclusion being a *proposition,*—that which I call the *Explication.* So far they both agree in being mediate, but they differ in four points. The first, that the medium of the Comparative syllogism is a term; of the Explicative a proposition. The second, that the medium of the Comparative is one; of the Explicative more than one. The third, that in the Comparative the medium is always the same; in the Explicative, it varies according to the various conclusion. The fourth, that in the Comparative the medium

[a] A better statement of the three different processes of Reasoning.

I. Given a proposition; commutative;—what are the inferences which its commutations afford?

II. Given two or more propositions; related and conditionally;—what are the inferences which the relative propositions, explicated under these conditions, afford?

III. Given three notions; two related, and at least one positively, to a third;—what are the inferences afforded in the relations to each other, which this comparison of the two notions to the third determines?

[b] [Kant, *Logik,* § 75. Fischer, *Logik,* c. v. §§ 99, 100, p. 137. Weiss, *Logik,* §§ 210, 251. Bouterwek, *Lehrbuch der philosophischen Vorkenntnisse,* § 100, p. 158, 2d ed. 1820. Herbart, *Lehrbuch zur Einleitung in die Philosophie,* § 64, p. 87, 1834.]

never enters the conclusion; whereas, in the Explicative, the same proposition is reciprocally medium or conclusion.

V. Logicians, in general, have held the Explicative class to be composite syllogisms, as compared with the Categoric; whilst a few have held them to be more simple. This dispute arises from each party taking a partial or one-sided view of the two classes. In one point of view, the Explicative are the more complex, the Comparative the more simple. In another point of view, the reverse holds good.

Our Hypothetical and Disjunctive Syllogisms may be reduced to the class of Explicative or Conditional. The Hypotheticals should be called, as they were by Boethius and others, *Conjunctive*, in contrast to the co-ordinate species of *Disjunctive*. Hypothetical, as a name of the species, ought to be abandoned.

The Conjunctives are conditional, inasmuch as negation or affirmation is not absolutely asserted, but left alternative, and the quality of one proposition is made dependent on another. They are, however, not properly stated. The first proposition,—that containing the condition,—which I would call the *Explicand*, should be thus enounced: "As B, so A;"—or, "As B is, so is A;" or, "As C is B, so is D A." Then follows the proposition containing the explication, which I would call the *Explicative;* and, finally, the proposition embodying the result, which I would call the *Explicate.*—They are called *Conjunctives* from their conjoining two convertible propositions in a mutual dependence, of which either may be made antecedent or consequent of the other.

Disjunctive Syllogisms are conditional, inasmuch as a notion is not absolutely asserted as subject or predicate of another or others, but alternatively conjoined with some part, but only with some part, of a given plurality of notions, the affirmation of it with one part involving its negation with the others. The first proposition, containing the condition, I would call the *Explicand*, and so forth as in the Conjunctives.—They are properly called *Disjunctives.*

DISTRIBUTION OF REASONINGS.

(Nov. 1848.)—Inference may be thus distributed, and more fully and accurately than I have seen. It is either, (I.) Im-

mediate, that is, without a middle term or medium of comparison ; or (II.) Mediate, with such a medium.*

Both the Immediate and the Mediate are subdivided, inasmuch as the reasoning is determined (A) to one, or (B) to one or
other conclusion. (It is manifest that this latter division may
constitute the principal, and that *immediate* and *mediate* may
constitute subaltern classes.)

All inference, I may observe in the outset, is hypothetic, and
what have been called *Hypothetical Syllogisms* are not more
hypothetic than others.

I. A—Immediate Peremptory Inference, determined [to] one
conclusion, contains under it the following species :—β . . .

I. B—Immediate Alternative Inference contains under it
these five species,—

1°. Given one proposition, the alternative of affirmation and
negation. As—A *either is or is not ; but A is ; therefore A is not
not.* Or, A *is or is not* B ; but A *is* B ; *therefore,* A *is not not-*B.

This species is anonymous, having been ignored by the logi-

* [Cf. Fonseca, *Instit. Dial.*, L. vi.
c. 1, [1st ed.] 1564. Eustachius, *Summa
Philosophiæ Quadripartita, Dialectica,*
P. III. tract. i., p. 112.] ["Quoniam
argumentatio est quædam consequentia,
(latius enim patet consequentia quam
argumentatio), prius de consequentia,
quam de argumentatione dicendum est.
Consequentia igitur, sive consecutio,
est oratio in qua ex aliquo aliquid col-
ligitur ; ut, *Omnis homo est animal,
igitur aliquis homo est animal.*"—Ed.]
[Whether Immediate Inference really
immediate, me, on the affirmative, E.
Reinhold, *Logik,* § 106 ; on the negative, Wolf, *Phil. Rat.,* § 461. Krug,
Logik, § 94, p. 237. Schulze, *Logik,*
§§ 85-90, (§ 80, 5th ed.) Cf. Maimon,
Versuch einer neuen Logik, Sect. v. § 2,
p. 74 et seq. F. Fischer, *Logik,* p. 104,
et seq. Bachmann, *Logik,* § 105, p.
154 et seq. Reimarus, *Vernunftlehre,*
§ 159 et seq. (1765). Bolzano, *Wissenschaftslehre, Logik,* vol. ii. § 255 et seq.
Twesten, *Logik, insbesondere die Analytik,* § 77, p. 64. Rö'sling, *Die Lehren*

der reinen *Logik,* § 130, p. 291. Scheibler, *Op. Log., De Proprii. Consec-
tions,* p. 492 et seq.]

β [Kinds of Immediate Inference.—
I. Subalternation. II. Conversion. III.
Opposition—(a) of Contradiction—(b)
of Contrariety—(c) of Subcontrariety.
IV. Equipollence. V. Modality. VI.
Contraposition. VII. Correlation. VIII.
Identity.

Fonseca (IV), (I), (II). Eustachius
(I), (IV), (II), (VIII). Wolf, (IV),
(VII), (III), a, b, c, (II). Stattler, (I),
(IV), (II), (III). Kant, (I), (III), a,
b, c, (II), (VI). E. Reinhold, (I), (II),
(VI), (VII). Rösling, (I), (IV), (II),
(III), a, b, c, (V). Krug, (IV), (I),
(III, a, b, c, (II), (V). G. E. Schulze,
(IV), (I), (III), (II). S. Maimon, (I),
(III), (II), (VI). Bachmann, (IV), (I),
(III), a, b, c, (II), (VI), (V). Platner,
(I), (II), (III), (IV). F. Fischer, (V),
(I), (III), (II), (VI). Reimarus, IV.,
(I), (III), a, b, (II. Twesten, I), (V),
(III), (IV), (II), (VI).] [See above,
pp. 237, 238.—Ed.]

cians; but it requires to be taken into account to explain the various steps of the process.

2°, Given one proposition, the alternative between different predicates. This is the common Disjunctive Syllogism.

3°, The previous propositions conjoined, given one proposition, &c. As, A *either is or is not either* B *or* C *or* D: *but* A *is* B; *therefore it is not not* B, *it is not* C, *it is not* D.

Alias, A *is either* B *or non*-B, *or* C *or non*-C, *or* D *or non*-D; *but* A *is* B; *therefore it is not non*-B, *and it is non*-C, *and it is non*-D.

4°, Given two propositions, second dependent on the first, and in the first the alternative of affirmation and negation. This is the Hypothetical Syllogism of the logicians. It is, however, no more hypothetical than any other form of reasoning; the so-called hypothetical conjunction of the two radical propositions being only an elliptical form of stating the alternation in the one, and the dependence on that alternation in the other. For example,—*If* A *is* B, B *is* C; this merely states that A *either is or is not* B, and that B *is or is not* C, *according as* A *is or is not* B. In short—*As* A *is or is not* B, *so* B *is or is not* C.

(Errors,—1, This is not a mediate inference.

2, This is not more composite than the categorical.

3, The second proposition is not more dependent upon the first, than the first upon the second.)

5°, Given two propositions, one alternative of affirmation and negation, and another of various predicates; the Hypothetico-disjunctive or Dilemmatic Syllogism of the logicians.

II. A—Mediate Peremptory Inference. This is the common Categorical Syllogism. Three propositions, three actual terms, one primary conclusion, or two convertible equally and conjunctly valid.

II. B—Mediate Alternative Syllogism. Three propositions, three possible terms, and conclusions varying according . . .

. . . .

2°, The Disjunctive Categorical.

. . . .

4°, The Hypothetical Categorical.

5°, Hypothetico-Disjunctive Categorical.

. . . .

HYPOTHETICAL SYLLOGISM.—CANON.

(Oct. 1848.)—Canon—Two or more propositions thought as indetermined in quality, but as in quality mutually dependent, the determination of quality in the one infers a determination of the corresponding quality in the other.

This canon embodies and simplifies the whole mystery of Hypothetical Syllogisms, which have been strangely implicated, mutilated, and confused by the logicians.

1°, What are called Hypothetical Propositions and Syllogisms are no more hypothetical than others. They are only hypothetical as elliptical. When we say, " If A is, then B is," we mean to say that the proposition, " A is or is not," and the proposition, " B is or is not," are mutually dependent,—that as the one so the other. *If* here only means taking for the nonce one of the qualities to the exclusion of the other. I, therefore, express in my notation the connection of the antecedent and consequent of a hypothetical proposition, thus:—

$$(A \ x \ \text{———} \ x \ \text{—+—} \ ) = (B \ x \ \text{———} \ x \ \text{—+—})$$

2°, The interdependent propositions are erroneously called *Antecedent* and *Consequent.* Either is antecedent, either is consequent, as we choose to make them. Neither is absolutely so. This error arose from not expressing overtly the quantity of the subject of the second proposition. For example, " If man is, then animal is." In this proposition, as thus stated, the negation of the first does not infer the negation of the second. For *man* not existing, *animal* might be realised as a consequent of *dog, horse,* &c. But let us consider what we mean. We do not mean *all animal,* but *some* only, and that *some* determined by the attribute of *rationality* or such other. Now, this same *some animal* depends on *man,* and *man* on it; expressing, therefore, what we mean in the proposition thus:—" If all man is, then some animal is,"—we then see the mutual dependence and convertibility of the two propositions.[a] For to say that *no animal is,* is not to explicate but to change the terms,

---

[a] Cf. Titius, *Ars Cogitandi*, c. xii. § 26 et seq. : " In specie falsum quoque arbitror, quod Syllogismi Conditionalem duas habent figuras, quæ his maniam. tur regulis, (1) *posito antecedente, ponitur consequens, non vero, remoto antecedente...*

3°, The interdependent propositions may be dependent through their counter qualities, and not merely through the same. For example:—"As our hemisphere is or is not illuminated, so the other is not or is; but the other is not illuminated; therefore ours is." Another:—"If A is, then B is not; but B is; therefore A is not."

DISJUNCTIVE AND HYPOTHETICAL SYLLOGISMS PROPER.

Aristotle ignores these forms, and he was right.[a] His followers, Theophrastus and Eudemus, with the Stoics, introduced them into Logic as co-ordinate with the regular Syllogism; and their views have been followed, with the addition of new errors, up to the present hour. In fact, all that has been said of them has been wrong.

1°, They are not composite by contrast to the regular syllogism, but more simple.

2°, If inferences at all, they are immediate and not mediate.

3°, But they are not argumentations but preparations (expli-

dentis removetur consequens, (2) remoto consequente removetur antecedens, non autem, posito consequente, ponitur antecedens, . . . Videamus specialius; contra primam regulam sic peccatur;

Si Chinenses sunt Mahometani, sunt infideles,

At non sunt Mahometani,

Ergo non sunt infideles,

nam conclusio hic est absurda! Verum si prædicatum conclusionis sumatur particulariter, nulla est absurditas, si autem generaliter, tum evadunt quatuor termini. Eodem exemplo secunda regula etiam illustratur, sed assumamus aliud ex Weisio, P. i. L. 2.

Si miles est doctus, novit libros (nempe sicut eruditi solent,)

Sed novit libros (scil. ut alii homines etiam indocti nosse solent,)

Ergo miles est doctus.

Hæc conclusio itidem pro falsa habetur! Sed jam indicavimus in addita parenthesi veram causam, nempe quatuor terminos, quodsi antem medius

terminus eodem sensu accipiatur, ac in syllogismo formaliter proposito quam minor probari, tum conclusio erit verissima, idque virtute præmissarum. Omnis igitur error exinde habet originem, quod quantitatem prædicati vel non intelligant, vel non observant; si igitur hunc lapsum evitas, objecta exempla omnia, qualia etiam Weisius d. l. commemorat, facile dilues."—ED.

[a] Cf. Titius, *Ars Cogitandi*, c. xii. § 7: "Syllogismus Disjunctivus est enthymema, sine majore, bis, oratione disjuncta et positiva, propositum." § 17: "Conditionalis seu Hypotheticus nihil aliud est quam enthymema vel sine majore, vel minore, bis, prima scil. vice conditionaliter, secunda pure, propositum." § 20: "Sequitur nullum peculiare concludendi fundamentum vel formam circa Syllogismos Conditionales occurrere, nam argumentationes imperfectas, adeoque materiam syllogismorum regularium, illi continent."—ED.

cations) for argumentation.[a] They do not deal with the quæsitum,—do not settle it; they only put the question in the state
required for the syllogistic process; this, indeed, they are frequently used to supersede, as placing the matter in a light which
makes denial or doubt impossible; and their own process is so
evident that they might, except for the sake of a logical, an articulate, development of all the steps of thought, be safely omitted,
as is the case with the quæsitum itself. For example:—

1. Hypothetical (so-called) Syllogism. Let the quæsitum or
problem be, to take the simplest instance,—"Does animal exist?"
This question is thus hypothetically prepared—"If Man is, Animal is;" but [as is conceded] Man is; therefore, Animal is."
But here the question, though prepared, is not solved; for the
opponent may deny the consequent, admitting the antecedent.
It, therefore, is incumbent to show that the existence of Animal
follows that of Man, which is done by a categorical syllogism."

Animal, ——— : Man : ———— Existent.

2. Disjunctive (so-called) Syllogism. Problem—" Is John
mortal?" Disjunctive syllogism—"John is either mortal or immortal; but he is not immortal; ergo, [and this, consequently,
is admitted as a necessary alternative], he is mortal." But the
[alternative antecedent] may be denied, and the alternative consequent falls to the ground. It is, therefore, necessary to show
either that he is not immortal, or,—the necessary alternative,—
that he is mortal, which is done by a categorical syllogism.

John ———, Man : ——|—— : Immortal,

John ———, Man : ———, Mortal,

HYPOTHETICAL INFERENCE.

Inasmuch as a notion is thought, it is thought either as existing
or as non-existing; and it cannot be thought as existing unless it
be thought to exist in this or that mode of being, which, conse-

a This I say, for, notwithstanding
what M. St Hilaire so ably states in
refutation of my paradox, I must adhere to it as undisproved.—See his
Translation of the Organon, vol. iv.
p. 56.

quently, affords it a ground, condition, or reason of existence.
This is merely the law of Reason and Consequent; and the hypothetical inference is only the limitation of a supposed notion to
a certain mode of being, by which, if posited, its existence is
affirmed; if sublated, its existence is denied. For example:—
" If A is, it is B; but A is," &c.

Again, we may think the existence of B (consequently of A B)
as dependent upon C, and C as dependent upon D, and so forth.
We, accordingly, may reason, " If A is B, and B is C, and C is
D," &c.

DISJUNCTIVE SYLLOGISM PROPER.

(October 1848.)—Inasmuch as a notion is thought, it is thought
as determined by one or other, and only by one or other, of any
two contradictory attributes; and inasmuch as two notions are
thought as contradictory, the one or the other, and only the one
or the other, is thought as a determining attribute of any other
notion. This is merely the law of Excluded Middle. The disjunctive inference is the limitation of a subject notion to the one
or to the other of two predicates, thought as contradictories; the
affirmation of the one inferring the negation of the other, and
vice versâ. As:—" A is either B or not B," &c. Though, for the
sake of brevity, we say " A is either B or C or D," each of these
must be conceived as the contradictory of every other; as, B = |
C | D, and so on with the others.

HYPOTHETICALS (CONJUNCTIVE AND DISJUNCTIVE SYLLOGISM).

(April 80, 1849)—These syllogisms appear to be only modifications or corruptions of certain immediate inferences; for they have
only two terms, and obtain a third proposition only by placing
the general rule of inference, (stating of course, the possible alternatives,) disguised, it is true, as the major premise. It is manifest
that we might prefix the general rule to every mediate inference;
in which case a syllogism would have four propositions; or, at
least, both premises merged in one complex proposition, thus—

If A and C be either subject or predicate [of the same term ?] they are
both subject or predicate of each other;
But B is the subject of A and predicate of B [C ?];
Therefore A is the predicate of C.*

* There seems to be an error here in the Author's MS. It is obvious that a

Thus, also, a common hypothetical should have only *two* propositions. Let us take the immediate inference, prefixing its rule, and we have, in all essentials, the cognate hypothetical syllogism.

1.—Conjunctive Hypothetical.

All B is (some or all) A ;	All men are (some) animals ;
Some or all B exists ;	(All or some) men exist ;
Therefore, some A exists.	Therefore, some animals exist.

Here it is evident that the first proposition merely contains the general rule, upon which all immediate inference of inclusion proceeds ; to wit, that, the subjective part being, the subjective whole is, &c.

Now, what is this but the Hypothetical Conjunctive?

If B is, A is ;	If man is, animal is ;
But B is ;	But man is ;
Therefore, A is.	Therefore, animal is.

2.—Hypothetical Disjunctives.

B is either A or not-A ;	Man is either animal or non-animal ;
But B is A ;	But man is animal ;
Therefore, B is not not-A.	Therefore, is not non-animal.

Stating this hypothetically, we may, of course, resolve the formal contradictory into the material contrary. But this is wholly extralogical.

HYPOTHETICAL AND DISJUNCTIVE SYLLOGISMS.

(1848 or 1849.)—The whole antecedent must be granted ; and there cannot be two propositions inferred. In Categorical Syllogisms, the antecedent is composed of the major and minor premises, and there is only one simple conclusion, (though this may, in the second and third figures, vary). So in Hypothetical and Disjunctive Syllogisms the whole antecedent is the two clauses of the first proposition ; and the whole inference is the first and

mediate inference may be expressed in the form of a hypothetical syllogism. Thus: "If B is A, and C is B, then C is A ; but B is A, and C is B ; therefore, C is A." This is apparently what the Author means to express in a somewhat different form. —Ed.

second clauses of the second proposition, erroneously divided into minor proposition and conclusion.

(January 1850.)—The Medium or Explicative may be indefinitely various, according to the complexity of the Explicand ; and so may the Explicate. The explicative and the explicate change places in different explications. There is, in fact, no proper explicative or explicate.

(January 1850.)—In Disjunctives there is always at least double the number of syllogisms (positive and negative) of the disjunct members ; and in all syllogisms where the disjunct members are above two, as there is thus afforded the possibility of disjunctive explicates, there is another half to be added. Thus, if there be two disjunct members, as A—x B C, there are four syllogisms, but all of an absolute explicate. But if there be three disjunct members, as A—x B C D, in that case there are six absolute explicates, three positive and three negative, and, moreover, three disjunctivo-positive explicates, after a negative explicative, and so on.

HYPOTHETICAL SYLLOGISM.—CANONS.

(February 1850.)

I. For *Breadth*,—The extensive whole or class being universally posited or sublated, every subjacent part is posited or sublated; or for *Depth*,—All the comprehensive wholes being posited or sublated, the comprehended parts are universally posited or sublated.

II. For *Breadth*,—Any subjacent part being posited or sublated, the extensive whole or class is partially posited or sublated ; or for *Depth*,—Any comprehensive whole being posited or sublated, the comprehended parts (or part) are, *pro tanto*, posited or sublated,—Conversion and Restriction.

III. If one contradictory be posited or sublated, the other is sublated or posited,—Contradiction.

IV.—If some or a part only of a notion be posited or sublated, all the rest (all other some) is sublated or posited,—Integration.

V. If the same under one correlation be posited or sublated, so under the other,—Equipollence.

VI. Law of Mediate Inference,[a]—Syllogism.

Mem.—The *some* in the explicand is, (as in the Conversion of

---

[a] See above, p. 290.—ED.

propositions), to be taken in the explicative as the *same some*. There is thus an inference equally from consequent to antecedent, as from antecedent to consequent.[a]

HYPOTHETICALS OR ALTERNATIVES.

CONJUNCTIVE, (HYPOTHETICALS EMPHATICALLY), AND DISJUNCTIVE, (ALTERNATIVES EMPHATICALLY.)

(August 1852.)

Quantification—*Any*.

Affirmative—*Any*, (*Anything*, *Aught*), contains under it every positive quantification,—*All* or *Every*,—*Some at least*,—*Some only*,—*This*, *These*. (Best.)

Negative—*Not any*, *None*, *No*, (*Nothing*, *Naught*), is equivalent to the most exclusive of the negations, *All not*, *All*, or *Every not*, *Not one*, and goes beyond the following, which are only partial negations,—*Not all*, *Not some*, *Some not*. (Worst.)

Affirmative—*Any*, a highest genus and best ; not so Negative —*Not any*,—a lowest species, and worst. Therefore can restrict, —subalternate in the former, not in the latter.

1	2
⎴—*Any*, (*all* or *every*,—*some*).	*Some not*, or *not some*, or *not all*— *some only*, (def.)
Pure affirmative.	
	Mixed affirmative and negative.

3
⎴*All* or *every not*, *not one*, *not any*.
Pure negative.

If any (every) M be an (some) A, and any (every) A an (some) S, then is any (every) M an S ; or, *vice versâ*, If no (not any) A be any S, and any M some A, then is no M any S.

∴ (On one alternative), Some M being an (some) A, and all A an (some) S, some M is some S.

(On the other), No A being any S, and every M some A, no M is any S.

If, (on *any possibility*), any M is, some A is ; or, v. v., If no A is, no M is.

∴ (In *this actuality*), (on one alternative), some M being, some A is ; (on the other), no A being, no M is.

Possible M :, ——— , A or A : ——— : M. Supposition of universal Possibility. In any case.

Actual M, ——— , A or A : ——— : A. Assertion of particular Actuality. In this case.

From *Possible*, we can descend to *Actual* ; from *Any*, to *Some* ; but *Not any* being lowest or worst, we can go [no] lower.

The *Possible* indifferent to Affirmation or Negation, it contains both implicitly. But when we descend to the *Actual*, (and *Potential* ?), the two qualities emerge. This explains much in both kinds of Hypotheticals or Alternatives,—the Conjunctives and Disjunctives.

Higher classes, — *Possible, Actual,* — *Semper, quandocunque, tunc, nunc*— *Ubicunque, ubique, ibi, hic*—*Any, all, some*—*In all, every, any, case, in this case*—*Conceivable, real.*

RULES OF HYPOTHETICAL SYLLOGISMS.

1. **Universal Rule of Restriction.**—What is thought of all is thought of some,—what is thought of the whole higher notion (genus), is thought of all and each of the lower notions (species or individual).

2. **General Rule of both Hypotheticals.**—What is thought (implicitly) of all the Possible (genus), is thought (explicitly) of all and each the Actual (species).

3. **Special Rule of Conjunctives.**—What is thought as consequent on every Possible, is thought as consequent on every Actual, antecedent.

4. **Special Rule of Disjunctives.**—What is thought as only Possible, (alternatively), is thought as only Actual, (alternatively).

5. Most Special Rule of Conjunctives. . . . . . : . . . .

6. Most Special Rule of Disjunctives. . . . .

HYPOTHETICALS—EXAMPLES UNQUANTIFIED.

(Higher to Lower.)

AFFIRMATIVE.	NEGATIVE.
If the genus is, the species is.	If the genus is not, the species is not.
If the stronger can, the weaker can.	If the stronger cannot, the weaker cannot.

(Lower to Higher.)

If the species is, the genus is.
If the weaker can, the stronger can.

If the species is not, the genus is not.
If the weaker cannot, the stronger cannot.

(Equal to Equal.)

If triangle, so trilateral.

Such poet Homer, such poet Virgil.

Where (when) the carcase is, there (then) are the flies.

If Socrates be the son of Sophroniscus, Sophroniscus is the father of Socrates.

If equals be added to equals, the wholes are equal.

If A be father of B, B is son of A ;

∴ A being father of B, B is son of A ;

∴ B not being son of A, A is not father of B.

If the angles be proportional to the sides of a △ ;

∴ An equiangular will be an equilateral △.

If " wheresoever the carcase is, there will the eagles be gathered together." (Matth. xxiv. 28).

∴ If here the carcase is, here, &c.

These examples not all *negation*.

A.)—CONJUNCTIVE HYPOTHETICALS.

1). If A be D, it is △ ; ∴ $\begin{cases} \text{A, being D, is △.} \\ \text{A, not being △, is not D.} \end{cases}$

In other words—A is either D △, or not △ D.

Identity and Contradiction.

2). If B be A, it is not non-A ; ∴ $\begin{cases} \text{B, being A, is not non-A.} \\ \text{B, being non-A, is not A.} \end{cases}$

In other words—B is either A or none-A.

Excluded Middle.

3). If B be not A, it is non-A ; ∴ $\begin{cases} \text{B, not being A, is non-A.} \\ \text{B, being non-A, is not A.} \end{cases}$

In other words—B is either not A or not non-A.

Excluded Middle.

4). If E be not D, it is not △ ; ∴ $\begin{cases} \text{E, not being D, is not △.} \\ \text{E, being △, is D.} \end{cases}$

In other words—E is either not D △, or △ D.

Contradiction and Identity.

B.)—DISJUNCTIVE HYPOTHETICALS.

If B be either A or non-A ; ∴ { B, being A, is non-A.
{ B, being non-A, is not A.

Excluded Middle.

" *If* " means *suppose that,—in case that,—on the supposition—hypothesis—under the condition—under the thought that,—it being supposed possible* ;

∴ &c., means *then,—therefore,—in that case,* &c., &c.—*actually either.*

Only, properly, in both Conjunctives and Disjunctives, two contradictory alternatives. For contrary alternatives only material, not formal, and,'in point of fact, *either* A *or* B *or* C means A *or non-*A, B *or non-*B, C *or non-*C.

The minor premise, on the common doctrine, a mere materiality. Formally,—logically, it is a mere differencing of the conclusion, which is by formal alternative afforded.

1). In Hypotheticals, (Conjunctive and Disjunctive), two (or three) hypotheses. The *first* is in the original supposition of *possibility*. (If B be A, it is not non-A—If B be either A or non-A). The *second* (*second* and *third*) is in the alternative suppositions of *actuality* (∴ either if B be A, it is not non-A, or if B be non-A, it is not A.—∴ If B be A, it is not non-A, or if B be non-A, it is not A). (Possibly,—by possible supposition) *If man is, animal is*; ∴ (actually) *Man being, animal is*; (or) *animal not being, man is not.*

1*). Possibility—a genus indifferent to negative and affirmative. These two *species* of Possibility, to wit two Actuals,—an actual yes and an actual no. The total formal conclusion is, therefore, of two contradictories. This explains why, in Conjunctive and Disjunctive Hypotheticals, there are two alternative consequents, and only one antecedent.

2). In Hypotheticals (Conjunctive and Disjunctive) a division of genus in the first supposition into two contradictory species. The inference, therefore, one of subalternation or restriction.

3). In Hypotheticals (Conjunctive and Disjunctive) two alternative contradictory conclusions—the *form* giving no preference between the two, the *matter* only determining, (other immediate inferences have only one determinate conclusion, and all mediate syllogism has virtually only one). Formally, therefore, we cannot categorically, determinately, assert, and assert exclusively, either alternative, and make a minor separate from the conclusion. This only materially possible; for we know not, by the laws of thought, whether a certain alternative is, knowing only that one of two alternatives must be. Formally, therefore, only an immediate inference, and that alternative double.

4). Hypothetical, (Conjunctive and Disjunctive), reasoning more marking out,—predetermining, how a thing is to be proved than proving it.

5). Thus, three classes of inference: 1°, Simple Immediate Inference.—2°, Complex Immediate Inference, (Hypotheticals Conjunctive and Disjunctive).—3°, Syllogisms Proper, Mediate Inference.

6). If we quantify the terms, even the formal inference breaks down.

7). The only difference between the first proposition and the two latter, is the restriction or subalternation. These last should, therefore, be reduced to one, and made a conclusion or restriction. The genera and species are of the most common and notorious kinds, as *Possible* and *Actual*,—*Wherever*, *Here*, &c.—*Whenever*, *Now*,—*All* or *Every*,—*Some*, *This*, &c. The commonness and notoriety of this subordination is the cause why it has not been signalised ; and if signalised, and overtly expressed, Hypotheticals might be turned into Categoricals. It is better, however, to leave them as immediate inferences. For it would be found awkward and round-about to oppose, for example, the Possible to the Actual, as determining a difference of terms. (See Molinæus, *Elem. Log.*, L. i. tr. iii. p. 95, and Pacius, *In Org., De Syll. Hyp.*, p. 533.) The example of the *Cadaver* there given, shows the approximation to the ordinary Hypotheticals. They may stand, in fact, either for Categoricals or Hypotheticals.

8). Disjunctives—(*Possibly*) A is either B or non-B ; ∴ (*Actually*) A is either, &c.

9). The doctrine in regard to the Universal Quantity, and the Affirmative Quality (see Krug, *Logik*, §§ 57, 83, 86, pp. 171, 261,

275) of the suppositive proposition in Conjunctive and Disjunctive
Hypotheticals, is solved by my theory of *Possibility*. In it is
virtually said, (whatever quantity and quality be the clauses)—
"*on any possible supposition*." (On the Quality v. Krug, *Logik*, §
57, p. 172. Pacius, *In. Org.*, p. 533. Molinæus, *Elem. Log.*, *l. c.*)

10). *Possibly,—problematically* includes as species the actual
affirmative, and the actual negative. It will thus be superfluous
to enounce a negative in opposition to an affirmative alternative ;
for thus the possible would be brought down to the actual ; and
the whole syllogism be mere tautological repetition.

11). The quantified terms, if introduced, must either be made
determinate, to suit the Hypotheticals, or must ruin their inference. For example—*If all or some Man be some Animal*, we
must be able to say, *But some Animal is not, therefore Man
(any or some) is not*. But here *some Animal*, except definitised
into the *same some Animal*, would not warrant the required inference. And so in regard to other quantifications, which the logicians have found it necessary to annul.

12). The minor proposition may be either categorical or hypothetical. (See Krug, *Logik*, § 88, p. 264. Heerebord, *Instit.
Logicar. Synopsis*, L. ii. c. 12, pp. 266, 267.) In my way of stating it :—" If Man is, Animal is, ∴ If Man is (or Man being),
Animal is."

13). Of notions in the relation of sub-and-superordination, (as,
in opposite ways, Depth and Breadth, Containing and Contained),
absolutely and *relatively*, the lower being *affirmed*, the higher are
(partially) affirmed ; and the higher being (totally) *denied*, the
lower are (totally) denied. A, E, I, O, U, Y may represent the
descending series.

The first proposition is conditional, complex, and alternative ;
we should expect that the second should be so likewise. But this
is only satisfied on my plan ; whereas, in the common, there is a
second and a third, each categorical, simple, and determinate.

The subalternation is frequently double, or even triple, to wit,
1°, From the Possible to the Actual. 2°, (for example) From *every-*

where to *here*, or *this place*, or the place by name. 3°, From *all* to *some*, &c.—in fact, this inference may be of various kinds.

The μετάληψις of Aristotle may mean the determination,—the subalternation ; the κατὰ ποιότητα may refer to the specification of a particular quality or proportion under the generic ; and the πρόσληψις of Theophrastus (for the reading in Aristotle should be corrected) may correspond to the κατὰ ποιότητα.

———

There is no necessary connection, formally considered, between the antecedent and consequent notions of the Hypothetical major. There is, consequently, no possibility of an abstract notation ; their dependence is merely supposed, if not material. Hence the logical rule,—*Propositio conditionalis nihil ponit in esse*. (See Krug, *Logik*, § 57, p. 166). But on the formal supposition,—on the case thought, what are the rules ? . . . . .

———

We should distinguish in Hypotheticals between a propositional antecedent and consequent, and a syllogistic A and C ; and each of the latter is one proposition, containing an A and C.

The antecedent in an inference should be that which enables us formally to draw the conclusion. Show in Categoricals and in Immediate Inferences. On this principle, the conclusion in a Hypothetical will contain what is commonly called the minor proposition with the conclusion proper ; but it will not be one and determinate, but alternative.

If there were no alternation, the inference would follow immediately from the fundamental proposition ; and there being an alternative only makes the conclusion alternatively double, but does not make a mediate inference.

To make one alternative determinate is extralogical ; for it is true only as materially proved. 1°, The splitting, therefore, of the conclusive proposition into two,—a minor and a conclusion proper, is wholly material and extralogical ; so also, 2°, Is the multiplying of one reasoning into two, and the dividing between them of the alternative conclusion.

———

Errors of logicians, touching Hypothetical and Disjunctive Reasonings :—

1°, That [they] did [not] see they were more immediate inferences.

2°, Most moderns that both Hypothetical.

3°, That both alternative reasonings in one syllogism.

4°, Mistook a part of the alternative conclusion for a minor premise.

5°, Made this a distinct part, (minor premise), by introducing material considerations into a theory of form.

6°, Did not see what was the nature of the immediate inference in both,—how they resembled and how they differed.

B.—HISTORICAL NOTICES.

(CONJUNCTIVE AND DISJUNCTIVE.)

I.—ARISTOTLE.

(August 1852.)

Aristotle, (*Anal. Pr.* L. i. c. 32, § 5, p. 262, Pacii,) describes the process of the Hypothetic Syllogism, (that called by Alexander δἰ ὅλων), but denies it to be a Syllogism. Therefore his syllogisms from Hypothesis are something different. This has not been noticed by Mansel, Waitz, ...

Thus literally:—" Again, if *Man* existing, it be necessary that *Animal* exist, and if *Animal*, that *Substance*; Man existing, it is necessary that Substance exist. As yet there is, however, no syllogistic process; for the propositions do not stand in the relation we have stated. But, in such like cases, we are deceived by reason of the necessity of something resulting from what has been laid down; whilst, at the same time, the syllogism is of things necessary. But the Necessary is more extensive than the Syllogism; for though all syllogism be indeed necessary, all necessary is not syllogism."—Why not 1 1°, No middle; 2°, No quality,—affirmation or negation:—problematic, not assertory;—hypothetical not syllogistic; 3°, No quantity. Compare also *An. Pr.* L. i. c. 24.

Aristotle, (*Anal. Post.*, L. i. c. 2, § 15, p. 418; c. 10, §§ 8, 9, p. 438), makes *Thesis* or Position the genus opposed to *Axiom*, and containing under it, as species, 1°, *Hypothesis* or Supposition; and, 2°, *Definition*. Hypothesis is that thesis which assumes one or other alternative of a contradiction. Definition is that thesis which neither affirms nor denies—Hypothetical, in Aristotle's sense, is thus that which affirms or denies one alternative or other,—which is not indifferent to yes or no,—which is possibly not either, and, consequently, includes both. Hypotheticals, as involving a positive and negative alternative, are thus, in Aristotle's sense, rightly named, if divided ; but, in Aristotle's sense, as complete, they are neither propositions nor syllogisms, as not affirming one alternative to the exclusion of the other.[a]

II.—AMMONIUS HERMIÆ.

I. Ammonius Hermiæ, on *Aristotle Of Enouncement*, Introduction, f. 3. ed. Ald. 1546. f. 1. ed. Ald. 1503.—After distinguishing the five species of Speech, according to the Peripatetics,— the *Vocative*, the *Imperative*, the *Interrogative*, the *Optative*, and the *Enunciative* or *Assertive*,—having further stated the corresponding division by the Stoics, and having finally shown that Aristotle, in this book, limited the discussion to the last kind, that alone being recipient of truth and falsehood, he thus proceeds :— " Again, of *Assertive* speech, (ἀποφαντικοῦ λόγου), there are two species ; the one called *Categoric* [or *Predicative*], the other *Hypothetic* [or *Suppositive*]. The Categoric denotes, that *something does or does not belong to something ; as when we say, Socrates is walking, Socrates is not walking ;* for we predicate *walking of Socrates,* sometimes affirmatively, sometimes negatively. The Hypothetic denotes, that *something being, something* [else] *is or is not, or something not being, something* [else] *is not or is:*

a [Whether the *Syllogisms ex Hypothesi* of Aristotle are correspondent to the ordinary Hypothetical Syllogism.

For the affirmative, see Pacius, *Com. in Org. An. Prior.,* L. i. cc. 23, 29, 44, pp. 153, 177, 194. St Hilaire, *Translation of Organon,* vol. ii. pp. 107, 139, 178.

For the negative, see Piccartus, *In Org. An. Prior.,* L. i. cc. 40, 41, 42, p. 500. Neldelius, *De Usu Org. Arist.,* P. iii. c. 2, pp. 38, 45, (1607.) Keckermann, *Opera,* pp. 766, 757. Scheibler, *Opera Logica Tract. Syll.,* P. iv. c. x.

tit. 2, p. 548. Burgersdicius, *Instit. Log.,* L. ii. cc. 12, 14, pp. 253, 270, 273. Ritter, *Gesch. der Phil.,* iii. p. 96. (Eng. Tr., p. 80), Ramus, *Scholæ Dial.,* L. vii. cc. 12, 13, pp. 492, 503. Molinæus, *Elementa Logica,* p. 95 et seq. Waitz, *Org.,* i. pp. 427, 433. Cf. Alexander, *In An. Prior.,* ff. 85, 109. Philoponus, *In An. Prior.,* ff. 60 a, 60 b, 87 b, 88. Anonymus, *De Syllogismo,* f. 44 b. Magentinus, *In An. Prior.,* f. 17 b. Ammonius, *In de Interp.,* 8 b. Blemmidas, *Epit. Log.,* c. 36.]

As when we say, *If man be, animal also is,—If he be man, he is not stone,—If it be not day, it is night,—If it be not day, the sun has not risen.*

"The Categoric is the only species of Assertive speech treated of by Aristotle, as that alone perfect in itself, and of utility in demonstration; whereas Hypothetic syllogisms, usurping [usually] without demonstration the [minor] proposition, called the *Transumption,* or *Assumption,* and sometimes even a [major premise] Conjunctive or Disjunctive, requiring proof, draw their persuasion from hypotheses, should any one [I read εἰ τις for ἥτις,] concede their primary suppositions. If, then, to the establishment of such suppositions, we should employ a second hypothetic syllogism,—in that case, we should require a further establishment for confirmation of the suppositions involved in it; for this third a fourth would again be necessary; and so on to infinity, should we attempt by hypotheses to confirm hypotheses. But to render the demonstration complete and final, it is manifest that there is needed a categoric syllogism to prove the point in question, without any foregone supposition. Hence it is, that Categoric [reasonings] are styled *Syllogisms* absolutely; whereas Hypothetic [reasonings] of every kind are always denominated *Syllogisms from hypothesis,* and never Syllogisms simply. Add to this, that Hypothetic enouncements are made up of Categoric. For they express the consequence or opposition (ἀκολουθίαν ἢ διάστασιν) of one Categoric proposition and another, uniting them with each other, by either the Conjunctive or Disjunctive particle, (συμπλεκτικῷ ἢ διαζευκτικῷ συνδέσμῳ), in order to show that they constitute together a single enouncement. For these reasons, therefore, Aristotle has only considered, in detail, the Categoric species of Assertive speech."

III.—ANONYMOUS SCHOLION.[a]

"In Hypothetic Syllogisms, the first [I] are those of two terms, [a] Conjunctive, or [b] Disjunctive, (ὅροι συνημμένοι ἢ διαλελυμένοι); then follow [II] the two [classes of] syllogisms with three, and these conjunctive, terms.

a In Waitz, *Org.,* i. pp. 9, 10.

[I. a.] "There are four syllogisms through the Return (ἡ ἐπάνοδος) on the prior (ὁ πρότερος, ὁ πρῶτος) [or antecedent clause of the hypothetical proposition], and four through it on the posterior (ὁ δεύτερος, ὁ ἔσχατος). For the terms are taken either both affirmatively, or both negatively. And the return upon the prior is ponent (κατὰ θέσιν), upon the posterior tollent (κατὰ ἀναίρεσιν.) For example [the return upon the prior] :—

(1). If A is, B is ; (Return) but A is ; (Conclusion, συμπέρασμα) therefore B is.
　(2). If A is, B is not ; but A is ; therefore B is not.
　(3). If A is not, B is ; but A is not ; therefore B is.
　(4). If A is not, B is not ; but A is not ; therefore B is not.

" The return upon the posterior :—
　(1). If A is, B is ; but B is not ; therefore A is not.
　(2). If A is, B is not ; but B is ; therefore A is not.
　(3). If A is not, B is ; but B is not ; therefore A is.
　(4). If A is not, B is not ; but B is ; therefore A too is.

[b.] "Following those of conjunctive, are syllogisms of disjunctive, terms. In these, the return is upon either [clause] indifferently. For example : If it must be that either A is or B is ; [in the one case,] B is not, therefore A is ; or, [in the other,] A is not, therefore B is.

[II.] "Of three conjunctive terms, there are [in the figures taken together] eight syllogisms, through a return on the prior, and eight [sixteen] [a] through a return on the posterior [clause]. For the three terms are correlated (συντίθενται), either all affirmatively, or some ; and here either the third alone, or the third and second, or the second alone, negatively. Again, either all are negatively correlated, or some ; and here either the third alone, or the third and second, or the second alone, affirmatively. In this manner the correlation [in each figure] is eightfold ; taking for exemplification only a single mood [in the several figures] :—

　　　　　　If A is, B is ;
　　　　　　If B is, C is ;
　　　　　　If A is, therefore C is.

a It would seem that the Author here, and in the last sentence, discounts altogether the first figure, puzzled, apparently, to which premise, (the minor placed first, according to the common practice of the Greeks, or the major prior, in Aristotelic theory), he should accord the designation of first.

This is of the first figure. For the middle collative term
(ὁ συνάγων ὅρος μέσος) is twice taken, being the consequent
(ὁ λῆγων) in the former conjunctive [premise] (τὸ πρότερον
συνημμένον), the antecedent (ὁ ἡγούμενος) in the latter.
Wherefore, these syllogisms are indemonstrable,* not requiring
reduction (ἡ ἀνάλυσις) for demonstration. The other moods
of the first figure are, as has been said, similarly circumstanced.

" The second figure is that in which the collative term [or
middle] (ὁ συνάγων) holds the same relation to each of the collated [or extreme] terms, inasmuch as it stands the antecedent
of both the conjunctive [premises], except that in the one it is
affirmative ; in the other, negative. Wherefore, when reduced to
the first figure, they demonstrate, as is seen, through the instance
of a single mood composed of affirmative collated terms. As :

> If A is, B is ;
> If A is not, C is ;
> If B is not, therefore C is.

This is reduced to the first figure in the following manner.
Whether it has the collated terms, both affirmative, or both negative, or both dissimilar to the reciprocally placed collative term,
there is taken in the reduction the opposite [and converse] of the
prior conjunctive [premise], and the latter is applied, in order
that the opposite of the consequent in the former conjunctive
[premise] may find a place in the foresaid mood. As :

> If B is not, A is not ;
> If A is not, C is ;
> If B is not, therefore C is.

This it behoved to show.

" The third figure is that in which the collative term holds
the same relation to each of the collated terms, being the consequent in either conjunctive [premise] affirmatively and negatively, as in the example of a single mood again consisting of
affirmative collated terms. Thus :

> If A is, B is ;
> If C is, B is not ;
> If A is, therefore C is not.

---

a Vide Apuleius. [De Dogm. Plat., 836.—Ed.]
lii. p. 37. Kln. Cf. Discussions, p.

The reduction of this to the first figure is thus effected. The opposite [and converse] of the second conjunctive [premise] is taken along with the first conjunctive [premise], and the antecedent of the former is applied to the opposite of the latter's consequent; as in the foresaid mood, thus:

> If A is, B is;
> If B is, C is not;
> If A is, therefore C is not.

"All this requires to be shown concretely. As in the first figure, [first mood:]

> If day is, light is;
> If light is, visible objects are seen;
> If day is, therefore visible objects are seen.

Second figure, first mood:

> If day is, light is;
> If day is not, the sun is under the earth;
> If light is not, the sun is [therefore] under the earth.

Reduction:

> If light is not, day is not;
> If day is not, the sun is under the earth;
> If light, therefore, is not, the sun is under the earth.

Third figure, first mood:

> If day is, light is;
> If things visible are unseen, light is not;
> If day, therefore, is, things visible are not unseen.

"There are eight moods of the second figure, and eight of the third; two composed of affirmatives, two of negatives, four of dissimilars with a similar or dissimilar collative.

"End of Aristotle's Analytica."

Relative to the translation from the Greek interpolator on Hypothetical Syllogisms, in Waitz, (Org., i. pp. 9, 10); and in particular to the beginning of [II.]

Better thus:—In all the Figures:—the Quality of the syllogism is either *Pure*,—and here two, viz., one affirmative and one negative; or *Mixed*,—and here six, viz., three in which affirmation, and three in which negation, has the preponderance.

The following are thus arranged :—

		First Figure.	Second Figure.	Third Figure.
	All **A**	If A is, B is; If B is, C is; ∴ If A is, C is.	If B is, A is; If B is, C is; ∴ If A is, C is.	If A is, B is; If C is, B is; ∴ If A is, C is.
	1, 2 **B**	If A is, B is; If B is, C is not; ∴ If A is, C is not.	If B is, A is; If B is, C is not; ∴ If A is, C is not.	If A is, B is; If C is not, B is; ∴ If A is, C is not.
Affirmation of parts preponderant.	**1, 3** **C**	If A is, B is not; If B is not, C is; ∴ If A is, C is.	If B is not, A is; If B is not, C is; ∴ If A is, C is.	If A is, B is not; If C is, B is not; ∴ If A is, C is.
	2, 3 **D**	If A is not, B is; If B is, C is; ∴ If A is not, C is.	If B is, A is not; If B is, C is; ∴ If A is not, C is.	If A is not, B is; If C is, B is; ∴ If A is not, C is.
	All **E**	If A is not, B is not; If B is not, C is not; ∴ If A is not, C is not.	If B is not, A is not; If B is not, C is not; ∴ If B is not, C is not.	If A is not, B is not; If C is not, B is not; ∴ If A is not, C is not.
Negation of parts preponderant; but affirmative in general.	**1, 2** **F**	If A is not, B is not; If B is not, C is; ∴ If A is not, C is.	If B is not, A is not; If B is not, C is; ∴ If A is not, C is.	If A is not, B is not; If C is, B is not; ∴ If A is not, C is.
	1, 3 **G**	If A is not, B is; If B is, C is not; ∴ If A is not, C is not.	If B is, A is not; If B is, C is not; ∴ If A is not, C is not.	If A is not, B is; If C is not, B is; ∴ If A is not, C is not.
	2, 3 **H**	If A is, B is not; If B is not, C is not; ∴ If A is, C is not.	If B is not, A is; If B is not, C is not; ∴ If A is, C is not.	If A is, B is not; If C is not, B is not; ∴ If A is, C is not.

These eight syllogisms are all affirmative, the negation not being attached to the principal copula.[a] If, therefore, the negation be attached to one or other premise, there will be sixteen negative syllogisms, in all twenty-four. The negatives are, however, awkward and useless.—(See Lovaniensea, p. 301.)

But each of these twenty-four syllogisms can receive twelve different forms of predesignation, corresponding to the twelve moods of the simple categorical; according to which they are arranged and numbered. It is hardly necessary to notice that

a See Lovaniensea, *In Arist. Dial., Tract. de Hypotheticis Syllogismis*, p. 299.

the order of the premises is in Comprehension, after the Greek fashion of the scholiast.

	i.	ii.	iii.	iv.	v.	vi.	vii.	viii.	ix.	x.	xi.	xii.
Γ A	:	,	.	:	:	,	:	.	.	:	.	,
M B	: :	: :	: :	: :	, :	: ,	, :	: ,	, :	, :	, :	: ,
C C	:	,	:	,	:	:	,	:	:	,	,	,

This is exemplified in the Syllogism E of the preceding table, thus:

1. If all A is not, all B is not; if all D is not, all C is not; ∴ If all A is not, all B is not.
2. If some A is not, all B is not; if all B is not, some C is not; ∴ If some A is not, some C is not.
3. If some A is not, all B is not; if all B is not, all C is not; ∴ If some A is not, all C is not.
4. If all A is not, all B is not; if all B is not, some C is not; ∴ if all A is not, some C is not.
5. If all A is not, some B is not; if all B is not, all C is not; ∴ If all A is not, all C is not.
6. If some A is not, all B is not; if some B is not, all C is not; ∴ if some A is not, all C is not.
7. If all A is not, some B is not; if all B is not, some C is not; ∴ If all A is not, some C is not.
8. If some A is not, all B is not; if some B is not, all C is not; ∴ If some A is not, all C is not.
9. If some A is not, some B is not; if all B is not, all C is not; ∴ If some A is not, all C is not.
10. If all A is not, all B is not; if some B is not, some C is not; ∴ If all A is not, some C is not.
11. If some A is not, some D is not; if all D is not, some C is not; ∴ If some A is not, some C is not.
12. If some A is not, all D is not; if some D is not, some C is not; ∴ if some A is not, some C is not.

There will thus be 96 (12 × 8) affirmative syllogisms of this kind, and 192 negatives; in all 288.

X.

SORITES.

(See above, Vol. I. p. 385.)

(Without order.)

All logicians have overlooked the Sorites of Second and Third Figures.

In Sorites of the Second or Third Figure, every term forms a syllogism with every other through the one middle term: in Sorites of the First Figure, every Second term at most forms a syllogism with every other, through its relative middle term.

No subordination in Sorites of Second or Third Figure, *ergo*, no one dominant conclusion.

Alias—In First Figure, there being a subordination of notions, there may be a Sorites with different middles, (all, however, in a common dependency). In Second and Third Figures, there being no subordination of terms, the only Sorites competent is that by repetition of the same middle. In First Figure, there is a new middle term for every new progress of the Sorites; in Second and Third, only one middle term for any number of extremes.

In First Figure, a syllogism only between every second term of the Sorites, the intermediate term constituting the middle term. In the others, every two propositions of the common middle term form a syllogism.

Alias—There being no subordination in Second and Third Figures between the extremes, there, consequently, are—

1°, No relations between extremes, except through the middle term.

2°, There is only one possible middle term; any number of others.

3°, Every two of the terms, with the middle term, may form a syllogism.

4°, No order.

———

Before concluding this subject, I would correct and amplify the doctrine in regard to the Sorites.[a]

---

[a] Interpolation in *Lectures.* See above, Vol. I. p. 385.—En.

1°, I would state that, by the quantification of the Predicate, (of which we are hereafter to treat, in reference to reasoning in general), there are two kinds of Sorites: the one descending from whole to part, or ascending from part to whole; the other proceeding from whole to whole; of which last it is now alone requisite to speak. It is manifest, that if we can find two notions wholly equal to a third notion, these notions will be wholly equal to each other. Thus, if all trilateral figure be identical with all triangular figure, and all triangular figure with all figure the sum of whose internal angles is equal to two right angles, then all figure the sum of whose internal angles is equal to two right angles, and all trilateral figure, will also be identical, reciprocating or absolutely convertible. We have thus a simple syllogism of absolute equation. On the same principle, if A and B, B and C, C and D, are absolutely equivalent, so also will be A and D. We may thus, in like manner, it is evident, have a Sorites of absolute equivalence. It is not, indeed, very easy always to find four or more terms or notions thus simply convertible. In geometry, we may carry out the concrete syllogism just stated, by adding the three following propositions;—*All figure, the sum of whose internal angles is equal to two right angles, is all figure which can be bisected through only one angle;—All figure which can be bisected through only one angle, is all figure which, bisected through an angle and a side, gives two triangles; and All figure which, thus bisected, gives two triangles, is all figure which, bisected through two sides, gives a triangle and a quadrangle,* and so forth. In theology, perhaps, however, these series are more frequently to be found than in the other sciences. The following twelve equivalent concepts constitute at once a good example of such a Sorites, and at the same time exhibit a compendious view of the whole Calvinistic doctrine. These are:—1. *Elected*—2. *Redeemed*—3. *Called* —4. *Graced with true repentance*—5. *With true faith*—6. *With true personal assurance*—7. *Pardoned*—8. *Justified*—9. *Sanctified*—10. *Endowed with perseverance*—11. *Saved.*—12. *Glorified.* This series could indeed be amplified, but I have purposely restricted it to twelve. Now, as *All the elect are all the redeemed, all the redeemed all the called, all the called all the [truly] penitent, all the [truly] penitent all the [truly] believing, all the [truly] believing all the [truly] assured, all the [truly] assured*

all the pardoned, all the pardoned all the justified, all the justi-
fied all the sanctified, all the sanctified all the persererant, all the
perseverant all the saved, all the saved all the glorified, all the
glorified all the blest with life eternal; it follows of necessity,
that *all the blest with life eternal are all the elect.* To turn this
affirmative into a negative Sorites, we have only to say, either
at the beginning,—*None of the reprobate are any of the elect,*
and, consequently, infer, at the end, that *none of the blessed with*
eternal life are any of the reprobate; or at the end,—*None of the*
blessed with eternal life are any of the punished, and, consequently,
infer that *none of the punished are any of the elect.* Perhaps the
best formula for this kind of Sorites is to be found in the letters
a, b, c. This will afford us a Sorites of six terms, viz., a, b, c—
a, c, b—b, a, c—b, c, a—c, a, b—c, b, a—which are all virtually
identical in their contents. If there be required a formula for
a longer Sorites, we may take the letters a, b, c, d, which will
afford us twenty-four terms. Perhaps the best formula for a
descending or ascending Sorites is, for example, a, b, c, d, e, f,—
a, b, c, d, e,—a, b, c, d,—a, b, c,—a, b,—a.

I.—COMPREHENSIVE SORITES—PROGRESSIVE AND REGRESSIVE.

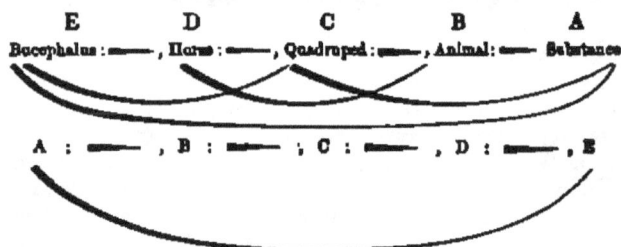

II.—EXTENSIVE SORITES.

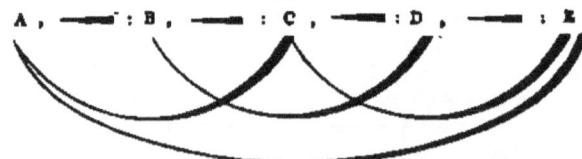

XI.

SYLLOGISM

A.—ITS ENOUNCEMENT—ANALYTIC AND SYNTHETIC— ORDER OF PREMISES.

(See above, Vol. I. p. 305.)

(a) ENOUNCEMENT OF SYLLOGISM.

(Nov. 1848.)—There are two orders of enouncing the Syllogism, both natural, and the neglect of these, added to the not taking into account the Problem or Question, has been the ground why the doctrine of syllogism has been attacked as involving a *petitio principii*, or as a mere tautology. Thus, Buffier cites the definition, *the art of confessing in the conclusion what has been already avowed in the premises.*[a] This objection has never been put down.

The foundation of all syllogism is the Problem. But this may be answered either Analytically or Synthetically.

I. *Analytically* (which has been wholly overlooked) thus:— Problem or quæsitum, " Is Γ C ?" Answer, " Γ is C; for Γ is M, and M is C." This is in the reasoning of Depth. More explicitly :—" Does Γ contain it in C ?" " Γ contains it in C; for Γ contains in it M, and M contains in it C." But it is wholly indifferent whether we cast it in the reasoning of Breadth. For example :—" Does C contain under it Γ ?" " C contains under it Γ; for C contains under it M, and M contains under it Γ."[b]

Here all is natural; and there is no hitch, no transition, in the order of progressive statement. The whole reasoning forms an organic unity; all the parts of it being present to the mind at once, there is no before and no after. But it is the condition of a verbal enouncement, that one part should precede and follow

---

[a] *Seconde Logique*, Art. lii. § 128.— ED.

[b] Plato, in a letter to Dionysius, (*Epist.* 3), reverses the common order of Syllogism, placing the conclusion first (*what he thinks there is some sense in the deed*); then the minor, (*what good men so think*) ; lastly the major, (*that the presentiments of divine men are of highest authority*). *Platonis Opera*, Bekker, ix. p. 74. Cf. Melanchthon, *Dialectica*, L. lii., *De Figuratione*, p. 98, ed. 1542.

another. Here, accordingly, the proposition in which the reasoning is absolved or realised, and which, from the ordinary mode of enouncement, has been styled the *Conclusion*, is stated first; and the grounds or reasons on which it rests, which, from the same circumstance, have been called the *Premises* or *Antecedent*, are stated last. This order is Analytic. We proceed from the effect to the cause,—from the principiatum to the principia. And it is evident that this may be done indifferently either in Depth or Breadth; the only difference being that in the counter quantities the grounds or premises naturally change their order.

II. *Synthetically;*—the only order contemplated by the logicians as natural, but on erroneous grounds. On the contrary, if one order is to be accounted natural at the expense of the other, it is not that which has thus been exclusively considered. For—

1°, It is full of hitches. There is one great hitch in the separation of the conclusion from the question; though this latter is merely the former proposition in an assertive, instead of an interrogative, form. There is also at least one subordinate hitch in the evolution of the reasoning.

2°, The exclusive consideration of this form has been the cause or the occasion of much misconception, idle disputation, and groundless objection.

(On the two methods; tumultuary observations, to be better arranged, and corrected.)

1°, In the first or analytic order, what is principal in reality and in interest, is placed first, that is, the Answer or Assertion, called on the other order the *Conclusion*.

2°, In this order all is natural; there is no hitch, no saltus, no abrupt transition; all slides smoothly from first to last.

a) The question slides into its answer, interrogation demands and receives assertion.

b) Assertion requires a reason and prepares us to expect it; and this is given immediately in what, from the other order, has been called the *Antecedent* or *Premises*.

c) Then the first term, either in Breadth or Depth, is taken first in the ground or reason, and compared with M; then M is compared with the other. As in Breadth:—"Does C contain under it Γ?" "C contains Γ; for C contains under it M, and M contains under it Γ."—In Depth :—Does Γ contain in it C?" "Γ contains in it C; for Γ contains in it M, and M contains in it C." This is the first Figure.—Second Figure, using common language:—"Is Γ C?" "Γ is C (and C is Γ); for Γ and C are both the same M." Here the two extremes taken together are compared with M.—In the Third Figure M is compared with both extremes:—"Is Γ C?" "Γ is C (and M is Γ); for the same M is both Γ and C."

3°, In this order there is nothing pleonastic, nothing anticipated.

4°, Nothing begged.

5°, In this method the process is simple. Thought is one; but to be enounced it must be analysed into a many. This order gives that necessary analysis, and nothing more.

6°, In this order, when assertive, answer is limited by question ; good reason why, in Second and Third Figures, *one* answer should be given.

7°, This order is the one generally used by the mathematicians. (See Twesten, *Logik*, § 117, p. 105, and below, p. 413. Plato also.)

8°, If the Quæsitum be stated as it ought to be, this order follows of course ; and the neglect of the quæsitum has followed from the prevalence of the other. If the quæsitum be stated in using the common form, we must almost of course interpolate a "yes" or a "no" before proceeding to the premises in the common method; and, in that case, the conclusion is only a superfluous recapitulation.

In the Synthetic, or common order, all is contrary. (The numbers correspond.)

1°, In this order, what is first in reality and interest, and in and for the sake of which the whole reasoning exists, comes last; till the conclusion is given we know not, (at least we ought not to know), how the question is answered.

2°, In this order all is unnatural and contorted by hitches and abrupt transitions. There is no connection between the question and what prepares the answer,—the premise. (Show in detail.)

3°, In this order all is pleonastic and anticipative. The premises stated, we already know the conclusion. This, indeed, in books of Logic, is virtually admitted,—the conclusion being commonly expressed by a "therefore," &c. Ancient doctrine of Enthymeme, (Ulpian, &c.), unknown to our modern logicians; among their other blunders on the Enthymeme. On the common doctrine, Logic,—Syllogistic,—is too truly defined the art of confessing in the conclusion what had been already avowed in the premises.

. 4°, On this order the objection of *petitio principii* stands hitherto unrefuted, if not unrefutable, against Logic.ª

5°, In this order the process is complex. The simple thought is first mentally analysed, if it proceed, as it ought, from the quæsitum; but this analysis is not expressed. Then the elements are recomposed, and this recomposition affords the synthetic announcement of the syllogism,—the syllogism being thus the superfluous regress of a foregone analysis. Aristotle's analytic is thus truly a synthetic; it overtly reconstructs the elements which had been attained by a covert analysis.ᵝ

6°, In this method, the problem hanging loose from the syllogism, and, in fact, being usually neglected, it does not determine in the Second and Third Figures one of the two alternative conclusions, which, *ex facie syllogismi*, are competent in them. The premises *only* being, there is no reason why one of the conclusions should be drawn to the preference of the other. *Mem.* Counter-practice old and new. The logicians ought not, however, to have ignored this double conclusion.

7°, See corresponding number.

8°, See corresponding number.⁷

(b) ORDER OF PREMISES.

Aristotle places the middle term in the first Figure between the

a [Stewart (*Elements*, vol. ii. ch. 3, § 2, *Works*, vol. iii. p. 202, *et alibi*) makes this objection. Refuted by Galluppi *Lez. di Logica e di Metafisica*, Lez. i. p. 242 *et seq.*]

β [Aristotle's *Analytics* are in synthetic order; they proceed from the simple to the compound; the elements they commence with are gained by a foregone analysis, which is not expressed. They are as synthetic as a grammar commencing with the letters.

The meaning of the term is the doctrine showing how to analyse or reduce reasonings to syllogisms; syllogisms to figure; figure to mood; second and third figures to first; syllogisms to propositions and terms; propositions to terms; for of all these analysis is said. See *Pacii Organon, An. Prior.*, i. cc. 2, 32, 42, 44, 45, [pp. 128, 281, 273, 275, 278, 280.]

γ Compare *Discussions*, p. 652.—ED.

extremes, and the major extreme first ;—in the second Figure
before the extremes and the major extreme, next to it ;—in the
third Figure after the extremes, and the minor extreme next to it.

In his mode of enouncement this relative order is naturally
kept; for he expresses the predicate first and the subject last,
thus : A *is in all* B, or A *is predicated of all* B, instead of saying *All* B *is* A.

But when logicians came to enounce propositions and syllogisms
in conformity to common language, the subject being usually first,
they had one or other of two difficulties to encounter, and submit
they must to either; for they must either displace the middle
term from its intermediate position in the first Figure, to say
nothing of reversing its order in the second and third ; or, if they
kept it in an intermediate position in the first Figure, (in the
second and third the Aristotelic order could not be kept), it
behoved them to enounce the minor premise first.

And this alternative actually determined two opposite procedures,—a difference which, though generally distinguishing the
logicians of different ages and countries into two great classes,
has been wholly overlooked. All, it must be borne in mind, regard the syllogism in Figure exclusively, and as figured only in
Extension.

The former difficulty and its avoidance determined the older
order of enouncement, that is, constrained logicians to state the
minor premise first in the first Figure; and, to avoid the discrepancy, they of course did the same for uniformity in the second
and third.

The latter difficulty and its avoidance determined the more
modern order of enouncement, that is, constrained logicians to
surrender the position of the middle term as middle, in following
the order of the major premise first in all the Figures.

Philoponus on the First Book of the *Prior Analytics*, c. iv. § 4,
(Pacian Division), f. 20 a, ed. Trincavelli.—"This definition appears
to be of the extremes and of the middle term; but is not. It
behoves, in addition, to interpolate in thought an '*only ;*' and
thus will it be rightly enounced, as if he had said :—' *But the ex-*
'*tremes are both that which is only in another, and that in which*
'*another only is.*'   For if A is [predicated] of all B, and B is [pre-

dicated] of all C, it is necessary that A should be predicated of
all C. This is the first syllogistic mood. Two universal affirmatives, inferring an universal conclusion. For if B is in all C,
consequently C is a part of B; but again D is a part of A; consequently, A is in all C, inasmuch as C is a part of B. But what
is here said will appear more clearly from a concrete example—
Substance of all animal; animal of all man; (there follows),
substance of all man. And backwards (ἀνάπαλιν), *All man
animal; all animal substance; all man therefore substance.* In
regard to this figure, it is plain how we ought to take the terms
of the first mood. The first [major] is most generic; the second
[middle] is a subaltern genus; and the third [minor] is a species
more special than the middle. But a conclusion is here always
necessary. Thus, following the synthetic order, that is, if we
start from the major term, *substance* begins, beginning also the
conclusion. *Substance of all animal,* (*substance* stands first);
animal of all man; (finally, the conclusion commences with
substance),—*substance of all man.* But if [on the analytic order]
we depart from the minor term, as from *man,* in this case the
conclusion will, in like manner, begin therewith: *All man animal; all animal substance; all man substance.*"

This is the only philosophic view of the matter. His syllogisms really analytic (=in Depth.)

Analytic and *Synthetic* ambiguous. Better,—order of *Breadth,*
and *Depth.*[a]

a [Instances and authorities for the
enouncement of Syllogism, with the
Minor Premiss stated first :—

A. ANCIENTS.

*Greeks :—*Gregory of Nyssa, *Opera,*
t. ii. p. 612, in his 12 (not 10) Syllogisms against Manicheans, varies.
These very corrupt. Joannes Damas-
cenus, (*Dialectica,* c. 64, *Opera,* ed.
Lequien, Paris, 1712, t. i. pp. 65, 66),
gives two Syllogisms, one with minor
first. Alcinous, *De Doct. Plat.,* L. i.
cc. 5 and 6. Aristotle often places
minor first. See Zabarella, *Opera Logica De Quarta Figura,* p. 124. Vallius, *Logica,* t. ii. pp. 72, 76. Aristo-

tle and Alexander not regular in stating major propositions. See in First
Figure, *An. Pr.* i. c. 4. Aristotle used
the "*whole*" only of the predicate.
See Zabarella, *Tabula, In An. Prior.,*
p. 149. (But see above, p. 307.) Boethius, *Opera,* pp. 562, 583. Aristotle,
An. Pr. i. c. 1, *sub fine,* ubi Alexander,
f. 9 a. Philoponus, f. 17 a, f. 11 b.
Alexander Aph., *In An. Pr.* i. ff. 9 a,
15 b. Philoponus, *In An. Pr.* i., ff. 11
b, 20 a, explains the practice of Greek
Peripatetics in this matter. See also
ff. 17 a, 18 a; and 11 a, 21 a—these
in i. Fig.—in ii. Fig. 23 b. The same,
In Physica, i. c. 1, f. 2. Themistius,
In An. Post. ii. c. 4. Anonymus,

B.—FIGURE—UNFIGURED AND FIGURED SYLLOGISM.

(1853.)—(a) CONTRAST AND COMPARISON OF THE VARIOUS KINDS OF FORMAL SYLLOGISM—DIFFERENCE OF FIGURE ACCIDENTAL.

A). *Unfigured Syllogism—One* form of syllogism: for here there is abolished, 1°, The difference of Breadth and Depth, for

De Syllogismo, I. 43 a. Gregorius Anepouymos, *Compend. Philosophiæ Syntagma*, L. v. cc. 1, 6, pp. 58, 70. Georgius Diaconus Pachymeria, *Epit. Log.*, tit. iv. cc. 1—4. Sextus Empiricus, *Pyrrh. Hypotypos.*, L. ii. cc. 13, 14, pp. 103, 110. Clemens Alex., *Strom.*, L. viii. *Opera*, p. 784, (ed. Sylburgii.) Blemmides, *Epitome Logica*, c. 31, p. 210. Gregorius Trapezuntius, *Dialectica De Syll.*, p. 30 : "Prima (Figura) est in qua medius terminus subjicitur in majore, et in minore prædicatur: quemvis contra fieri et soleat et possit." A Greek, he wrote in Italy for the Latins ; but refers here to the practice of his countrymen.

Latins :—Cicero, *De Fin.* iii. 8; iv. 18; *Tusc. Disp.* iii. 7; v. 15; *Opera Phil.* pp. 885, 903, 981, 1029, ed. Verburgii. Macrobius, *Opera*, p. 181, Zennii. Seneca, *Epist.* 85, p. 303. Apuleius, *De Habit. Doct. Plat.*, L. iii. p. 36, ed. Elmenhorst. Isidorus, in *Gothofr. Auctores*, p. 878. Cassiodorus, *Dialectica*, *Opera*, p. 556, Genev. 1650, gives alternative, but in Psalm xxxi. v. 16, gives a syllogism with minor first. Martianus Capella, *De Septem Artibus Liberalibus*, allows both forms for first Figure ; generally makes the minor first (see below, p. 433). Boethius, (origo mali), v. *Opera*, p. 594 *et seq.*

Orientals.—a. *Mahommedans :*—Averroes (enouncing as we) in all the Figures has minor first. (See below, p. 433.)

b. *Jews :* — Rabbi Simeon [truly Maimonides] (in Hebrew,) *Logica*, per S. Munsterum, cc. 6, 7, Basil, 1527.

B. MODERNS.

Modern anticipations of the doctrine that the Minor Premise should precede the Major. Valla, *Dialectica*, I. 60 l., &c. *Opera*, pp. 733, 736. Joannes Neomagus, *In Trapezuntium*, f. 38 b, (only adduces examples.) Caramuel, *Rat. et Realis Philosophia Logica*, Disp. ix. xvi. Aquinas, *Opusc.* 47. (Camerarius, *Disp. Phil.*, P. i. qu. 13, p. 117.) Alstedius, *Encyclopædia*, p. 437. Gassendi, *Opera*, ii. p. 413; i. p. 107. Camerarius, *Disp. Phil.*, P. i. qu. 13, p. 117. Leibnitz, *Opera*, T. ii. Pars i. p. 356, *Dissert. de Arte Combinatoria*, (1666), ed. Dutens, who refers to Ramus, Gassendi, Alcinous, &c. Cf. *Nouveaux Essais*, L. iv. § 6, p. 454, ed. Rasp; and Locke's *Essay*, ibid. Buffier, *Logique*, § 68. Camerius, *Dialectica*, Tract. v. *De Syll. Cat.* p. 158, (first ed. 1532). J. C. E. *Nova Detecta Veritas*, &c., see Beusch, *Systema Logicum*, § 547, p. 626. Chauvin, *Lexicon Philosophicum*, v. *Figura*. Hobbes, *Computatio*, c. iv. prefixes the minor, (see Hallam, *Lit. of Europe*, vol. iii. c. 3, p. 309, ed. 1839.) Lambert, *Neues Organon*, i. 136, § 225. Bachmann, *Logik*, § 133, pp. 202, 226. Hollmann, *Logica*, § 454. Esser, *Logik*, § 107, p. 210. Krug, *Logik*, § 114, p. 408. Denze, *System der Logik*, c. v. p. 210 *et seq.* Stapulensis, in Sergeant's *Method to Science*, p. 127. Facciolati, (though he errs himself), *Rudimenta Logicæ*, p.

the terms are both Subject or both Predicate, and may be either indifferently; 2°, All order of the terms, for these may be enounced first or second indifferently ; 3°, All difference of major or minor term or proposition, all duplicity of syllogism ; 4°, All difference of direct and indirect conclusion.

B). *Figured Syllogism—Two* forms of syllogism by different orders of terms :—

First Figure.—Here the two forms of syllogism are possible, each with its major and minor terms, each with its direct or immediate, its indirect or mediate, conclusion. These two various forms of syllogism are essentially one and the same, differing only accidentally in the order of enouncement, inasmuch as they severally depart from one or from the other of the counter, but correlative, quantities of Depth and Breadth, as from the containing whole. But in fact, we *may* enounce each order of syllogism, [in] either quantity, the one is the more natural. . . . . .

Second and Third Figures.—In each of these figures there are possible the two varieties of syllogism ; but not, as in the first figure, are these different forms variable by a counter quantity, and with a determinate major and minor term ; for in each the extremes and the middle term (there opposed) are necessarily in the same quantity, being either always Subject or always Predi-

88, P. lii. c. 3, note 4, where Boethius, Sextus Empiricus, Alcinous, &c. Ch. Mayne, *Essay on Natural Notions*, p. 172 *et seq.* Lamy, *Acta Erud.*, 1708, p. 67.

Who have erred in this subject,— making our order of enunciation the natural and usual. Vives, *Censura Veri, Opera*, t. i. p. 606. J. G. Vossius, *De Nat. Art. Liberal.*, *Logica*, c. viii. § 9. J. A. Fabricius, *Ad. Sext. Emp.* 103. Facciolati, *Rudimenta Logica*, p. 86. Waitz, *In Org. Comm.*, pp. 380, 386.

That Reasoning in Comprehensive Quantity most natural. Wolf, *Phil. Rat.* § 399, p. 327. Reusch, *Systema Logicum*, § 547. Schulze, *Logik*, § 77 of old (1817) § 72 of last (1831), edition, holds that *dictum de omni*, &c., evolved out of *nota notæ*, for mere subordination syllogisms. Hauschius, in

Acta Erud. 1728, p. 470. Lamy (B.) in *Acta Erud.* 1708, p. 67. Oldfield, *Essay on Reason*, p. 240. Valla, *Dialectica*, L. iii. c. 45. Hoffbauer, *Analytik der Urtheile und Schlüsse*, § 152, p. 198. Mayne's *Rational Notions*, p. 123 *et seq.* Mariotte, *Logique*, Part ii. disn. iii. p. 161. Paris, 1678. Chladenius, *Phil. Def.*, p. 18, (in Wolf, *Phil. Rat.*, §581.) Castillon, *Mem. de Berlin*, 1802. Hallam, *Lit. of Europe*, vol. iii. p. 309. Thomson (W.), *Outlines of the Laws of Thought*, p. 30. In reference to the above, the mathematicians usually begin with what is commonly called the Minor Premise, (as A = B, B = C, *therefore* A = C) ; and frequently they state the Conclusion first, (as A = B, *for* A = M, and M = B), or &c. ; see Wolf, *Phil. Rat.*, § 551, Twesten, *Logik*, § 117, p. 105, and Lambert, *Neues Org.*, I. § 225.]

cale in the jugation. They differ only as the one extreme, or
the other, (what is indifferent), is arbitrarily made the Subject
or Predicate in the conclusion. Indirect or Mediate conclusions
in these figures are impossible; for the indirect or mediate conclusion of the one syllogism is in fact the direct conclusion of
the other.

Thus difference of Figure accidental.

——————

If rule true, it will follow that it is of no consequence
whether:—

1°, The middle one or any other of the three terms be, in any
proposition, subject or predicate, if only either. Hence difference of *Figure* of no account in varying the syllogism. Thus,
(retaining the subordination of terms) convert major proposition
in Extension of first Figure, and you have second Figure; convert minor proposition, and you have third Figure; convert both
premises, and you have fourth Figure.

2°, Whether one of the extremes, one or other of the premises,
stand first or second, be, in fact, major or minor term of a proposition; all that is required is, that the terms and their quantities
should remain the same, and that they should always bear to
each other a relation of subject and predicate. Thus, (if [in] any
of the Figures,) the major and minor terms and propositions
interchange relation of subordination, when, in the first Figure,
you convert and transpose; and when [in] the other three
Figures (fourth ?), you simply transpose the premises.

Indifferent (in first Figure) which premise precedes or follows.
For of two one not before the other in nature. But not indifferent in either whole, which term should be subject and predicate
of conclusion.[a]

(b) DOUBLE CONCLUSION IN SECOND AND THIRD FIGURES.

My doctrine is as follows:—

In the *Unfigured Syllogism* there is no contrast of terms, the
notions compared not being to each other subject and predicate;
consequently, the conclusion is here necessarily one and only one.

a Compare *Discussions*, p. 652.—ED.

In the *Figured Syllogism* we must discriminate the Figures.

In the First Figure where the middle term is subject of the one extreme and predicate of the other, there is of course a determinate major extreme and premise, and a determinate minor extreme and premise; consequently, also, one proximate or direct, and one remote or indirect, conclusion,—the latter by a conversion of the former.

In the Second and Third Figures all this is reversed. In these there is no major and minor extreme and premise, both extremes being either subjects or predicates of the middle; consequently, in the inference, as either extreme may be indifferently subject or predicate of the other, there are two indifferent conclusions, that is, conclusions neither of which is more direct or indirect than the other.

This doctrine is opposed to that of Aristotle and the logicians, who recognise in the Second and Third Figures a major and minor extreme and premise, with one determinate conclusion.

The whole question in regard to the duplicity or simplicity of the conclusion in these latter figures depends upon the distinction in them of a major and a minor term; and it must be peremptorily decided in opposition to the universal doctrine, unless it can be shown that in these figures this distinction actually subsists. This was felt by the logicians; accordingly they applied themselves with zeal to establish this distinction. But it would appear, from the very multiplicity of their opinions, that none proved satisfactory; and this general presumption is shown to be correct by the examination of these opinions in detail,—an examination which evinces that of these opinions there is no one which ought to satisfy an inquiring mind.

In all, there are six or five different grounds on which it has been attempted to establish the discrimination of a major and minor term in the Second and Third Figures. All are mutually subversive; each is incompetent. Each following the first is in fact a virtual acknowledgment that the reason on which Aristotle proceeded in this establishment, is at once ambiguous and insufficient.—I shall enumerate these opinions as nearly as possible in chronological order.

1. *That the major is the extreme which lies in the Second Figure nearer to, in the Third Figure farther from, the middle.*—

This is Aristotle's definition, (*An. Pr.*, L. i., cc. 5, 6). At best it is ambiguous, and has, accordingly, been taken in different senses by following logicians; and in treating of them it will be seen that in none, except an arbitrary sense, can the one extreme, in these figures, be considered to lie nearer to the middle term than the other. I exclude the supposition that Aristotle spoke in reference to some scheme of mechanical notation.

2. *That the major term in the antecedent is that which is predicate in the conclusion.*—This doctrine dates from a remote antiquity. It is rejected by Alexander; but, adopted by Ammonius and Philoponus, (ff. 17 b, 18 a, ed. Trinc.), has been generally recognised by subsequent logicians. Its recognition is now almost universal. Yet, critically considered, it explains nothing. Educing the law out of the fact, and not deducing the fact from the law, it does not even attempt to show why one being, either extreme may not be, predicate of the conclusion. It is merely an empirical,—merely an arbitrary, assertion. The Aphrodisian, after refuting the doctrine, when the terms are indefinite (preindesignate), justly says :—" Nor is the case different when the terms are definite [predesignate]. For the conclusion shows as predicate the term given as major in the premises ; so that the conclusion is not itself demonstrative of the major; on the contrary, the being taken in the premises as major, is the cause why a term is also taken as predicate in the conclusion."—(*An. Pr.*, f. 24 a, ed. Ald.)

3. *That the proximity of an extreme to the middle term, in Logic, is to be decided by the relative proximity in nature, to the middle notion of the notions compared.*—This, which is the interpretation of Aristotle by Herminus, is one of the oldest upon record, being detailed and refuted at great length by the Aphrodisian, (ff. 23 b, 24 a). To determine the natural proximity required is often difficult in affirmative, and always impossible in negative, syllogism ; and, besides the objections of Alexander, it is wholly material and extralogical. It is needless to dwell on this opinion, which, obscure in itself, seems altogether unknown to our modern logicians.

4. *That the major term in the syllogism is the predicate of the problem or question.*—This is the doctrine maintained by Alexander, (f. 24 b); but it is doubtful whether at first or second hand. It has been adopted by Averroes, Zabarella, and sundry of

the acuter logicians in modern times. It is incompetent, however, to establish the discrimination. Material, it presupposes an intention of the reasoner; does not appear *ex facie syllogismi*; and, at best, only shows which of two possible quæsita,—which of two possible conclusions,—has been actually carried out. For it assumes, that of the two extremes either might have been major in the antecedent, and predicate in the conclusion. If Alexander had applied the same subtlety in canvassing his own opinion, which he did in criticising those of others, he would not have given the authority of his name to so untenable doctrine.

5. *That the major extreme is that contained in the major premise, and the major premise that in the order of enouncement first.*—This doctrine seems indicated by Scotus, (*An. Pr.*, L. i., qu. xxiv. §§ 5, 6) ; and is held explicitly by certain of his followers. This also is wholly incompetent. For the order of the premises, as the subtle Doctor himself observes, (*Ib.*, qu. xxiii. § 6), is altogether indifferent to the validity of the consequence ; and if this external accident be admitted, we should have Greek majors and minors turned, presto, into Latin minors and majors.

6. *That the major extreme is that contained in the major premise, and the major premise that itself most general.*—All opposite practice originates in abuse. This opinion, which coincides with that of Herminus (No. 3), in making the logical relation of terms dependent on the natural relation of notions, I find advanced in 1614, in the *Disputationes* of an ingenious and independent philosopher, the Spanish Jesuit, Petrus Hurtado de Mendoza, (*Disp. Log. et Met.*, L, Disp. x. §§ 50-55). It is, however, too singular, and manifestly too untenable, to require refutation. As material, it is illogical ; as formal, if allowed, it would at best serve only for the discrimination of certain moods ; but it cannot be allowed, for it would only subvert the old without being adequate to the establishment of aught new. It shows, however, how unsatisfactory were the previous theories, when such a doctrine could be proposed by so acute a reasoner, in substitution. This opinion has remained unnoticed by posterior logicians.

The dominant result from this historical enumeration is, that, in the Second and Third Figures, there is no major or minor term, therefore no major or minor premise, therefore two indifferent conclusions.

This important truth, however natural and even manifest it may seem when fully developed, has but few and obscure vaticinations of its recognition during the progress of the science. Three only have I met with.

The first I find in the Aphrodisian, (f. 24 b); for his expressions might seem to indicate that the opinion of there being no major and minor term in the Second figure, (nor, by analogy, in the Third,) was a doctrine actually held by some early Greek logicians. It would be curious to know if these were the "ancients," assailed by Ammonius, for maintaining an overt quantification of the predicate. The words of Alexander are:—"Nor, however, can it be said, that in the present figure there is no major. For this at least is determinate,—that its major must be universal; and,'if there be [in it] any syllogistic combination, that premise is the major which contains the major term." (F. 24 a.) Demurring to this refutation, it is, however, evidence sufficient of the opinion to which it is opposed. This, as it is the oldest, is, indeed, the only authority for any deliberate doctrine on the point.

The second indication dates from the middle of the fifteenth century, and is contained in the *Dialectica* of the celebrated Laurentius Valla (L. iii. c. θ [51]). Valla abolishes the third figure, and his opinion on the question is limited to his observations on the second, in treating of *Cesare* and *Camestres*, which, after a host of previous logicians, he considers to be a single mood. There is nothing remarkable in his statement : "Neque distinctæ sunt propositio et assumptio, ut altera major sit, altera minor, sed quodammodo pares ; ideoque sicut neutra vindicat sibi primum aut secundum locum, ita utraque jus habet in utraque conclusione. Verum istis placuit, ut id quod secundo loco poneretur, vendicaret sibi conclusionem : quod verum esset nisi semper gemina esset conclusio. Sed earum dicamus alteram ad id quod primo loco, alteram ad id quod secundo loco positum est, referri." We, therefore, await the development of his doctrine by relation to the other moods, *Festino* and *Baroco*, which thus auspiciously begins :— "Idem contingit in reliquis duobus : qui tamen sunt magis distincti." We are, however, condemned to disappointment. For, by a common error, excusable enough in this impetuous writer, he has confounded singulars (definites) with particulars (indefinites);

APPENDIX. 419

and thus the examples which he adduces of these moods are, in
fact, only examples of *Cesare* and *Camestres*. The same error
had also been previously committed (L. iii. c. 4). The whole,
therefore, of Valla's doctrine, which is exclusively founded on
these examples, must go for nothing; for we cannot presume, on
such a ground, that he admits more than the four common moods,
identifying, indeed, the two first, by admitting in them of a
double conclusion. We cannot, certainly, infer, that he ever
thought of recognising a particular,—an indefinite, predicate in
a negative proposition.

The third and last indication which I can adduce is that from
the *Method to Science* of John Sergeant, who has in this as in
his other books, (too successfully), concealed his name under the
initials "J. S." He was a Catholic priest, and, from 1665, an
active religious controversialist; whilst, as a philosopher, in his
Idea Philosophiæ Cartesianæ, a criticism of Descartes, in his
Solid Philosophy, a criticism of Locke,ª in his *Metaphysica*, and in
the present work he manifests remarkable eloquence, ingenuity,
and independence, mingled, no doubt, with many untenable, not
to say ridiculous, paradoxes. His works, however, contain genius
more than enough to have saved them, in any other country, from
the total oblivion into which they have fallen in this,—where, indeed, they probably never were appreciated. His *Method to Science*, (a treatise on Logic), was published in 1696, with a "Preface,
dedicatory to the learned students of both our Universities," extending to sixty-two pages. But, alas! neither this nor any other
of his philosophical books is to be found in the Bodleian.

In the Third Book of his *Method*, which treats of Discourse,
after speaking of the First, or, as he calls it, "only right figure of a
syllogism," we have the following observations on the Second and
Third :—" § 14. Wherefore the other two figures, [he does not recognise the Fourth], are unnatural and monstrous. For, since
nature has shown us that what conjoins two notions ought to be
placed in the middle between them ; it is *against nature and rea-*

ª Sergeant is an intelligent antagonist of both these philosophers, and I
have elsewhere had occasion to quote
him as the first and one of the ablest
critics of the *Essay on Human Under-*
standing. In certain views he antici-
pates Kant; and Pope has evidently
taken from his brother Catholic the
hint of some of his most celebrated
thoughts.

son to place it either *above* them *both*, as is done in that they call the *second figure*, or *under* them *both*, as is done in that figure they call the *third.*

"§ 15. Hence no determinate conclusion can follow, in either of the last figures, from the disposal of the parts in the syllogisms. For since, as appears (§ 13), the extreme which is predicated of the middle term in the *major*, has thence a title to be the predicate in the *conclusion*, because it is above the middle term, which is the *predicate*, or *above* the *other* extreme in the *minor;* it follows, that if the middle term be *twice above* or *twice below* the other two terms in the premises, that reason ceases ; and so it is left indifferent which of the other terms is to be subject or predicate in the conclusion; and the indeterminate conclusion follows, not from the artificial *form* of the syllogism, but merely from the *material* identity of all the three terms ; or from this, that their notions are found in the same *Ens.* Wherefore, from these premises, [in the Second figure],

> Some laudable thing is [all] virtue,
> [All] courtesy is a virtue ;

or, from these, [in the Third],

> [All] virtue is [some] laudable,
> Some virtue is [all] courtesy ;

the conclusion might either be,

> Therefore, [all] courtesy is [some] laudable,
> Or, Some laudable thing is [all] courtesy.

"So that, to argue on that fashion, or to make use of these awkward figures, is not to know certainly the end or conclusion we aim at, but to shoot our bolt at no determinate mark, since no determinate conclusion can in that case follow." (P. 232).

Extremes, it is said, meet. Sergeant would abolish the Second and Third figures, as petitory and unnatural, as merely material corruptions of the one formal first. I, on the contrary, regard all the figures as equally necessary, natural, and formal. But we agree in this : both hold that, in the Second and Third figures, there is a twofold and indifferent conclusion ; howbeit, the one makes this a monstrosity of the syllogistic *matter*, the other, a beauty of the syllogistic *form*. Therefore, though I view Sergeant as wrong in his premises, and "shooting his bolt at no determinate mark," I must needs allow that he has, by chance, hit the bull's eye. I have inserted, within square brackets, the quantifications required to restore and show out the formality of his examples; on my scheme of notation they stand as follows:—

C, ———— : M, ———— : Γ   C, ———— : M, ———— : Γ'

C.—HISTORICAL NOTICES REGARDING FIGURE OF SYLLOGISM.

I.—ARISTOTLE.

Aristotle; Figures and Terms of Syllogism, *Prior Analytics* B. I.

FIRST FIGURE.—Ch. iv.

§ 2. "When three terms [or notions] hold this mutual relation, —that the last is in the whole middle, whilst the middle is or is not in the whole first,—of these extremes there results of necessity a perfect syllogism." [a]

§ 3. "By *middle* term, [B (B)], I mean that which itself is in another and another in it; and which in position also stands intermediate. I call *extremes* both that which is itself in another [the minor], and that in which another is [the major]. For if A be predicated of all B, and B of all C, A will necessarily be predicated of all C."

---

[a] Ch. iv. § 2.—This definition of the First Figure, (founded on the rules De Omni and de Nullo), applies only to the universal moods, but, of these, only to those legitimate and useful,— Barbara and Celarent. It, therefore, seems inadequate, but not superfluous.

Aristotle uses the phrase, "to be in all or in the *whole*," both with reference to *extension*,—for the lower notion B, as contained under the all or whole of the higher notion A: and with reference to comprehension, — for the higher notion A as contained in the all or whole of the lower notion B. In the former sense, which with Aristotle is the more usual, and, in fact, the only one contemplated by the logicians, there is also to be observed a distinction between the inhesion and the predication of the attribute.

§ 10. "I call that the *major* extreme [A (A)] in which the middle is; the *minor* [Γ (C)] that which lies under the middle."

```
[Thus, A ——— A.
       B ——— B.
       Γ ——— C.]
```

SECOND FIGURE.—Ch. v.

§ 1. " When the same [predicate notion] inheres in all of the one and in none of the other, or in all or in none of both [the subject notions],—This I denominate the *Second Figure*."

§ 2. " The *middle* [M (M)] in this [figure] I call that which is predicated of both [notions]; the *extremes*, the [notions] of which the middle is said. The *major* extreme [N (N)] is that towards the middle; the *minor* [Ξ (O)], that from the middle more remote.

§ 3. " The middle is placed out [from between] the extremes, first in position."

```
[So, M ——— M
     N ——— N
     Ξ ——— O]
```

THIRD FIGURE.—Ch. vi.

§ 1. " When in the same [subject notion] one [predicate notion] inheres in all, another in none of it, or when both inhere in all or in none of it, such *Figure* I call the *Third*.

§ 2. " In this [figure] I name the *middle*, that of which both [the other terms] are predicated; the *extremes*, the predicates themselves. The *major* extreme [Π (P)] is that farther from, the *minor* [P (Q)] that nearer to, the middle.

§ 3. " The *middle* [Σ (R)] is placed out [from between] the extremes, last in position."

```
[As, Π ——— P
     P ——— Q
     Σ ——— R]
```

*　　*　　*　　*　　*　　*　　*

General Theory of Figure.—*Prior Analytics*, B. i. c. 23, § 7.

' If, then, it be necessary [in reasoning] to take some [term] common [or intermediate] to both [extreme terms]; this is possible in three ways. For we predicate either [the extreme] A of [the middle] C, and [the middle] C of [the extreme] D; or [the middle] C of both [extremes]; or both [extremes] of [the middle] C. These are the [three] Figures of which we have spoken; and it is manifest, that through one or other of the Figures every syllogism must be realised." a

II. AND III.—ALEXANDER AND HERMINUS.

Alexander, *In An. Pr.*, f. 23 b.

Second Figure, c. v. § 2, Aristotle.—"'The middle extreme is that which lies towards the middle.'

"But it is a question, whether in the Second Figure there be by nature any major and minor extreme, and if there be, by what criterion it may be known. For if we can indifferently connect with the middle term whichsoever extreme we choose, this we may always call the major: and as negative conclusions only are drawn in this figure, universal negatives being also mutually convertible, it follows, that in universal negatives the one term has no better title to be styled major than the other, seeing that the major term is what is predicated, whilst both are here indifferently predicable of each other. In universal affirmatives, indeed, the predicate is major, because it has a wider extent; and for this reason, such propositions are not [simply] convertible; so that here there is by nature a major term which is not to be found in universal negatives.

"Herminus is of opinion that, in the Second Figure,—

[1°.] "If both the extremes, of which the middle is predicated, be homogeneous [or of the same genus], the major term is that most proximate to the genus common to the two. For example:—If the extremes be *bird* and *man; bird* lying nearer to the common genus [*animal*] than *man*, as in its first division, *bird* is thus the major extreme; and, in general, of homogeneous terms, that holding such a relation to the common genus is the major.

a Aristotle here varies the notation by letters of the three syllogistic terms, making C (Γ) stand for the middle term, A and B for the two extremes. This he did, perhaps, to prevent it being supposed, (what his previous nota- tion might appear to indicate,) that the middle term was a notion in the First Figure, necessarily intermediate be- tween the two extremes, in the Second superior, in the Third inferior, to them.

[2°.] " But if the terms be equally distant from the common genus, as *horse* and *man*, we ought to regard the middle predicated of them, and consider of which [term] it is predicated through [that term] itself, and of which through some other predicate; and compare that through which it is predicated of another with that through which it is predicated of [the term] itself. And if that through which [the middle] is predicated of another, (viz., the one extreme), be nearer [than the other extreme] to the common genus, that [extreme] of which [for τούτων οὗ, I read τοῦτον οὗ], the middle is [mediately] predicated, from its closer propinquity to the common genus, rightly obtains the title of *major*. For example: If the extremes be *horse* and *man*, *rational* being predicated of them,—negatively of *horse*, affirmatively of *man*; seeing that *rational* is not of itself denied of *horse*, but because *horse* is *irrational*, whereas *rational* is of itself affirmed of *man*, *horse* is nearer than *man* to their common genus *animal*; *horse* will, therefore, be the major extreme, though *man* be no further removed than *horse* from its proper genus. And this, because that through which the predicate [*i.e.* the middle] is predicated of this last, as being *irrational*, is greater; for *rational* is not denied of *horse* qua *horse*, whilst it is affirmed of *man* qua *man*.

[3°.] "But if the extremes be not homogeneous, but under different genera, that is to be considered the major term, which of the two holds the nearer of its own genus. For instance: If aught be predicated of *colour* and *man*, *colour* is the major extreme; for *colour* stands closer to *quality*, than *man* to *substance*; as *man* is an individual [or most special] species, but not *colour*.

[4°.] " Finally, if each be equally remote from its proper genus, we must consider the middle, and inquire of which term it is predicated through [that term] itself, and of which through something else; and if that through which the middle is predicated of another [*i.e.* one extreme], be nearer to its proper genus, and if through that the middle be actually predicated of this term, this term is to be deemed the major. For example: If the terms be *white* and *man*, the one being an individual species in *quality*, the other in *substance*; and if *rational* be affirmatively predicated of *man*, negatively of *white*; the affirmation is made in regard to *man* as *man*, whereas the negation is made of *white*, not as *white*, but as *inanimate*. But since *inanimate*, through which *rational* is denied of *white*, is more common, more universal, and more proximate to *substance inanimate* than *man* to [*substance*] *animate*, on that account, *white* is the major term in preference to *man*. [So far Herminus.]

" But to reason thus, and to endeavour to demonstrate a major term by nature, in the Second Figure, is a speculation which may be curious, but is not true. [I read πρὸς τῷ περιουργίαν ἔχειν.]

[1°.] " For, in the first place, if we consider the given terms, not in themselves, but in relation to others, in which the predicated term does not inhere ; the major term will be always found in the negative proposition. For, in this case, the major is always equal to the middle term ; since whether it be thus or thus taken from the commencement, or be so made by him who denies it, the negative major will still stand in this relation to the middle term. For the middle does not inhere, where it is not supposed to inhere. Wherefore, its repugnant opposite inheres in the subject, but the repugnant opposite of the middle is equal to the middle. And this, either through the middle itself, or through another notion of wider extent ; as when *rational* is denied of something through *inanimate*. For there is here an equalisation through *irrational*, through which *rational* is negatively predicated of *horse*. For either the middle is equal to this of which it is denied, or [I read, ἤ for ὁ] it is less ; as when, through *inanimate*, *rational* is denied of aught. For *inanimate* is equal to *animate*, under which is *rational*, a notion greater than that other of which it is affirmed. For since the affirmative predicate is greater than its subject, of which the middle is denied or not affirmed ; and since the reason why the middle is denied, is equal to or greater than the middle itself, which middle, again, in an affirmative proposition, is greater than its subject ;—on these accounts, a negative proposition is always greater than an affirmative. Nevertheless, Aristotle himself says that a negation is to be placed in the minor [proposition] ; for the second syllogism in this figure [Camestres] has as its minor premise an universal negative.

[2°.] " Further, why in the case of negatives alone should explanation or inquiry be competent, in regard to the reason of the negative predication, seeing that in the case of affirmatives the reason is equally an object of inquiry? For *rational* is predicated of *man*, of itself, indeed, but not primarily, that is, not inasmuch as he is *man*, but inasmuch as he is *rational*; so that if *rational* [be denied] of *horse* through *irrational*, still these are both branches of the same division. By this method, assuredly, no major can be ever found. Wherefore, we ought not, in this way, to attempt a discrimination of the major of affirmative syllogisms in the Second Figure. For in this figure affirmation and negation are equally compatible with the major term; so that whatsoever term has by the forementioned method been found major, the same, taken either as major or minor, will effectuate a syllogistic jugation; which being competent, there is no longer any major [or minor] in this figure. For the problem is to find not a major term absolutely, but one of this figure." [So much touching Herminus.]

[3°.] " Nor, on the other hand, as is thought by some, is that unconditionally to be called the major term, which stands predicate in the conclusion. For neither is this manifest. If left indefinite [praeindesignate], the same term will hold a different relation, though a conversion of the universal negative; so that what is now the major, may be anon the minor; we may, in fact, be said to constitute the same term both major and minor. Naturally there is in negative propositions no major notion, nor, from the conclusion, ought we to mark out the major at all. Nor is the case different when the term is defined [predesignate]. For the conclusion shows, as predicate, the term given as major in the premises ; so that the conclusion is not itself demonstrative of the major; on the contrary, the being taken in the premises as major is the cause why a term is also taken as predicate in the conclusion.

" Nor, however, can it be said that in this figure there is no major. For this at least is determinate,—that its major must be universal; and, if there be [in it] any syllogistic combination, that premise is the major which contains the major term.

[4°.] " But, in the Second Figure, which of the terms is to be deemed the major? That is to be deemed the major, and to be placed first,which in the problem[question or quaesitum]we intend to demonstrate, and which we regard as predicate. For every one who reasons, first of all determines with himself, what it is he

would prove; and to this end he applies his stock of suitable propositions; for no one stumbles by chance on a conclusion. The
notion, therefore, proposed as predicate in the problem to be
proved, is to be constituted the major term; for although the proposition be converted, and the notion thereby become the subject,
still in what we proposed to prove, it [actually] was, and [there-
fore virtually] remains, the predicate. Hence, even if there be
drawn another conclusion, we convert it; so that, to us who prove
and syllogise and order terms, that always stands as the major.
For major and minor are not, in negative syllogisms, regulated by
their own nature, but by the intention [of the reasoner] to conclude. Thus it is manifest, that what is the predicate in the problem, is also the predicate in the conclusion."

Alexander on *Prior Analytics*, L. i. c. vi., f. 30 a, ed. Ald.

(Third Figure.) . . . "This is the Third Figure, and holds
the last place because nothing universal is inferred in it, and because sophistical syllogisms chiefly affect this figure with their
indefinite and particular conclusions. But the sophistical are the
last of all syllogisms. . . . Add to this, that while both the
Second and Third Figures take their origin from the First, of the
two the Third is engendered of the inferior premise. For the
minor, *qua* minor, is the inferior premise, and holds reasonably a
secondary place, [the conversion of the minor proposition of the
First figure giving the Third figure].

F. 30 b. (Darapti.) "The first syzygy in this figure is of two
universal affirmatives [Darapti.] But it may be asked—Why,
whilst in the second figure there are two syllogistic conjugations,
having one of the premises an universal affirmative, the other an
universal negative, (from having, now their major, now their
minor, as an universal negative proposition converted);—why, in
the third figure, there is not, in like manner, two syllogistic combinations of two universal affirmatives, since of these, either the
major or the minor proposition is convertible? Is it that in the
second figure, from the propositions being of diverse form [quality],
the commutation of a universal negative into something else by
conversion is necessary, this being now the major, now the minor,
and it not being in our power to convert which we will? In the
third figure, on the other hand, there being two universal affirmatives, the position [relation] of the propositions, (for they are simi-

lar in character and position), is not the cause of one being now
converted, now another; the cause lying in us, not in the jugation. Wherefore, the one or other being similarly convertible,
inasmuch as the position [relation] of the two propositions is the
same; the one which affords the more important probation is
selected, and hereby is determined the syllogistic jugation. Moreover, the differences of syllogisms [moods] in each figure are
effected by the differences among their jugations, not by those
among their probations. Thus that the combination of propositions is syllogistic [or valid], is proved by conversion and *reductio
ad impossibile*, also by exposition. But from this circumstance
there does not emerge a plurality of syllogisms [moods]. For the
different probations [are not valid from such plurality, but] from
the unity of the jugation from which they are inferred, so that
one jugation of two universal affirmatives may constitute, in the
third figure, a single syllogism [mood], howbeit the probations
are different; inasmuch as now the one, now the other, of the
propositions can be converted."

IV.—PHILOPONUS.

Philoponus (or rather Ammonius) on Aristotle, *An. Pr.*, L. i.
c. 4, § i. f. 17 a, ed. Trincavelli, 1536.

"The Predicate is always better than the Subject, because the
predicate is, for the most part, more extensive (ἐπὶ πλέον) than the
subject, and because the subject is analogous to the matter, the
predicate to the form; for the matter is the subject of the forms.
But when the middle term is predicated of the two extremes, or is
the subject of both; in this case, it is not properly intermediate.
But, howbeit, though in position external to the middle, it is still
preferable to be the predicate than to be the subject. On this
ground, that is called the first figure, the middle term of which
preserves its legitimate order, being subject of the one extreme,
and predicate of the other. The second figure is that in which the
middle is predicated of both extremes, and in which it occupies
the *better* position of those remaining. Finally, the third figure
is that in which the middle term is subjected to the two extremes;
here obtaining only the *lowest* position. Wherefore, in the first
figure the middle term is delineated on a level with the extremes;

whereas in the second it is placed *above*, and in the third *below*, them." "

Philoponus (or rather Ammonius) on Aristotle, *An. Pr.*, f. 17 a, ed. Trincavelli, 1536.

Syllogistic Figures in general.—" We must premise what is the Major Proposition of the Syllogism, and what the Minor. But to understand this, we must previously be aware what are the Major and Minor Terms. And it is possible to define these, both, in common, as applicable to all the three figures and, in special, with reference to the first alone. In the latter relation, that is, regarding specially the first figure, *the Major term is that which constitutes the Predicate, the Minor that which constitutes the Subject, of the Middle.* So far as limited to the first figure. But since in neither of the other figures do the extremes reciprocally stand in any definite (?) relation to the middle term; it is manifest that this determination is inapplicable to them. We must, therefore, employ a rule common to all the three figures; to wit, that the *major term is that predicated, the minor that subjected, in the conclusion.* Thus, *the Major Proposition is the one containing the Major Term; the Minor Proposition the one containing the Minor Term.* Examples : Of the First Figure,—*Man* [is] *animal; animal, substance; therefore, man, substance.* . . . . Of the Second,—*Animal* [is predicated] *of all man; animal of no stone; man, therefore, of no stone.* . . . Of the Third,—*Some stone is white; all stone is inanimate; consequently, some white is inanimate."* . . .

First Figure.—F. 19 b, *et seq.*; Aristotle, *l. c.* § 3. " ' But I call

a Ammonius, or Philoponus, here manifestly refers to the diagrams representing the three figures, and accommodated to Aristotle's three sets of letters, noting the three terms in each of these; thus :—

Whether these diagrams ascend higher

than Ammonius does not appear; for they are probably not the constructions referred to by Aristotle ; and none are given by the Aphrodisian in his original text, though liberally supplied by his Latin translator. The diagrams of Ammonius were long generally employed. By Neomagus 1533 (*In Trapezuntii Dialect.*, f. 35), they are most erroneously referred to Faber Stapulensis. [See further, *Discussions*, p. 670.—ED.]

that the middle term which itself is in another, and another in it; and which in position lies intermediate.'

"This definition of the middle term is not common to the three figures, but limited to the middle of the first figure only. For, &c. . . . . But, if there be a certain difference in species between the middle terms of the three figures, they have likewise something in common; to wit, that the middle term is found twice in the premises, throughout the three figures; which also in position is middle. For Aristotle wishes in the Diagraph (ἐν αὐτῇ τῇ καταγραφῇ) to preserve the order of intermediacy, so that, placing the three terms in a *straight line*, we assign the middle place to the middle term." [?]

Aristotle, *l. c.* § 4. "'But [I call] the *extremes* both that which is in another, and that in which another is. For if A be predicated of all B, and B of all C, it is necessary that A should also be predicated of all C. We have previously said what we mean by the expression [predicated] *of all.*'

"It may seem perhaps that this is a [perfect] definition of the extremes and of the middle term. But it is not. For it behoves us to sub-understand, in addition, the word *only;* and thus the definition will rightly run,—But [I call] the extremes, both that which is in another [minor], and that in which another only is [major]. For if A be predicated of all B, and B of all C, it is necessary that A be predicated of all C.

"This, the first syllogistic mood, is of two affirmative universals, collecting an affirmative conclusion. For if B inhere in all C, C is, consequently, a part of B. But B is a part of A; A, therefore, also inheres in all C, C being a part of B. The reasoning will be plainer in material examples—as *substance* [is predicated] *of all animal; animal of all man;* and there is inferred *substance of all man;* and conversely, *all man* [is] *animal; all animal, substance; therefore, all man, substance.*

"But it is manifest how, in this figure, the terms of the first mood [Barbara] ought to be taken. The first is the most general, and the second the subaltern genus; whilst the third is a species more special than the middle. The conclusion ought always to be drawn. Thus, if, proceeding synthetically, we commence by the major term [and proposition], *substance* begins; wherefore it also leads the way in the conclusion. [There is predicated] *substance of all animal* (here *substance* commences); *animal of all man*; whilst the conclusion again commences with *substance—substance of all man*. But if we start from the minor term [and proposition], as from *man*, with this also the conclusion will commence: *all man* [is] *animal*; *all animal, substance*; *all man, substance*.

" Aristotle takes the terms A, B, C; and from the relation of the letters, he manifests to us the order of the first figure. The major term he calls A, because A stands first in order; the minor term C; and the middle term B, as B, in its order, follows A, and precedes C.

," It is plain that the terms may possibly be coadequate [and therefore reciprocating]; as *receptive of science—risible—man*; for *all man is risible; all risible is receptive of science; therefore, all man is receptive of science*."

F. 23 b, Aristotle ch. 5, § 2. Second figure. "'The major extreme is that which lies nearer to the middle; the minor that which lies farther from the middle.'

" In place of more akin and more proximate to the middle; not in position, but in dignity. For since, of the terms, the middle is twice predicated, while (in the conclusion) the major is once, but the minor not even once predicated; [consequently], that which is once predicated will be the more proximate to that which is twice predicated, that is, to the middle, than that which is not even once predicated. Wherefore, we shall hear him [Aristotle], in the Third Figure, calling the minor the term more proximate to the middle on account of their affinity, for they are both subjecta, while he calls the major term the more remote. Perhaps, also, he wishes that in the diagraph ($\tau\tilde{\eta}$ $\kappa\alpha\tau\alpha\gamma\rho\alpha\phi\tilde{\eta}$), the major term should be placed closer to the middle, and the minor farther off. But the major extreme in this figure, the two premises being universal, exists not by nature but by position, for the first of the extremes which you meet with as a subject in the second figure, this is the minor extreme, the other is the major. So in the example—*All man animal; no plant animal; therefore, no man plant*. In like manner, if we take the commencement from *plant*, this becomes the minor term, and *man* the major; as *no plant animal; all man animal; no plant, therefore, man*. Consequently, the major and minor terms exist in these examples by position, not by nature. If, indeed, one or other of the proposi-

tions be particular, the major and the minor terms are then determined ; for we hold that in this figure the universal is the major."

Aristotle.—§ 3. "'The middle is placed external to, [not be-
tween], the extremes, and first in position.'

" The middle term passes out of what is properly the middle
position ; it is also placed ont of or external to the extremes; but
either above these or below.   But if it be placed above, so as to
be predicated of both, it is called first in position ; if below, so as
to be subjected, it is called second.   Wherefore, here as predicate
of both premises, he styles the middle term the first ; for if it
be placed above, it is first in position, and, in being apart from
the extremes, it is placed without them."

F. 27 b, Aristotle, ch. 6, § 2.   Third Figure.   "'The major
extreme is that more remote from, the minor is that more proximate to, the middle.'

" The major term in this figure is twice predicated, of the middle
and in the conclusion ; but the minor once only, and that of the
middle, for it is subjected to the major in the conclusion; the
middle is alone subjected, never predicated.   When he, therefore,
says that the major term is more remote from the middle, he means
the term always predicate is in affinity more remote from that
which is never predicate, but always subject.   And that which is
never subject is the major and more proximate term ; that again,
which is now subject, now predicate, is the minor."

V.—MARTIANUS CAPELLA.[a]

Martianus Capella, *De Septem Artibus Liberalibus*, L. iv. *De
Dialectica*, in capite, *Quid sit Predicativus Syllogismus*, p. 127,
ed. Grotii ; p. 83, ed. Basil. 1532.

" Hujus generis tres formæ [figuræ] sunt.

" Prima est, in qua declarativa [prædicatum] particula superioris sumpti, sequentia efficitur subjectiva [subjectum] ; aut subjectiva superioris, declarativa sequentis.   Declarativa superioris
fit subjectiva sequentia, ut *Omnis voluptas bonum est ; omne
bonum utile est ; omnis igitur voluptas utilis est.*   Subjectiva
superioris fit declarativa sequentis, si hoc modo velis conver-
tere : *Omne bonum utile est ; omnis voluptas bonum est ; omnis
igitur voluptas utilis est.*"

a Flourished A.C. 457, Passow ; 474, Tennemann.

In First Form or Figure, notices the four direct and five indirect moods,—*reflexim;* and in the Second and Third, the usual number of moods.[a]

In Second Figure—"Hic reflexione si utaris, alius modus non efficitur, quoniam de utrisque subjectivis fit illatio." He seems to hold that two direct conclusions are competent in Second and Third Figures.

In Second Figure, he enounces generally (four times) as thus:— "*Omne justum honestum; nullum turpe honestum; nullum igitur justum turpe;* but sometimes (once) thus:—"*Nullum igitur turpe justum.*"

In Third Form or Figure generally (six times) thus as—"*Omne justum honestum; omne justum bonum; quoddam igitur honestum bonum;*" but sometimes (once) as,—"*Quoddam igitur bonum honestum.*"

VI.—ISIDORUS.

Isidorus, *Originum,* L. ii. c. 28. *De Syllogismis Dialecticis.* (*Opera,* p. 20 (1617); in *Gothofredi Auctores,* p. 878.)

"Formulæ Categoricorum, id est, Prædicativorum Syllogismorum sunt tres.

"Primæ formulæ modi sunt novem. Primus modus est qui conducit, id est, qui colligit ex universalibus dedicativis dedicativum universale directim: ut, 'Omne justum honestum; omne honestum bonum; ergo omne justum bonum.'" All in first figure, with minor first; in second and third figures, varies ; uses "*per reflexionem*" and "*reflexim*" indifferently; and through all moods of all figures follows Apuleius. "Has formulas Categoricorum Syllogismorum qui plene nosse desiderat, librum legat qui inscribitur *Perihermenias Apuleii,* et quæ subtilius sunt tractata cognoscet."

VII.—AVERROES.

Averroes, *In Anal. Prior.,* L. i.

C. v. On First Figure.—"If, therefore, the middle term be so

a Cassiodorus, in First Figure, gives both forms, "vel sic;" in Second and Third, though he gives also "an vel sic," they are examples, both in conversa, of Capella's general mode of enunciation. See *Dialect., Opera,* pp. 535, 556, Genev. 1650, and above, p. 412. (§. 520). Cf. Apuleius, *De Syllogismo Categorico, Op.,* p. 36, Elmen. (A.C. 160). Isidorus, of Seville, in *Gothofr. Auct.,* p. 578. (A.C. 600; died 636.)

ordered between the two extremes, that it be predicated of the minor and subjected to the major, (as, if we say *all* C *is* B and *all* B *is* A); it is plain that this order of syllogism is natural to us; and it is called by Aristotle the First Figure." And thus are stated all the examples in detail.

C. vi. Figure Second.—"And the proposition whose subject is the subject of the quæsitum is the minor proposition, but that whose subject is the predicate of the quæsitum is the major. *Let us then place first in order of enunciation the minor extreme, let the middle term then follow,* and the major come last, to the end that thus the major may be distinguished from the minor; *for in this figure the terms are not distinguished, unless by relation to the quæsitum."* So all the examples.

C. vii. Third Figure.—"That proposition in which lies the subject of the quæsitum is called the minor proposition, since the subject itself is called the minor term; that proposition which contains the predicate of the quæsitum is named the major. In the example, let the minor term be C, the middle B, and the major A, and their order be that we first enounce the middle, then the minor, and last of all the major." And so the examples.

VIII.—MELANCHTHON.

Melanchthon, *Erotemata Dialecticas,* L. iii. p. 175.

"Demonstration why there are necessarily three [and only three] *Figures.*

" Every argumentation which admits the syllogistic form, (for of such form Induction and Example are not recipient, [?]) proceeds either [1°], From genus to species universally with an universal conclusion, or [2°], From species to genus with a particular conclusion, or [3°] A distraction of two species takes place, or [4°], There is a concatenation of a plurality of causes and effects. Nor are there more modes of argumentation, if we judge with skill.

" The process from genus to species engenders the First Figure. And that the consequence is valid from the genus with an universal sign both affirmatively and negatively to the species,— this is naturally manifest.

" The process from species to genus with a particular conclusion engenders the Third Figure. And it is evident that, the species posited, the genus is posited.

" The distraction of species engenders the Second Figure. And the reason of the consequence is clear, because disparate species are necessarily sundered. These may be judged of by common sense, without any lengthened teaching. Both are manifest,— that the figures are rightly distributed, and that the consequences are indubitably valid."

IX.—ARNAULD.

Arnauld, *L'Art de Penser*, (*Port Royal Logic*), P. iii. ch. 11, p. 235.—"General principle of syllogisms :—*That one of the premises should contain the conclusion, and the other show that it does so contain it.*"—[So Purchot, *Instit. Phil.*, Vol. I. P. iii. ch.1.]

Ch. v. p. 215.—" Foundation of First Figure.

" Principle of affirmative moods :—*That what agrees with a notion taken universally, agrees also with all of which this notion is affirmed; in other words, with all that is the subject of this notion, or is comprised within its sphere.*" [Or, more shortly, (says Purchot, c. vi.), *Whatever is predicated of the superior is predicated of the inferior.*]

" Principle of the negative moods :— *What is denied of a notion taken universally, is denied of all whereof this notion is affirmed.*" [Purchot— *What is repugnant to the superior, is repugnant also to the inferior,* ch. vi. p. 217.]

" Foundation of the Second Figure. "

" Principle of the syllogisms in Cesare and Festino :—*That what is denied of a universal notion, is denied also of whatever this notion is affirmed, that is to say, of all its subjects.*

" Principle of the syllogisms of Camestres, Baroco :—*All that is contained under the extension of a universal notion, agrees with none of the subjects whereof that notion has been denied, seeing that the attribute of a negative proposition is taken in its whole extension.*"

Ch. vii. p. 220.—" Foundation of the Third Figure.

" Principle of the affirmative moods :— *When two terms may*

" Purchot says this Figure rests upon a single principle—*Two things are* **and the same, but something agrees with the one, which is repugnant to the other.**

be affirmed of the same thing, they may also be affirmed of each other, taken particularly." [So Purchot nearly.]

"Principle of the negative moods:—*When of two terms, the one may be denied, and the other affirmed, of the same thing, they may be particularly denied of each other."* [So Purchot nearly.]

No foundation of principle given for the Fourth Figure.

X.—GROSSER.

Samuel Grosser, *Pharus Intellectus,* 1697, P. iii., S. i., Mem. 3, c. 2, p. 137. (Probably from Weiss, see Pref.)

"The foundation of the first figure is the Dictum de Omni et Nullo; for whatever is universally affirmed or denied of a universal subject, that is also affirmed or denied of all and each contained under that subject.

"The foundation of the second figure is Contrariety; for the predicates of contrary things are contrary.

"The foundation of the third figure is the agreement of the extremes in any third; for what agree with any third agree with each other, and may be joined or separated in the same proposition, inasmuch as they are in agreement or confliction in relation to any third thing."

Illustrates the three figures by three triangles, p. 132. In the first we ascend to the apex on one side, and descend on the other; in the second we ascend at both sides; in the third we descend on both sides.

XI.—LAMBERT.

Lambert, *Neues Organon,* Vol. I., § 225.—See Melanchthon, (above, p. 434.)

Relation of Figures. "We further remark that the first discoverer of Syllogisms and their Figures was, in his arrangement of their propositions, determined by some arbitrary circumstance; his views and selections at least were not founded on aught natural and necessary (§ 196). *He places, to wit, that premise after the other, which contains among its terms the subject of the conclusion,* probably in order to introduce into all the figures a common law. To that law, however, we do not restrict ourselves either in speech or in writing. The mathematicians, who perhaps draw the greatest number of formal syllogisms with the fewest paralogisms, commence to take the first figure, for example, not with the major but with the minor proposition, because not only in this figure is such premise always the more obtrusive, but also because its subject is the proper matter of discourse. Frequently the major premise is only quoted, or it is absolutely omitted, whensoever it is of itself obvious to the reader, or is easily discoverable from the minor and conclusion. The conclusion inferred is then, in like manner, constituted into the minor proposition of a new syllogism, wherewith a new major is connected. This natural arrangement of the syllogisms of the First Figure, rests, consequently, altogether on the principle,—*That we can assert of the Subject of an affirmative proposition, whatever we may know of its Predicate; or what may be said of the attribute of a thing is valid of the thing itself.* And this is what the syllogisms of the First Figure have peculiar to themselves. It is also so expressed: *What is true of the Genus is true also of each of its Species.*"

§ 226. "On the other hand, in the Second and Third Figures there is no talk of species and genera. *The second Figure denies the subjects of each other, because they are diverse in their attributes;* and every difference of attribute is here effectual. We, consequently, use this figure principally in the case *where two things ought not to be intercommuted or confounded.* This becomes necessarily impossible, so soon as we discover in the thing A something which does not exist in the thing B. We may, consequently, say that *syllogisms of the second figure lead us to distinguish things, and prevent us from confounding notions.* And it will be also found, that, in these cases, we always use them.

§ 227. "The Third Figure affords Examples and Exceptions ; and, in this Figure, we adduce all *Exempla in contrarium.* The two formulæ are as follows :—

" 1. There are B which are C ; for M is B and C.

" 2. There are B which are not C ; for M is B and not C.

" In this manner we draw syllogisms of the Third Figure, for the most part, in the form of copulative propositions (§ 135) ; because we are not wont twice to repeat the subject, or to make thereof two propositions. Sometimes one proposition is wholly omitted, when, to wit, it is self-manifest.

" In the Fourth Figure, as in the First, species and genera appear, only with this difference, that in the moods, *Baralip, Dibatis, Fesapo, Fresison*, the inference is from the species to the genus ; whereas in *Calentes* there is denied of the species what was denied of the genus. For where the genus is not, neither are there any of its species. This last mood we, therefore, use when we conclude negatively *a minori ad majus*, seeing that the genus precedes, and is more frequently presented than, any of its species."

§ 229. "The syllogisms of the four Figures are thus distinguished in relation to their employment, in the following respects :—

" 1. The First Figure ascribes to the thing what we know of its attribute. It concludes from the Genus to the Species.

" 2. The Second Figure leads to the discrimination of things, and relieves perplexity in our notions.

" 3. The Third Figure affords Examples and Exceptions in propositions which appear general.

" 4. The Fourth Figure finds Species in a Genus in *Baralip* and *Dibatis ;* it shows that the species does not exhaust the genus in *Fesapo, Fresison ;* and it denies the species of what was denied of the genus in *Calentes*."

§ 230. "This determination of the difference of the four Figures is, absolutely speaking, only manifested when we employ them after a natural fashion, and without any thought of a selection. For, as the syllogisms of every figure admit of being transmuted into those of the first, and partly also into those of any other, if we rightly convert, or interchange, or turn into propositions of equal value, their premises ; consequently, in this point of view, no difference subsists between them. But whether we in every case should perform such commutations in order to bring a syllogism under a favourite figure, or to assure ourselves of its correctness,—this is a wholly different question. The latter is manifestly futile. For, in the commutation, we must always undertake a conversion of the premises, and a converted proposition is assuredly not always of equal evidence with that which we had to convert, while, at the same time, we are not so well accustomed to it. For example, the proposition, *Some stones attract iron*, every one will admit, because *The magnet is a stone*, and *attracts iron*. This syllogism is in the Third Figure. In the First, by conversion of one of its premises, it would run thus :—

Major,—*All magnets attract iron ;*
Minor,—*Some stones are magnets ;*
Conclusion,—*Some stones attract iron.*

Here we are unaccustomed to the minor proposition, while it appears as if we must pass all stones under review, in order to pick out magnets from among them. On the other hand, that *the magnet is a stone,* is a proposition which far more naturally suggests itself, and demands no consideration. In like manner :—*A circle is no square ; for the circle is round,—the square not.* This proof [in the Third Figure] is as follows, when cast in the First :—

> *What is not round is no circle ;*
> *A square is not round ;*
> *Consequently, &c.*

Here the major proposition is converted by means of a *terminus infinitus,* and its truth is manifested to us only through the consciousness that *all circles are round.* For, independently of this proposition, should we not hesitate,—there being innumerable things which are *not* round,—whether the circle were one of those which belonged to this category ? We think not ; because we are aware."

§ 231. "It is thus apparent that we use every syllogistic figure there, where the propositions, as each figure requires them, are more familiar and more current. The difference of the figures rests, therefore, not only on their form, but extends itself, by relation to their employment, also to things themselves, so that we use each figure where its use is more natural : *The First for finding out or proving the Attributes of a thing ; the Second for finding out or proving the Difference of things ; the Third for finding out and proving Examples and Exceptions ; the Fourth for finding out and excluding the Species of a Genus.*"

§ 232. " Further, whether the three last Figures are less evident than the first, is a question which has been denied [affirmed (?)] on this account, that the First Figure only rests immediately on the *Dictum de Omni et Nullo* [§ 220], whilst the others have hitherto, by a circuit, been educed therefrom. We have already remarked [§ 221], that this circuit, through our mode of notation, is wholly superseded. We need, therefore, only translate its principle into the vernacular, and we shall find that the *Dictum de Omni et Nullo*

is on that account applicable to the First Figure, because its truth
is based on the nature of the propositions. From this principle,
therefore, the First Figure and its moods admit of an immediate
deduction; it is thus only a question whether the other figures are
incapable [capable (?)] of such immediate deduction, or whether
it is necessary previously to derive them through the first figure?
Our mode of notation shows that the latter is an [unnecessary] circuit, because every variety of syllogism admits for itself a various
notation, and because, in that case, the premises are taken for what
they actually are. Consequently, every figure, like the first, has its
own probation,—a probation drawn exclusively from the natures of
the propositions. The whole matter is reduced to this,— *Whether
a notion wholly or in part is, or wholly or in part is not, under
a second; and whether, again, this second wholly or in part is, or
wholly or in part is not, under a third.* All else proceeds only on
the interchange of equivalent modes of expression,—the figured,
namely, and those which are not figured. And this interchange
we may style *translating*, since the figured modes of expression
may be regarded as a special language, serving the purpose of a
notation. We have above (§ 220), after all the syllogistic modes
were discovered and denoted, adduced the *Dictum de Omni et
Nullo*, but only historically, since our manner of determining the
syllogistic moods is immediately founded on the nature of the propositions, from which this *Dictum* is only a consequence. Moreover, this consequence is special, resting as it does on the notions
of *Species* and *Genera*. Wherefore, its validity only extends so
far as propositions can be recalled to these notions; as, for example, in the First Figure. In the Second, the notion of *Difference* emerges; and in the Third, the notion of *Example*. If we,
therefore, would have special *dicta* for the several Figures, in that
case it would follow, and, at the same time, become manifest that
the middle term of a syllogism, considered for itself, expresses,
in the First Figure, a *principle [of Ascription or Procreation]*;
in the Second, *Difference*; in the Third, an *Example*; and in
the Fourth, the *principle of Reciprocity.*

" 1. For the First Figure. *Dictum de Omni et Nullo.* What is
true of all A, is true of every A.

" 2. For the Second Figure. *Dictum de Diverso.* Things which
are different, are not attributes of each other.

" 3. For the Third Figure. *Dictum de Exemplo.* When we find things A which are B, in that case some A are B.

" 4. For the Fourth Figure. *Dictum de Reciproco.* I. If no M is B; then no B is this or that M. II. If C is [or is not] this or that B; in that case some B are [or are not] C."

XII.—PLATNER.

Platner, *Philosophische Aphorismen*, 3d ed., 1793.—Part I., § 544, conformed to his *Lehrbuch der Logik und Metaphysik*, 1795, § 227. " The reason why the predicate belongs to the subject is in all possible syllogisms this,—because the subject stands in a relation of subordination with, [is either higher or lower than], a third notion to which the predicate belongs. Consequently, all inference proceeds on the following rule:—If the subject of the [concluding] judgment stand in a relation of subordination with a third notion, to which a certain predicate pertains; in that case, this predicate also pertains to the same judgment, affirmatively or negatively."

In his note on this Aphorism, Platner (*Lehrbuch*) admits— " My fundamental rule is only at fault in the second Aristotelic figure, which, however, is no genuine figure; because here, in the premises, the subject and predicate have changed places," &c. In the 2d edition of his *Aphorisms* (1784) he had adopted the principle of identity with the same third, as he has it: " *In what extension or proportion (Maasse) two notions are like or unlike to a third, in the same extension or proportion are they like or unlike each other.*" (§ 628.)

Philosophische Aphorismen, Part I., third edition, (1793,) § 568, compared with second, (1784,) § 672-676.—" Nevertheless, each of these grammatical figures of syllogism has its peculiar adaptation in language for the dialectical application of proof; and the assertion is without foundation, that the first is the most natural. Its use is only more appropriate, when we intend to show, —*that a predicate pertains* [or *does not pertain*] *to a subject in virtue of its class.* More naturally than in the first, do we show, in the second, *the difference of things apparently similar;* and in the third, *the similarity of apparently different things.* The fourth figure, [it is said in the second edition], on account of the position of its terms, is always unnatural in language."

Philosophische Aphorismen, Part I., last edition, 1793, § 561.
—" The principle of the first figure is the *Dictum de Omni et Nullo.*"

§ 564.—" Touching the other figure, [the third, for in this edition Platner abolishes, in a logical relation, the second], its special principle is the following rule :—*What belongs to the subordinate, that, since the subordinate is a part of the universal, belongs also in part (particularly) to the universal.*"

In the second edition, 1784, the second figure is recognised, and, with the third, obtains its special law.

§ 659.—" The principle of the second figure is :—*If two notions, wholly or in part, are opposite to a third, so are they also, wholly or in part, opposite to each other.*"

§ 664.—" The principle of the third figure is :—*What can be particularly affirmed or denied of a subaltern species, that also, in so far as such subaltern species is part of a genus, may be particularly affirmed or denied of the genus.*"

Philosophische Aphorimen. Part I., § 546. Note.—" In general, logicians treat the subject as if it were necessarily subordinated to the predicate. It may, however, on the contrary, be the higher notion, and the predicate thus be subordinated to it. This is the case in all particular propositions where the predicate is not an attribute of the genus, but an accident of the subject. For instance,—*Some creatures are animals ;* here the subject is the higher : *Some men are imperfect ;* here the higher is the predicate. We must not, therefore, in our syllogistic, thus enounce the fundamental rule of reasonings,—*If the subject be subordinated to a third notion,* but *with or in the relation of subordination with a third notion.*"

XIII.—Fries.

Fries, *System der Logik*, § 56.—" The species of categorical syllogisms are determined by the variety of relations in which three notions may stand to each other, so that a syllogism may be the result.

" These relations may be thought as three.

" Case I.—Three notions are reciprocally subordinated in gradation, so that the second is subordinated to the first, but superordinated to the third.

" Case II.—Two notions are subordinated to a third.

" Case III.—Two notions are superordinated to a third."

" When, in these cases, is a syllogism possible?

§ 57.—" In all the three cases, the syllogisms are equally valid, for they are founded on the general laws of the connection of notions.

" They all follow, to wit, from the relation of a whole sphere to its parts, which lies in the *Dictum de Omni et Nullo*. The principles for the three mentioned cases are thus:—

" For the first,—*The part* (C) *of the part* (B) *lies in the whole* (A), *and what* (A) *lies out of the whole* (B), *lies also out of the part* (C).

" For the second,—*What* (A *or some* A) *lies out of the whole* (B), *lies also out of its parts* (C).

" For the third,—*If a part* (B) *lie in two wholes* (A *and* C), *in that case these have a part in common; and if a part* (B) *lie in a whole* (C), *but out of another whole* (A), *in that case the first* (C) *has a part out of the other* (A).

" The first case alone coincides immediately with the perfect declaration of a syllogism,—that a case is therein determined by a rule. For the third case, therefore, our two declarations of a major premise—that *it is the rule*, and that *it contains the major term*,—do not coincide, seeing that here the minor term may be forthcoming in the rule. On this account, the arrangement of the first case is said to be the only *regular*, and the *others are reduced to it*. That this reduction is easily possible, we may in general convince ourselves, by reflecting that every syllogism requires a general rule as premise, and that the other cases are only distinguished from the first by the converted arrangement of the propositions. But as all propositions may be either purely converted or purely counterposed, consequently the two last cases can at most so far deviate from the first, that they are connected with the first case only through reversed (*gegentheilige*) notions.

§ 57 b. "The doctrine of the several species of categorical syllogisms, as regulated by the forms of their judgments, is at bottom an empty subtlety; for the result of all this circuity is only, that, in every categorical syllogism, a case is determined by a rule, and this

a [See Jordano Bruno (in Denzinger, § 237, p. 163.]
Logik, t. II. p. 269). Statller, *Logica*,

is already given in the law, that in every reasoning one premise must be universal. The scholastic logic treats of this doctrine only in so far as the species of syllogism are determined by the forms of judgment, and thereby only involves itself in long grammatical discussions. Aristotle has been falsely reproached for overlooking the fourth figure, he only having admitted three. For Aristotle proceeds, precisely as I have here done; only on the relation of notions in a syllogism, of which there are possible only our three cases. His error lies in this,—that he did not lay a general rule at the root of every figure, but, with a prolixity wholly useless, in determining the moods of the several figures, details each, even of the illegitimate, and demonstrates its illegitimacy. This prolixity has been too often imitated by other logicians, in the attempts at an evolution of the moods. Kant goes too far, in denouncing this whole doctrine as a mere grammatical subtlety. The distinction of the three cases is, however, a logical distinction; and his assertion, that the force of inference in the other two is wholly derived from that of the first case, is likewise not correct. I manifestly, however, conclude as easily in the third case,—' A part which lies in two wholes, is a part common to both,'—as in the first,—' The part of the part lies in the whole.' The third case presents, indeed, the readiest arrangement for reasonings from the particular to the general, i.e., for syllogisms in the second figure according to our terminology.

" The scholastic doctrine of the four syllogistic figures and nineteen moods of categorical syllogisms requires no lengthened illustration. If the figures are determined by the arrangement of notions in the premises, then the following combination is exhaustive. For the conclusion in all cases S——P [being supposed the same], the [terms or] notions stand :

1) According to our first case,   M——P
                                     S——M

2) With converted major premise, P——M
                                     S——M

3) With converted minor premise, M——P
                                     M——S

4) Both premises converted,   P——M
                                     M——S

" Should we therefore simply convert both premises in a syllo-

gism of the first figure, we are able to express it in all the figures.
Let the notions given be *fireproof, lead, metal*, there then follows
the conclusion—*Some metal is not fireproof*—from the premises:—

> In the First Figure — *No lead is fireproof;*
> *Some metal is lead;*
> In the Second Figure — *Nothing fireproof is lead;*
> *Some metal is lead;*
> In the Third Figure — *No lead is fireproof;*
> *All lead is metal;*
> In the Fourth Figure — *Nothing fireproof is lead;*
> *All lead is metal.*

"It is here apparent that the three first figures are our three
cases; but the fourth we did not employ, as it contains no peculiar relations or notions, but only under our first case supercor-
dinates, and then subordinates a middle term. This manner of
enunciating a syllogism is thus only possible, where we are competent, through conversions, to transmute the arrangement of the
first figure into that of the fourth. Now this happens: 1] If we
convert the conclusion S——P into P——S, since then the major
and the minor terms, as also the major and minor premises, change
names; or, 2] If both premises allow of an immediate conversion,
so that the one remains universal; for then the converted propositions contain the same thoughts as those given, and, consequently, establish the same conclusion."

[Objections to Fries' doctrine of figure—1°, Only applies to affirmatives; 2°, Only the arrangement of the results of a successful
comparison, and takes no heed of the comparisons that may have
been fruitless, (the illegitimate moods); 3°, Takes account of only
one subordination, for, in second and third cases, in each there is
a reciprocal subordination in Extension and Comprehension.]

XIV. AND XV.—KRUG AND BENEKE—THEIR DOCTRINES OF SYLLOGISM CRITICISED.

The authority of the two following philosophers, who conclude
this series, is rather negative than positive; inasmuch as they
both concur in proving, that the last attempts at a reformation
of the Syllogistic Theory proceed on a wholly different ground
from that on which, I think, this alone can be accomplished.

These two philosophers are Krug and Baneke; for, beside them, I am aware of no others by whom this has been attempted.

Krug was a disciple of the Kantian school, Kant's immediate successor in his Chair of Logic and Metaphysics at Kœnigsberg, and, subsequently, Professor of Philosophy in the University of Leipsic. He is distinguished, not only as a voluminous writer, but as a perspicuous and acute thinker; and his peculiar modification of the Kantian system, through a virtual return to the principle of Common Sense, is known, among the German theories, by the name of *Synthetism.* His *Logic*, (the first part of his *System of Theoretical Philosophy*), was published in 1806, and is one of the best, among the many excellent, treatises on that science, which we owe to the learning and ability of the Germans. (I have before me the fourth edition, that of 1833.) Krug propounded a new theory of syllogistic; but the novelty of his scheme is wholly external, and adds only fresh complication to the old confusion. It has, accordingly, found no favour among subsequent logicians.

Passing over the perverse ingenuity of the principles on which the whole doctrine is founded, it is enough to state, that Krug distributes the syllogistic moods into *eight* classes. Of these the *first*, (which, with some other logicians, he considers not as a figure at all, but as the pure, regular, and ordinary form of reasoning), corresponds to the First Figure of the Aristotelico-Scholastic distribution. The *other seven* classes, as so many impure, irregular, and extraordinary forms, constitute, (on the analogy of Rhetoric and Grammar), so many *figures.* Of these, the new is only the old *First Figure*, the minor premise, in extension, being stated before the major. Krug, like our other modern logicians, is not aware that this was the order in which the syllogism was regularly cast, in common language, by the Greeks, by the Arabians, by the Jews, and by the Latins prior to Boethius.* The old and new first figures are only a single figure, the syllogism being drawn in the counter orders of breadth and of depth. A mood in these orders, though externally varying, is intrinsically,—is schematically,— the same. Krug's distinction of his new first figure is, therefore, null. Thus, Damma is Barbara; Caleme is Celarent; Dirami is Darii; Firomo is Ferio. Nor is his discrimination of the other six better founded. His new (the old) *Second*, and his *Fifth* Figures, are also one. The latter is precisely the same with the former; *Fimeso* is *Festino*, and *Fomaco* is *Baroco*. In one case, (under *Camestres*), Krug adopts, as alone right, the conclusion rejected by the logicians. In this, he and they are, in fact, both wrong; though in opposite ways. Each mood, in the second (as in the third) figure, has two indifferent conclusions; and the special one-sided practice of the former is only useful, as gainsaying the general one-sided precept of the latter. The same objection applies to Krug's new (the old) *Third* in connection with his *Sixth* Figure. They are one; *Daroco* is *Bocardo*, *Fapimo* is *Felapton*, and *Fisemo* is *Ferison*. In two cases, (under *Disamis* and *Bocardo*), Krug has recognised the repudiated conclusion. Krug (§ 109) has, however, committed an error in regard to Bocardo. He gives, as its example, the following syllogism, in which, for brevity, I have filled up the quantifications:

"Some animals are not [any] viviparous;
All animals are [some] organised things;
Therefore, some organised things are not [any] viviparous."

In a note, he adds: "The conclusion should here be:—'Therefore, some things which are not viviparous are [some] organised.' And this is seen also by reduction. We have, however, followed the arbitrary precept of the logicians, that the extreme in the second proposition should stand subject in the conclusion; although it be here indifferent, which extreme becomes the subject. The conclusion is only changed into another quality." Only changed into another quality! Only an affirmative conclusion from a negative premise! The legitimate inference is:—

'Therefore, no viviparous is some organic;' or,
'Therefore, any viviparous is not some organic.'

Bachmann, (*Logik*, § 135), another eminent logician, has erred with Krug. A particular predicate in a negative proposition, seems indeed one of the last difficulties for reformed logic. Krug's new (the old) *Fourth* Figure bears a corresponding relation to his *Seventh*. He is right, certainly, in abolishing all the moods of the fourth figure, except *Fesapo* and *Fresiso*; and, from his point of view, he is hardly to be blamed for not abolishing these likewise, along with the correlative moods, *Fapesmo* and *Frisesmo*, and with them, his seventh figure. Finally, rejecting the scholastic doctrine of Reduction, he adopts, not without sundry perverse additions, Kant's plan of accomplishing the same end; so that Krug's conversive and contrapositive and transpositive interpolations, by which he brings back to propriety his sevenfold figured aberrations, are merely the substitution of one "false subtlety" for another. He, and Bachmann after him, renounce, however, "the crotchet of the Aristotelians," in making the extreme of the prior premise the predicate, always, of the conclusion, in the first and second figures; and, though both do this partially and from an erroneous point of view, their enunciation, such as it is, is still something.

Professor Beneke, of Berlin, is the last to whom I can refer, and in him we have, on the point in question, the final result of modern speculation. This acute and very original metaphysician stands the uncompromising champion of the philosophy of experience, against the counter doctrine of transcendentalism, in all its forms, now prevalent in Germany; and, among the other departments of mental science, he has cultivated the theory of reasoning, with great ability and success. In 1832 appeared his *Lehrbuch der Logik*, &c.; in 1839, his *Syllogismorum Analyticorum Origines et Ordo Naturalis*, &c.; and in 1842, his *System der Logik*, &c., in two volumes. In Logic, Beneke has devoted an especial share of attention to the theory and distribution of Syllogism; but it is precisely on this point, though always admiring the ingenuity of his reasonings, that I am compelled overtly to dissent from his conclusions.

The Syllogistic of Beneke is at once opposed, and correspondent, to that of Krug; there is an external difference, but, without imitation, an internal similarity. Instead of erroneously multiplying the syllogistic figures, like the Leipsic philosopher, the philosopher of Berlin ostensibly supersedes them altogether. Yet, when considered in essence and result, both theories agree, in being, and from the same side, severally, the one an amplification, the other an express doubling, of the nineteen scholastic moods. In this, both logicians were unaware, that the same had been, long ago, virtually accomplished in the progress of the science; neither considered, that the amplification he proposed was superficial, not to say mistaken; and that, instead of simplicity, it only tended to introduce an additional perplexity into the study. Beneke has the merit of more openly relieving the opposition of Breadth and Depth in the construction of the syllogism ; and Krug, though on erroneous grounds, that of partially renouncing the old error of the logicians in regard to the one syllogistic conclusion in the second and third figures. But, in his doctrine of moods, Beneke has, I think, gone wrong in two opposite ways : like Krug, in his arbitrary multiplication of these forms ; like logicians in general, in their arbitrary limitation.

In regard to the former : The counter quantities of breadth and depth do not discriminate two moods, but merely two ways of stating the same mood. Accordingly, we do not multiply the moods of the first figure, to which alone the principle applies, by casting them in the one dependency and in the other ; we only show, that in that figure every single mood may be enounced in a twofold order, more german, the one to the quantity of extension, the other to the quantity of intension. An adequate notation ought equally and at once, to indicate both.—But in reference to the second and third figures, the case is worse. For in them we have no such dependency at all between the extremes ; and to double their moods, on *this* principle, we must take, divide, and arbitrarily appropriate one of the two indifferent conclusions. But, as every single mood of these figures has a double conclusion, this division cannot be made to difference their plurality. If Professor Beneke would look (*instar omnium*) into Apuleius or Isidorus, or, better than either, into Blemmidas, he would find all his new moods, (not, of course, those in the fourth figure) stated by these, as by other ancient logicians ; who, however, dreamed not that the mere accidental difference of what they called an *analytic* and *synthetic* enouncement, determined any multiplication of the moods themselves.

In the latter respect: Dr Beneke has only followed his predecessors ; I therefore make no comment on the imperfection.—But, in accomplishing what he specially proposes, whilst we do not find any advancement of the science, we find the old confusion and intricacy replaced by another, perhaps worse. To say nothing of his non-abolition of the fourth figure, and of his positive failure in doubling its moods ; the whole process is carried on by a series of arbitrary technical operations, to supersede which must be the

aim of any one who would reconcile Logic with nature. His new
(but which in reality are old) amplifications are brought to bear
(I translate his titles) through " Commutations of the Premises,—
by Subalternation,—by Conversion,—by Contraposition ;" and "of
the Major,—of the Minor,"—in fact of both premises, (e.g. Fesapo
2, &c.) And so difficult are these processes, if not so uncertain the
author's language, that, after considerable study, I am still in doubt
of his meaning on more points than one. I am unable, for example,
to reconcile the following statements :—Dr Beneke repeatedly
denies, in conformity with the common doctrine, the universal
quantification of the predicate in affirmative propositions ; and
yet founds four moods upon this very quantification, in the conversion of a universal affirmative. This is one insolubility.—But there
arises another from these moods themselves (§ 28-31). For, if we
employ this quantification, we have moods certainly, but not of
the same figure with their nominal correlatives; whereas, if we do
not, simply rejecting the permission, all slides smoothly,—we have
the right moods in the right figure. This, again, I am unable to
solve.—Dr Beneke's duplication of the moods is also in sundry
cases only nominal; as is seen, for example, in Ferio 2, Fesapo 2,
and Fresiso 2, which are forms, all, and in all respects, identical.—
I must protest also against his violence to logical language. Thus,
he employs everywhere "non omne," "non omnia," "alle sind
nicht," &c., which is only a particular, (being a mere denial of
omnitude), for the absolute or universal negative, "nullum,"
"nulla," "kein ist," no, none, not any, &c., in opposition both to
principle, and to the practice of Aristotle and succeeding logicians.

[XVI.—TITIUS.

Gottlieb Gerhard Titius, Ars Cogitandi, sive Scientia Cogita-
tionum Cogitantium, Cogitationibus Necessaria Instructa et a
Peregrinis Liberata. Lipsiæ, 1723, (first edition, 1701).

Titius has been partially referred to by Sir W. Hamilton, as
having maintained the doctrine of a Quantified Predicate. See
above, p. 318. His theory of the Figure and Mood of Syllogism
is well deserving of notice,—proceeding, as it does, on the application of that doctrine. This theory is principally contained in the
following extracts from his Ars Cogitandi, which show how

closely he has approximated, on several fundamental points, to the doctrines of the *New Analytic*.[a]

Titius gives two canons of syllogism :—

I. Affirmative. "Quæcunque conveniunt in uno tertio, illa etiam, juxta mensuram illius convenientiæ, inter se conveniunt."

II. Negative. "Quæcunque pugnant in certo aliquo tertio, illa, juxta mensuram illius disconvenientiæ, etiam inter se pugnant." C. ix. §§ 30, 27.

The following relates to his doctrine of Figure and Mood, and to the special rules of Syllogism, as commonly accepted :—

C. x. § i. "Sic igitur omnium Syllogismorum formalis ratio in genuina medii termini et prædicati ac subjecti, Conclusionis collatione consistit ; eam si dicere velis *formam essentialem*, aut *figuram generalem* vel *communem*, non valde reluctabor.

§ ii. " Præter eam vero Peripatetici *Figuras* ex *peculiari medii termini situ* adstruunt, ea ratione ut *Primam* figuram dicant, in qua medius terminus in Majore est subjectum, in Minore prædicatum, *Secundam*, ubi idem bis prædicati, et *Tertiam*, ubi subjecti locum bis subit. Galenus adjecit *Quartam* primæ contrariam, in qua medius terminus in majore est prædicatum, in minore subjectum, quam pluribus etiam exposuit Autor *Art. Cog.* P. iii. c. 8.

§ iii. " Cæterum illæ figuræ tantum sunt *accidentales*, ab iisque vis concludendi non dependet. Quodsi tamen quis diversum medii termini situm attendendum esse putet, tum nec quarta figura negligenda esse videtur, licet eam Peripatetici nonnulli haut curandam existiment. *Vide* Ulman. *Synops. Log.* L. iii. c. 2, p. 164.

§ iv. " Interim *prima* cæteris magis naturalis ex eo videri potest, quod subjectum et prædicatum conclusionis in præmissis suam retineat qualitatem, cum in *secunda et tertia* alterum qualitatem suam exuere, in *quarta* vero utrumque eam deponere debeat.

§ v. " Postea in unaquaque figura, pro ratione quantitatis et qualitatis propositionum, peculiares *Modi* adstruuntur, ita quidem ut primæ figuræ *quatuor totidem* secundæ, tertiæ *sex* attribuantur, ex quibus etiam debite variatis quarta *quinque* accipiat, prout illa passim cum vocabulis memorialibus recenseri solent, ut illa quidem

huc transcribere opus non sit.   *Vide* Autor. *Art. Cogit.*, P. iii.
cc. 5, 6, 7, 8.

§ vi. " Non opus esse istis figuris et modis ad dijudicandam
Syllogismorum bonitatem, ex monito § 3, jam intelligi potest.
Quomodo tamen sine iis bonitas laudata intelligi queat, id forte
non adeo liquidum est. . . . . . .

§ vii. " Non diu hic quærenda sunt remedia : Observetur forma
essentialis seu figura communis, ac de veritate Syllogismi recte
judicabitur.   Applicatio autem hujus moniti non est difficilis,
nam primo respiciendum ad conclusionem, deinde ad medium
terminum, quo facto etiam judicari potest, an ejus et terminorum
conclusionis collatio in præmissis recte sit instituta nec ne.

§ ix. " De cætero uti anxie jam non inquiram, an omnis bene
concludendi ratio *numero modorum denario* circumscribatur,
quod quidem, juxta ἀκρίβειαν mathematicam demonstrasse
videri vult Autor *Art. Cog.* P. iii. c. 4, ita id haut admiserim, quod
illi *modi*, quos vulgo laudant, primæ, secundæ aut tertiæ figuræ
præcise sint assignandi, licet hoc itidem acumine mathematico
se demonstrasse putet dictus Autor. d. l. c. 5 *seqq.*

§ x. " Cum enim quævis propositio possit converti, modo quantitas prædicati probe observetur, hinc necessario sequitur, quod
quivis Syllogismus, adhibita propositionum conversione, in qnavis
figura possit proponi, ex quo non potest non æqualis modorum
numerus in unaquaque figura oriri, licet illi non ejusdem semper
sint quantitatis.

§ xi. " Operæ pretium non est prolixe per omnia Syllogismorum
singulis figuris adscriptorum exempla ire.   Sufficiat uno assert-
ionem illustrasse, v. gr. in prima figura, modo *Barbara* hic oc-
currit Syllogismus apud d. Autor. c. 5.

>   *O. Sapiens subjicitur voluntati Dei,*
>   *O. Honestus est sapiens,*
>   *E. O. honestus subjicitur voluntati Dei.*

§ xii. " Hanc in secunda figura ita proponere licet :

>   *Quidam, qui subjicitur voluntati Dei, est omnis sapiens,*
>   *Omnis honestus est sapiens,*
>   *E. Omnis honestus subjicitur voluntati Dei.*

Ratio concludendi manet eadem, *sapiens* enim et *is qui subjicitur voluntati Dei*, uniuntur in majore, dein *sapiens et Honestus* in minore, ergo in conclusione idea *sapientis* et *ejus qui voluntati Dei subjicitur*, quoque conveniunt.

§ xiii. " In tertia figura ita se habebit :

> *O. Sapiens subjicitur voluntati Dei,*
> *Q. Sapiens est omnis honestus,*
> *E. O. honestus subjicitur voluntati Dei :*

nec in hac concludendi ratione aliquid desiderari potest, nam medius terminus universaliter unitur cum conclusionis prædicato, deinde, quantum sufficit, conjungitur cum ejusdem subjecto, seu *omni honesto*, ergo subjectum et prædicatum se quoque mutuo admittent.

ı § xiv. " Cæterorum eadem est ratio, quod facile ostendi posset, nisi tricas illas vel scribere vel legere tædiosum foret. Ex his autem sequitur, quod *omnes regulæ speciales, quæ modis vulgaribus attemperatæ vulgo circumferuntur, falsæ sint*, quod speciatim ostendere liceat.

§ xv. " In universum triplici modo impingitur, vel enim *conclusio creditur absurda, quæ talis non est*, vel *vitium est in materia, ac altera præmissarum falsa*, vel *adsunt quatuor termini*, adeoque absurditas conclusionis, si aliqua subest, nunquam ab ea causa dependet, quam referunt regulæ.

§ xvi. " Sed videamus distinctius, (1) *Major in prima figura semper sit universalis* . . . . . .

§ xvii. " Inflectam huc exemplum minus controversum, quod Autor *Art. Cog.* P. iii. c. 7, in modo *Disamis* tertiæ figuræ, proponit :

> *Quidam impii in honore habentur in mundo,*
> *Quidam vituperandi sunt omnes impii,*
> *E. quidam vituperandi in honore habentur in mundo.*

§ xviii. " Hic habes primam figuram cum majore particulari, optime iterum concludentem, nam licet medius terminus particulariter sumatur in majore, ejus tamen ille est capacitatis, ut in eodem convenientia prædicati et subjecti ostendi queat, et nisi hoc esset, nec in tertia figura rite concluderetur.

§ xix. " Nec valde obsunt, quæ vulgo illustrandæ regulæ adducuntur. Ex sententia Weis. *in Log.* P. i. lib. ii. c. 2, § 4, male ita concluditur :

> *Q. animal volat,*
> *O. Leo est animal,*
> *E. Q. Leo volat.*

Verum si animal sumitur in minore sicut in majore, tum illa falsa est, si vero alio sensu, tum existunt quatuor termini; his ergo causis, non particularitati majoris, vitiosa conclusio tribuenda.

§ xx. " Nam alias ita bene concluditur:

> *Q. animal volat,*
> *O. avis est animal,* (illud quoddam,)
> *E. O. avis volat.*

Num licet medius terminus particularis sit, tantæ tamen est latitudinis ut cum utroque conclusionis termino possit uniri.

§ xxi. " Porro (2) *Minor semper sit affirmans.* Sed quid desiderari potest in hoc Syllogismo:

> *O. Homo est animal rationale,*
> *Leo non est homo,*
> *E. non est animal rationale ?*

et nonne illa ratio concludendi manifeste bona est, quæ subjectum et prædicatum, quæ in certo tertio non conveniunt, inter se quoquo pugnare contendit?

§ xxii. " Sed ais, mutemus paululum Syllogismum et absurditas conclusionis erit manifesta:

> *O. Homo est animal,*
> *Leo non est homo,*
> *E. Leo non est animal !*

Verum si terminus animalis in conclusione perinde sumitur, sicut suppositus fuit in majore, nempe *particulariter*, tum conclusio est verissima; si autem aliter accipiatur, tum evadunt quatuor termini, quibus adeo, non negationi minoris, absurditas conclusionis est imputanda, quæ observatio in omnibus exemplis quæ hic objici possunt et solent, locum habet.

. . . .

§ xxviii. " Sed revertamur ad regulas vulgares! Nimirum (3) *In secunda figura major sit universalis.* Verum cur non ita liceat concludere:

> *Quidam dives est Saxo,*
> *Quidam Germanus est omnis Saxo,*
> *E. Quidam Germanus est dives !*

quod argumentum Weis. L. ii. c. 4, § 2, intuitu tertiæ figuræ proponit.

§ xxix. " Argumenta, quæ fallere videntur, v. gr. quod Weisius L. ii. c. 3, § 3, profert :

> *Quidam homo est sapiens,*
> *Nullus stultus est sapiens,*
> *E. Nullus stultus est homo,*

et similia, responsione, § 22, data eliduntur; nimirum conclusio vel non est absurda, si recte intelligatur, vel adsunt quatuor termini, quibus adeo, non particularitati majoris, vitium est imputandum.

§ xxx. " Amplius (4) *Ex puris affirmativis in secunda figura nihil concluditur,* sed mirum foret, si illa concludendi ratio falleret, quæ fundamentum omnium Syllogismorum affirmativorum tam evidenter præ se fert ! Hoc argumentum utique formaliter bonum est :

> *Omnis sapiens sua sorte est contentus,*
> *Paulus sua sorte est contentus,*
> *E. Paulus est sapiens.*

§ xxxi. " Sed fallunt multa argumenta, v. gr. Weisio L. ii. c. 3, § 3, adductum :

> *Omnis lepus virit,*
> *Tu viris,*
> *E. Tu es lepus;*

verum non fallunt ob affirmationem præmissarum, sed qui vel minor falsa est, si scil. prædicatum accipiatur eodem sensu, quo in majore sumtum est, vel quia adsunt quatuor termini, si prædicatum minoris particulariter et alio sensu accipiatur.

§ xxxii. " Non possunt etiam vulgo diffiteri, quin ex puris affirmativis aliquando quid sequatur, verum id non vi *formæ* sed *materiæ* fieri causantur, *vide* Ulman., *Log.* L. iii. c. 3, § 4. Hæc vero est petitio principii ; nam quæ conveniunt in uno tertio, illa etiam inter se convenire debent, idque non fortuito, sed virtute unionis laudatæ, seu beneficio formæ.

. . . .

§ xxxiv. "In tertia figura (5) *Minor semper sit affirmans.* Ego tamen sic recte concludi posse arbitror :

> *Quoddam laudandum est omnis virtus,*
> *Nullum laudandum est quædam magnificentia,*
> *E. Quædam magnificentia non est virtus.*

§ xxxv. " Nec valde urgent exempla opposita Weisius L. ii., c. 4, § 2, hoc affert :

> *Omnis homo ambulat,*
> *Nullus homo est porcus,*
> *E. quidam porcus non ambulat ;*

nam recurrit responsio § 22 data, quæ vel conclusionem falsam non esse, vel causam falsitatis a quatuor terminis dependere ostendit, quæ etiam locum haberet, licet conclusionem universalem, *Nullus porcus ambulat,* assumas.

§ xxxvi. "Tandem (6) *In tertia figura conclusio semper sit particularis.* Verum Syllogismum cum conclusione universali jam exhibui § 13. In exemplis autem quæ vulgo afferuntur, v. gr.

> *Omnis senator est honoratus,*
> *Omnis senator est homo,* (quidam scil.),
> *E. omnis homo est honoratus.*

vide Weis. d. L. ii., c. 4, § 3, occurrunt quatuor termini, (nam homo, in minore particulariter, in conclusione universaliter sumitur), qui adeo veram absurdæ conclusionis causam, ac simul regulæ vulgaris falsitatem ostendunt.

§ xxxvii. "Illa autem omni, quæ contra vulgares regulas hactenus disputavimus, non eo pertinent, quasi rationem concludendi rejiciendis regulis hinc inde confectam commendemus, ita ut in demonstrationibus eadem uti, aut valde delectari, debeamus. Quin omni potius eo spectant, ut Peripateticos, qui formam Syllogismorum essentialem vel omnino non vel nimis frigide exponunt, in explicandis etiam eorum figuris accidentalibus, falli probarem.

. . . .

§ xxxix. "Atque ex hactenus dictis etiam intelligi potest, quæ nostra de *Reductione* sit sententia. Nimirum ex nostris hypothesibus illa nihil aliud est, quam *Syllogismorum per omnes quatuor figuras accidentales, salva semper conclusione, facta variatio.*

§ xl. "Pertinet igitur illa tantum ad *præmissas.* Syllogismus enim semper ut instrumentum veritatis inquirendæ considerari, adeoque quæstio probanda, quæ semper immobilis sit, nec, prout visum est, varietur, præsupponi debet.

§ xli. " Reductionis unica *Lex* est, ut simpliciter, juxta figuræ indolem, propositiones convertamus, quod sine ulla difficultate

procedit, dummodo quantitatem subjecti et prædicati debite con-
fideremus, ceu ex iis quæ de Conversione diximus satis liquet.

§ xlii. " Finis est, ut per ejusmodi variationem terminorum
unionem vel separationem eo accuratius intelligamus.   Hinc
omnis *utilitas* reductioni non est abjudicanda.   Si enim recte in-
stituatur, ingenium quantitati propositionum observandæ magis
magisque assuescit, ac inde etiam in peniliorem formæ essentialis intelligentiam provehitur.

§ xliii. " In *vulgari Reductione*, quæ in libellis logicis passim
exponitur, (*vide* Aut. *Art. Cog.* P. iii.,'o. 9,) quædam exempla re-
prehendi non debent, quando v. g. *Cesare* ad *Celarent* reducitur,
nam ibi simplici conversione alicujus propositionis defunguntur,
juxta legem, quam § 41, reductioni dedimus.

§ xliv. "Sed si ab illis exemplis abeas, parum vel nihil est,
quod in eadem laudari debeat, dum fere ex falsis hypothesibus
omnis reductio oritur, nam *conversio per contrapositionem* præ-
supponitur, quam tamen valde dubiam esse supra ostendimus,
præterea *peculiares modi* in singulis figuris adstruuntur, ac omnis
reductio ad *primam figuram* facienda esse existimatur, cum tamen
idem Syllogismus per omnes figuras variari queat.

§ xlv. " Ipsa vero reductio nullis legibus adstricta est.   Convertitur conclusio, transponuntur præmissæ, propositiones nega-
tivæ mutantur in affirmativas, atque ita quidvis tentatur, modo
figura intenta obtineatur.   Quo ipso puerilis error, quo Logica
pro arte concinnandi tres lineas, easque in varias formas mutandi,
habetur, satis elucet.   Inepta scientia est, quæ in verbis dispon-
endis, circumagendis, aut torquendis, unice occupatur.

§ xlvi. " Juxta hæc igitur vulgari modo reducere, maximam
partem nihil aliud est, quam errorem errore tegere, ingenia dis-
centium torquere, ac magno conatu magnas nugas agere, insciti-
amque professa opera ostendere."—ED.]

D.—SYLLOGISTIC MOODS.

(Vol. I. p. 401.)

(a) DIRECT AND INDIRECT MOODS.
(1) THEIR PRINCIPLE.—FIRST AND FOURTH FIGURES.

(See above, Vol. I. p. 423.)

Direct and Indirect Moods,—principle of.—That the two terms should hold the same relation to each other in the conclusion, that they severally held to the middle term in the premises. This determined by the Question. This constitutes direct, immediate, natural, orderly inference. When reversed, by Conversion, there emerges indirect, mediate, unnatural, irregular inference.

In the two last Figures, (Second and Third), the two terms hold the same relation to the middle term in the premises; ergo no indirect inference, but always two direct conclusions possible.

In the first Figure, as the two terms are subordinated to each other in the premises, one direct conclusion from premises, whether read in Extension or Comprehension, and, consequently, an indirect one also ;—the First Figure being first figure in Extensive quantity, the Fourth Figure being first figure in Comprehensive quantity. Direct and indirect moods in each.

1. Blunder about definition of major and minor terms by logicians, (for which Aristotle not responsible),[a] cause of fancy of a Fourth Figure constituted by indirect moods in comprehension.

2. That predicate could have no prefinition, and, therefore, though they allowed its converse, the direct inference was not suffered. This in Fapesmo, Frisesmo, (these alone, by some logicians, admitted in the First Figure), and Fesapo and Fresison in Fourth or Comprehensive First.[B]

---

[a] See Stahl, [*Notæ et Animadversiones in Compendium Dialecticum D. Conradi Hornelii, nunc primum ex Auctoris Autographo edita cura Caspari Pomeri Prof. Pub. Jenæ.* 1656, *Ad. L. iii. c. viii.*]

[B] [That fourth Figure differs from first only by transposition of Premises,—held by Derodon, *Logica Restituta,* p. 608. Camerarius, *Disputationes Philosophicæ,* Disp. i., qu. 13, p. 116. Cara-

muel, *Rat. et Real. Phil.,* Disp. xii. p. 45. Irenæus, *Integ. Phil. Elementa Logices,* Sect. iii. § 3, p. 29. Campanella, *Phil. Rat. Dialect.,* Lib. II. c. vi. art. xi. p. 391, and art. iv. p. 383, (1635). Ridiger, *De Sensu Veri et Falsi,* li. 8, § 38. Crusius, *Weg zur Gewissheit,* § 335, p. 606. Platner, *Philosophische Aphorismen,* i. § 554, p. 267.]

3. That major proposition, that which is placed first.

Fourth Figure.—The *First* Figure, and that alone, is capable of being enounced in two orders, those of Breadth and of Depth. It is exactly the same syllogism in either order ; and, while the order of Depth was usually employed by the Greeks, Orientals, and older Latins, that of Breadth has been the common, if not the exclusive, mode of enouncement among the western logicians, since the time of Boethius. In either form, there are thus *four* direct moods, and *five* indirect,—in all *nine* moods ; and if the Figure be held to comprise the moods of either form, it will have *eighteen* moods, as in fact is allowed by some logicians, and, among others, by Mendoza, (*Disp. Log. et Met.* T. I. pp. 515, 516). Martianus Capella, (*De Septem Artibus Liberalibus*, L. iv., *De Dialectica*, in cap. *Quid sit Prædicativus Syllogismus*, see above, p. 432), states and allows *either* form, but, like his contemporaries, Greek and Latin, he employs in his examples the order of *Depth.*

Now, mark the caprice of the logicians of the west subsequent to Boethius. Overlooking entirely the four direct moods in the order of Depth, which they did not employ, as the conclusion would, in these cases, have been opposed to their own order ; they seized upon the five *indirect* moods of the order of Depth, as this afforded a conclusion corresponding to their own, and constituted it, thus limited, into a Fourth Figure.

Did not make two forms of First Figure.

An indirect conclusion is in subject and predicate the reverse of a direct ; opposed, therefore, to the order of predication marked out by the premises which the direct conclusion exclusively follows.—An indirect conclusion, (what the logicians have not observed),[a] is an inference from the direct conclusion, and, therefore, one *mediate* from the premises.

a But see Contarenus, *De Quarta Figura Syllog.*, *Opera*, p. 235.—ED.

(2) Moods of Fourth Figure redressed.

(Early Paper—previous to 1844.　Later signs of quantity
substituted.—Ed.)

I. Bamalip,—only Barbara with transposed premises and converted conclusion.

> (2) *All irons are (some) metals;*
> (1) *All metals are (some) minerals;*
> *All irons are (some) minerals.*
> (By conversion)
> *Some minerals are (all) irons.*
>
> A : ———— , B : ———— , C
>
> (*Minerals,* ———— : (*Metals*), ———— : (*Irons*).
> (Redressed)

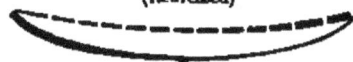

II. Calemes,—only Celarent with transposed premises and
converted conclusion.

> (2) *All snails are (some) mollusca ;*
> (1) *No molluscum is any insect ;*
> *No snail is any insect.*
> (By conversion)
> *No insect is any snail.*
>
> A : ———— , B : —+— : C
>
> (*Insect*): —+— : (*Molluscum*), ———— : (*Snail*)
> (Redressed)

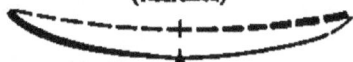

III.—Dimatis,—only Darii with transposed premises and
converted conclusion.

> (2) *Some stars are (some or all) planets ;*
> (1) *All planets are some things moving round sun ;*
> *Some stars are some things moving round sun ;*
> (By conversion)
> *Some things moving round sun are some stars.*

A, ——— , : B : ——— , C

(*Moving round Sun*), ——— : (*Planets*) : , ——— , (*Stars*)
(Redressed)

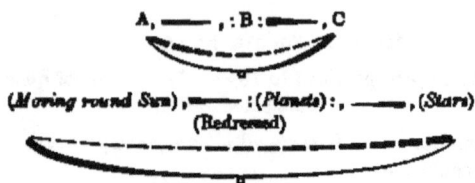

IV. Fesapo, [Felapos.][a]

    (2) *No artery is any vein ;*
    (1) *All veins are (some) bloodvessels ;*
    No artery is (some) bloodvessel.
        (By conversion)
    Some bloodvessel is no artery.

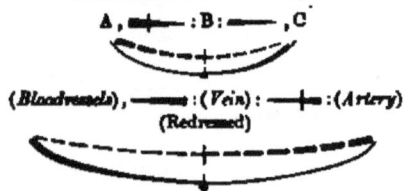

A, ——|— : B : ——— , C

(*Bloodvessels*), ——— : (*Vein*) : ——|— : (*Artery*)
(Redressed)

V. Fresison, [Frelilos].

    (2) *No muscle is any nerve ;*
    (1) *Some nerves are (some) expansion on hand ;*
    No muscle is (some) expansion on hand.
        (By conversion)
    Some expansion on hand is no muscle.

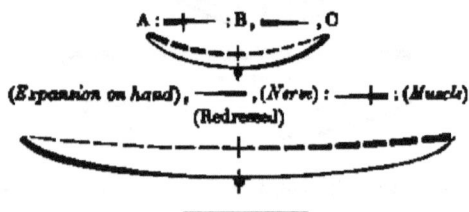

A : ——|— : B , ——— , C

(*Expansion on hand*), ——— , (*Nerve*) : ——|— : (*Muscle*)
(Redressed)

(March 1846.)—My universal law of Figured Syllogism excludes the Fourth Figure.— *What worse relation of subject and predicate subsists between either of two terms and a common third term with which one, at least, is positively related; that relation subsists between the two terms themselves.* Now, in Fourth Figure, this is violated; for the predicate and subject notions, relative to the middle term in the premises, are in the conclusion turned severally into their opposites by relation to each other. This cannot, however, in fact be; and, in reality, there is a silently suppressed conclusion, from which there is only given the converse, but the conversion itself ignored.

Fourth Figure. Reasons against.

1°, Could never directly, naturally, reach (a) Conclusion from premise, or (b) Premises from quæsitum.

2°, All other figures conversion of premises of First, but, by conversion of conclusion, (as it is), no new figure.

3°, All other figures have one conclusion, Fourth a converted one, often different.

————

(March 1850.)—Fourth Figure. The logicians who attempt to show the perversion in this Figure, by speaking of higher and lower notions, are extra-logical. Logic knows nothing of higher and lower out of its own terms; and any notion may be subject or predicate of any other by the restriction of its extension. Logic must show the perversion in this Figure *ex facie syllogismi*, or it must stand good.

On true reason, why no Fourth Figure, see Aristotle, *Anal. Pr.*, L. i. c. 23, § 8, and Pacius, in his *Commentary*.

————

(March 1850.)—*Fesapo* and *Fresiso* (also *Fapesmo, Frisesmo*) proceed on the immediate inference, unnoticed by logicians, that the quantities, apart from the terms, may, in propositions *InA* and *AnI*, be converted.

————

Averroes on *Prior Analytics*, L. I. c. 8.

" If we ask whether A be in C, and say that A is in C, because A is in B, and B in C; in this case, there is a natural syllogism by general confession; and this in the First Figure.

In like manner, if we say that A is not in C, because B is in C, and B is not in A, it is plain that we collect that conclusion by a natural process; and this is the Second Figure, which is frequently found employed by men in their ordinary discourse.

" In like manner, also, if we say that A is in C, because A and C are in B; that syllogism is also natural to us, and is the Third Figure.

" But if we say A is in C, because C is in B, and B in A, the reasoning is one which no one would naturally make; for the reason that the quæsitum (that is, C to be in A) does not hence follow,—the process being that in which we say A is in C, since A is in B, and B in C; and this is something which thought would not perform, unless in opposition to nature. From this it is manifest, that the Fourth Figure, of which Galen makes mention, is not a syllogism on which thought would naturally light," &c. Thereafter follows a digression against this figure. See also the same book, c. 23, and the *Epitome*, by Averroes, of the same, c. i.

(3) FOURTH FIGURE—AUTHORITIES FOR AND AGAINST.

Admitted by—

Ildefonsus de Peñafiel, *Cursus Philosophicus, Disp: Summul.*, D. iii. p. 39. G. Camerarius, *Disput. Philos.*, P. i. q. xiii., p. 116. *Port Royal Logic*, P. iii. c. 8, and c. 4. Ridiger, *De Sensu Veri et Falsi*, L. ii. c. 6, § 36. Hauschius in *Acta Erud.* p. 470 *et seq.* Lips. 1728. Noldius, *Logica Recognita*, c. xii. p. 277. Crakanthorpe, *Logica*, L. iii. c. xv. p. 194, (omitted, but defended). Lambert, *Neues Organon*, I. § 237 *et seq.* Hoffbauer, *Analytik der Urtheile und Schlüsse*, § 138. Twesten, *Logik, insbesondere die Analytik*, § 110. Leibnitz, *Opera*, ii. 357 ; v. 405 ; vi. 216, 217, ed. Dutens. Oddus de Oddis, (v. Contarenus, *Non Dari Quart. Fig. Syll., Opera Omnia*, p. 233, ed. Venet, 1589.)

Rejected by—

Averroes, *In An. Prior.* L. i. c. 8. Zabarella, *Opera Logica, De Quarta Fig. Syll.*, p. 102 *et seq.* Purchot, *Instit. Phil.* T. I. *Log.* P. iii. c. iii. p. 169. Molinaeus, *Elementa Logica*, L. i. c. viii. Facciolati, *Rudimenta Logica*, P. iii. c. iii. p. 85. Scaynus, *Paraphrasis in Organ.*, p. 574. Timpler, *Logicæ Systema*, L. iv. c. i. qu. 13, p. 543. Platner, *Philosophische Aphorismen*, I. p. 267. Burgersdicius, *Instit. Log.*, L. ii. c. vii. p. 165. Derodon, *Logica Restituta*, p. 606. Wolf, *Phil. Rat.*, § 343 *et seq.*, (ignored). Hollmann, *Logica*, § 453, p. 569. Goclenius, *Problemata Logica*, P. iv. p. 119. Keckermann, *Opera*, T. I. *Syst. Log.*, Lib. iii. c. 4, p. 745. Arriaga, *Cursus Philosophicus, In Summulas*, D. iii. § 5, p. 24. Aristotle, *An Prior.*, i. c. 23, § 8; c. 30, § 1, (omitted). Jo. Picus Mirandulanus, *Conclusiones, Opera*, p. 88. Melanchthon, in 1st edition of *Dialectic*, L. iii. *De Figuratione*, (1520), afterwards (1547) restored, (Heumanni, *Acta*, iii. 753). Cardinalis Caspar Contarenus, *Epistola ad Oddum de Oddis de Quart. Fig. Syll.*, *Opera*, p. 233 (1st ed. 1571). Trendelenburg, *Elementa Logica*, § 28, &c. Herbart, *Lehrbuch der Logik*, Einleit. 3, § 71. Hegel, *Encyclopædie*, § 187. Fries, *System der Logik*, § 57 b. Griepenkerl, *Lehrbuch der Logik*, § 29 *et seq.* Drobisch, *Logik*, § 77, p. 70. Wallis, *Institutio Logicæ*, L. iii. c. ix. p. 179.

(b) INDIRECT MOODS OF SECOND AND THIRD FIGURES.[a]

From	(II. Fig.)	
i.	Cesaro	*Reflexim;* (1, 2, 5, 8, 9.)[β] Cesares.
ii.	Camestres	*Reflexim;* (2, 5, 8, 9.) Camestre, Camestres.
		Faresmo, (only subaltern of Camestres); rejected (5), admitted (3, 6.)
iii.	Festino	Premises reversed ; (2, 3, 4, 5, 6, 7, 8, 9.) Firesmo, Frigesos.
iv.	Baroco	Premises reversed ; (2, 5, 7, 8, 9.) Bocardo, Moracos, Foranmeno.

	(III. Fig.)	
i.	Darapti	*Reflexim;* (1, 2, 3, 4, 10, 11.)
ii.	Felapton	Premises transposed; (4, 5, 6, 7, 8, 9, 11.) Fapemo, Fapelmon.
iii.	Disamis	*Reflexim;* (4, 7, 10, 11.)
iv.	Datisi	*Reflexim;* (4, 7, 10, 11.)
v.	Bocardo	Premises transposed ; (4, 7, 9, 11.) Baroco, Macopos, Danorooc.
vi.	Ferison	Premises transposed ; (4, 5, 6, 7, 8, 9, 11.) Frisesmo, Pisaros.

[a] The indirect Moods of the First Figure are universally admitted. to the authorities given on following page.—Ed.

[β] The numbers within brackets refer

	(II. Fig.)	
1.	Mart. Capella, (475)	Cesare, *referim.*
2.	Duns Scotus, (1300)	Cesare and Camestres, conclusions simply converted ; Festino and Baroco, premises transposed. Rejects (and rightly) what has since been called Faresmo, as a mere subaltern of Camestres. (*Super An. Pr.* L. i. qu. 23. See also Conimbricenses, *In Arist. Dial.,* ii. p. 362.)
3.	Lovanienses, (1535)	Faresmo, Firesmo. (*Comm.* p. 923, ed. 1547.)
4.	Pacius, (1584)	Firesmo. (On *An. Pr.* L. i. c. 7, and relative place of his *Comm. Anal.*)
5.	Conimbricenses, (1607)	Record that indirect moods from Cesare and Camestres ; and also Friseso, Bocardo were admitted by some " recentiores." (ii. p. 362.)
6.	Burgersdicius, (1626)	Faresmo, Firesmo.
7.	Caramuel, (1642)	Moracos, Frigesos.
8.	Scheibler, (1653)	Cesares, Camestres, Firesmo, Bocardo.
9.	Noldius, (1666)	Cesares, Camestre, Firesmo, Forameno ; (he has for the direct mood, Facrono, in place of Baroco).

	(III. Fig.)	
1.	Apuleius, (160)	Darapti, *referim.*
2.	Cassiodorus, (550)	Do.
3.	Isidorus, (610)	Do.
4.	Duns Scotus	Darapti, Disamis, and Datisi, their conclusions simply converted ; Felapton, Bocardo, Ferison, premises transposed. (*Super An. Pr.* L. i. qu. 24.)
5.	Lovanienses	Fapemo, Frisemo. (As above.)
6.	Pacius	Fapemo, Frisemo. (As above.)
7.	Conimbricenses	Record that some " recentiores " admit indirect moods from Darapti, Disamis, Datisi ; also Fapesmo, Frisesmo, and Baroco.
8.	Burgersdicius	Fapemo, Frisemo.
9.	Caramuel	Fapelmos, Macopos, Fiseros.
10.	Scheibler	Admits them from Disamis, Datisi, Darapti, but not from those which conclude particular negations.
11.	Noldius	Danoroc, (he has for Bocardo Ducamroc), Frisemo, Fapemo, and what are converted from Darapti, Disamis, and Datisi without names.

Indirect moods are impossible in the Second and Third Figures, for what are called indirect conclusions are only the direct conclusions. *Mem.*, that in the Second Figure Cesare and Camestres are virtually one; whilst in the Third Figure Darapti is virtually two,[a] as Disamis and Datisi are one.

For the particular quantification of the Predicate, useful illustrations, as in the First from Fapesmo, Frisesmo, or (in the pseudo Fourth), from Fesapo and Fresiso; so in the Second Figure from what have been called the *indirect moods* of Figure II.

FIGURE II.

1. Bocardo.
2. Firesmo.

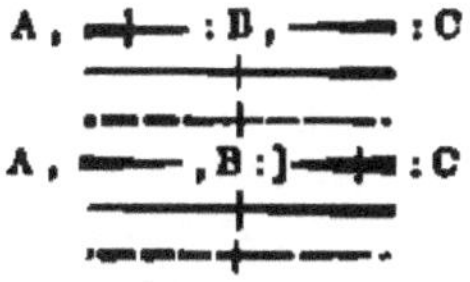

FIGURE III.

1. Baroco.
2. Fapemo.
2. Frisemo.

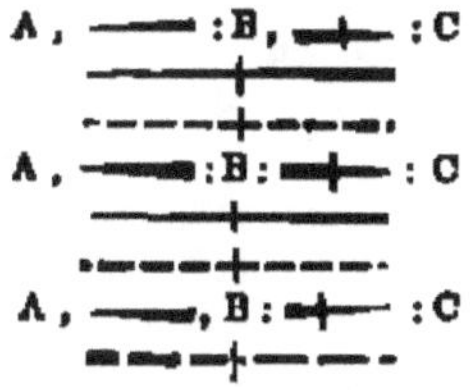

(1853.) Blunders of Logicians.—What have been called the Indirect Moods of the Second and Third Figures, arise only from the erroneously supposed transposition of the premises; and the Fourth Figure is made up of the really indirect moods of the First Figure, with the premises transposed.

(*c*) NEW MOODS—NOTES UPON TABLE OF SYLLOGISMS.[β]

Fig. I. vi.—Corvinus, (*Institutiones Philosophiæ Rationalis*, 1742, § 540), says:—"There sometimes appears to be an inference from pure particulars. For example,—*Some learned are* [*some*]

a This is maintained by Theophrastus.

β See below, Appendix XII.—ED.

ambitious men; some men are [all the] learned; therefore, some men are ambitious. But the minor proposition, although formally particular, involves, however, a universal, to wit, its converse,— *All the learned are [some] men,*—which is equipollent."—Why not, then, scientifically enounce, (as I have done), without conversion, what the thought of the convertend already really and vulgarly involved?

In all Figures.—I have been not undoubtful, whether the syllogisms of this class, in which the two premises, being the same, are mutually interchangeable, should be regarded as a single or as a double mood. Considered abstractly from all matter, the mood is single; for the two premises, however arranged, afford only a repetition of the same form. But so soon as the form is applied to any matter, be it even of a symbolical abstraction, the distinction of a double mood emerges, in the possible interchange of the now two distinguished premises. To the logicians this question was only presented in the case of Darapti(III. ii.); and on this they were divided. Aristotle (*An. Pr.* i. c. 6, § 6) contemplates only one mood; but his successor, Theophrastus, admitted two, (Apuleius, *De Hab. Doctr. Platonis,* L. iii. *Op.* p. 38, Elm.) Aristotle's opinion was overtly preferred by Alexander, (*ad locum,* f. 30, ed. Ald. quoted above, p. 427), and by Apuleius, (*l. c.*); whilst that of Theophrastus was adopted by Porphyry, in his lost commentary on the *Prior Analytics,* and, though not without hesitation, by Boethius, (*De Syll. Categ.* L. ii. *Op.* pp. 594, 598, 601, 604). The other Greek and Roman logicians silently follow the master; from whom, in more modern times, Valla (to say nothing of others) only differs, to reduce, on the counter-extreme, Cesare and Camestres, (II. ix. a, and x. b), and, he might have added, Disamis and Datisi, (III. iv. v.), to a single mood, (*De Dial.,* L. ii. c. 51). (For the observations of the Aphrodisian, see above, p. 423 *et seq.*)

To me it appears, on reflection, right to allow in Darapti only a single mood; because a second, simply arising through a first, and through a trnusposition, has, therefore, merely a secondary, correlative, and dependent existence. In this respect all is different with Cesare and Camestres, Disamis and Datisi. The principle here applies in my doctrine to the whole class of syllogisms with balanced middle and extremes.

Fig. II. xii. b.—David Derodon, (*Log. Rest., De Arg.,* c. ii. § 51), in canvassing the Special Rule of the Second Figure,—that the major premise should be universal, now approbates, now reprobates syllogisms of this mood; but wrong on both alternatives, for his admissions and rejections are equally erroneous. "Hic syllogismus *non valet* :—' Aliquod animal est [aliquod] rationale ; sed [ullus] asinus non est [ullus] rationalis ; ergo, [ullus] asinus non est [aliquod] animal.'" (P. 635.) The syllogism is valid; only it involves a principle which Derodon, with the logicians, would not allow,—that in negatives the predicate could be particular. (See p. 623.) Yet almost immediately thereafter, in assailing the rule, he says :—" At multi dantur syllogismi constantes majori particulari, qui tamen *sunt recti* ; ut, —' Aliquod animal non est [ullus] lapis ; sed [omnis] adamas est [aliquis] lapis ; ergo, [ullus] adamas non est [aliquod] animal.'" (This syllogism is, indeed, II. iii. a ; but he goes on :) "Item : ' Aliquod animal est [aliquod] rationale ; sed [ullus] lapis non est [ullus] rationalis ; ergo. [ullus] lapis non est [aliquod] animal.'" Now these two syllogisms are both bad, as inferring what Derodon thinks they do infer,—a negative conclusion with, of course, a distributed predicate, (p. 623); are both good, as inferring what I suppose them to infer,—a negative conclusion with an undistributed predicate.

Fig. III. viii. b.—Derodon, (*Ibid.,* § 54), in considering the Special Rule of the Third Figure,—that the minor premise should be affirmative,—alleges the following syllogism as "*vitious :* "— "'Omnis homo est [aliquod] animal; sed [ullus] homo non est [ullus] asinus ; ergo, [ullus] asinus non est [aliquod] animal.'" (P. 638.) It is a virtuous syllogism,—with a particular predicate (and not a universal, as our logician imagines,) in a negative conclusion.—Again, (omitting his reasoning, which is inept), he proceeds :—" Hic vero syllogismus *non est vitiosus, sed rectus* :— ' [Omnis] homo est [quidam] rationalis ; sed [ullus] homo non est [ullus] asinus [or Deus] ; ergo, [ullus] asinus [or Deus] non est [quidam] rationalis.'" This syllogism is indeed correct ; but not, as Derodon would have it, with a distributed predicate in the conclusion. That this conclusion is only true of the *asinus,* per accidens, is shown by the substitution of the term *Deus ;* this showing his illation to be formally absurd.

Fig. III. ii.—Derodon (*Ibid.*) says :—" Denique, conclusionem in tertia figura debere esse particularem, non universalem, statuunt communiter Philosophi ; unde hic syllogismus *non valet ;* —' Omnis homo est [quidam] rationalis ; sed omnis homo est [quoddam] animal ; ergo, omne [quoddam] animal est [quoddam] rationale.' Verum, licet conclusio sit *universalis,* syllogismus erit bonus, modo," &c., (p. 638.) The syllogism is, and must remain, vicious, if the subject and predicate of the conclusion be taken universally, whilst both are undistributed in the antecedent. But if taken, as they ought to be, in the conclusion particularly, the syllogism is good. Derodon, in his remarks, partly overlooks, partly mistakes, the vice.

Derodon, criticising the Special Rule of the First Figure,—that the major premise should be universal,—says, *inter alia :*—" At multi dantur syllogismi primæ figuræ constantes majori particulari, qui tamen *sunt recti :* ut,—' Aliquod animal est [aliquod] rationale ; sed homo est [aliquod] animal ; ergo, [!!] homo est [aliquis] rationalis :' item," &c., &c., (p. 627.) This syllogism is vicious ; the middle term, *animal,* being particular in both its quanifications, affords no inference."

XII.

LOGICAL NOTATION.

(See Vol. I. p. 305.)

(a) LAMBERT'S LINEAR NOTATION.β

This very defective,—indeed almost as bad as possible. It has accordingly remained unemployed by subsequent logicians ; and although I think linear diagrams do afford the best geometrical illustration of logical forms, I have found it necessary to adopt a

a See above, p. 324.—Ed.

β For Lambert's scheme of notation, see his *Neues Organon,* I. § 219; and for a criticism of the schemes of Lam-

bert and Euler, see S. Maimon, *Versuch einer neuen Logik,* Sect. iv. § 7, p. 64 et seq. Berlin, 1794.—Ed.

method opposite to Lambert's, in all that is peculiar to him. I have been unable to adopt, unable to improve, anything.

1°, Indefinite or particular notions can only be represented by the relation of two lines, and in two ways: 1°, One being greater than the other; 2°, One being partially out of relation to the other. Instead of this, Lambert professes to paint particularity by a dotted line, i.e., a line different by an accidental quality, not by an essential relation. But not even to this can he adhere, for the same notion, the same line, in different relations, is at once universal and particular. Accordingly, in Lambert's notation, the relation of particularity in notions is represented sometimes by a continuous, sometimes by a dotted line, or not represented at all. (See below, 1*, 1, 2, 3, 4, 5.)

2°, This inconsistency is seen at its climax in the case of the predicate in general affirmatives, where that term is particular. In Lambert's notation it, however, shows in general as distributed or universal; but in this he has no consistency. (See 1*, 1, 2, 3, 4.) But the case is even more absurd in negative propositions, where the predicate is really taken in its whole extent, and yet is, by the dotted line, determinately marked as particular. (See 4.)

3°, The relation of negativity, or exclusion, is professedly represented by Lambert in one line beyond, or at the side of, another. This requires room, and is clumsy, but is not positively erroneous:—it does express exclusion. But his affirmative propositions are denoted by two unconnected lines, one below the other. This is positively wrong; for here the notions are equally out of each other as in the lateral collocation. But even in this he is inconsistent; for he as often expresses the relation of negativity by lines in the relation of higher and lower. (See his whole scheme, and below, 1, 4.)

4°, He attempts to indicate the essential relation of the lines by the fortuitous annexation of letters, the mystery of which I have never fathomed.

5°, He has no order in the relation of his lines. The middle term is not always the middle line, and there is no order between the extremes. This could not indeed be from his method of notation; and except it be explained by the affixed letters, no one could discover in his lines the three compared notions in a syllogism, or guess at the conclusion inferred. (See 1—5.)

6°, From poverty the same diagram is employed to denote the most different moods in affirmative and negative. (Compare 2 and 3 with 4.)

7°, No order in the terms in the same figure.

8°, Incompleta. Lambert can represent ultra-total, &c., inclusion in affirmatives, but not ultra-total exclusion in negatives. Has the merit of noticing this relation.

9°, Lambert—but it is needless to proceed. What has been already said, shows that Lambert's scheme of linear notation is, in all its parts, a failure, being only a corruption of the good, and a blundering and incongruous jumble of the natural and conventional. The only marvel is, how so able a mathematician should have propounded two such worthless mathematical methods. But Lambert's geometrical is worse even than algebraic notation.

To vindicate what I have said, it will be enough to quote his notation of the moods of the Third Figure, (I. p. 133), which I shall number for the previous references.

III. Figure.

1.° Daraptl.
    . . . . C— — — c . . . .
    M — — m
    . . . . B— — — b . . . .

1. Felapton.
    M — — m C — — c
    B— — — b . . . .

2. Disamis.
    B— — — — b
    M — — — m
    . . . C . . . . . .

3. Datisi.
    C— — — — c
    M — ·— ·— m
    . . B . . . . . .

4. Bocardo.
    B— — — — — b
    M — — — m
    C . . . .

5. Ferison.
    M— — m C — — c
    . . B . . . . .

(b) Notation by Maass.

Professor Maass of Halle,[a] discontented, not unreasonably, with the geometrical notations of Lambert and Euler, has himself proposed another, compared with which those of his predecessors show as absolutely perfect. [It will be sufficient to despatch this scheme with a very few remarks. To use it is wholly impossible; and even the ingenious author himself has stated it towards the conclusion of his *Logic* (§ 495—512), in the course of which, it is not, (if I recollect aright,) honoured with a single reference. It is, however, curious as the only attempt made to illustrate Logic, not by the relations of geometrical quantities, but by the relations of geometrical relations,—angles.

1°, It is fundamentally wrong in principle. For example, Maass proposes to represent coinclusive notions, notions, therefore, to be thought as the same, by the angles of a triangle, which cannot possibly be imaged as united, for surely the identity of the concepts, " triangle," " trilateral," and " figure with angles equal to two right angles," is not illumined by awarding each to a separate corner of the figure. On the contrary, coexclusive notions he represents by angles in similar triangles and these can easily be conceived as superposed. The same may be said of co-ordinates. But, waving the objection that the different angles of a figure, as necessarily thought out of each other, are incapable of typifying, by their coincidence, notions to be thought as coinclusive,—it is further evident, that the angles of an equilateral triangle cannot naturally denote reciprocal or wholly identical notions, in contrast to others partially identical; for every angle of every triangle infers,—necessitates,—contains, if you will,—the whole of every other, equally as do the several angles of an equilateral triangle.

2°, But Maass is not consistent. He gives, for instance, a triangle (Fig. 12) to illustrate the subordination of one notion to another; and yet he represents the lower or contained notion by an obtuser, the higher or containing notion by an acuter, angle.

3°, The scheme is unmanifest,—in fact, nothing can be less ob-

[a] *Grundriss der Logik*, 1793. I quote from the fourth edition, 1821. I regret the necessity imposed on me of speaking in the way I do of Maass' scheme of notation ; for his *Logic* is one of the best compends published even in Germany.

trusive. It illustrates the obscure by the obscure, or, rather, it
obscures the clear. Requiring itself a painful study to comprehend its import, (if comprehended it be), instead of informing
the understanding through the eye, it at best only addresses the
eye through the understanding. Difficult,—we only regret that
it had not been impossible.

4°, It is clumsy, operose, complex, and superfluous. For, to
represent a notion denoted by a single angle, it is compelled to
give the redundance of a whole triangle ; and three repugnant
notions demand an apparatus of three several figures, and six
vacant angles. In fact, the only manifestation to which this
scheme of angles can pretend, is borrowed from the scheme of
figures which it proposes to supersede.

5°, It is wholly dependent upon the accidents of foreign aid.
To let it work at all, it calls into its assistance an indefinite
plurality of figures, a Greek and Latin alphabet, combinations of
letters straight and deflected, and an assortment of lines, thick
and thin, plain and dotted. I have counted one diagram of the
eighteen, and find that it is brought to bear through three varieties of line, four triangles, and eleven letters.

It is needless to enumerate its other faults,—its deficiencies,
excesses, ambiguities, &c.; *transeat in pace.*

(c) THE AUTHOR'S NOTATION.—NO. I. LINEAR.

The notation previously spoken of,[a] represents every various
syllogism in all the accidents of its external form. But as the
number of Moods in Syllogisms Analytic and Synthetic, Intensive
and Extensive, Unfigured and Figured, (and of this in all the
figures,) is the *same;* and as a reasoning, essentially identical,
may be carried through the *same numerical mood,* in every genus
and species of syllogism :—it seems, as we should wish it, that
there must be possible also, a notation precisely manifesting the
modal process, in all its essential *differences,* but, at the same
time, in its internal *identity,* abstract from every accidental variety
of external form. The anticipation and wish are realised ; and
realised with the utmost clearness and simplicity, in a notation
which fulfils, and *alone* fulfils, these conditions. This notation

[a] See Tabular Scheme at the end of the present volume.—ED.

I have long employed : and the two following are specimens. Herein, four common lines are all the requisites : three (horizontal) to denote the *terms;* one (two?—perpendicular) or the want of it, at the commencement of comparison, to express the *quality* of affirmation or of negation ; whilst *quantity* is marked by the relative length of a terminal line within, and its indefinite excurrence before, the limit of comparison. This notation can represent equally *total* and *ultra-total* distribution, in simple Syllogism and in Sorites ; it shows, at a glance, the *competence* or *incompetence* of any conclusion ; and every one can easily evolve it.

Of these: the former, with its converse, includes, Darii, Dabitis, Datisi, Disamis, Dimaris, &c.; whilst the latter, with its converse, includes Celarent, Cesare, Celanes, Camestres, Camelos, &c. But of these, those which are represented by the same diagram are, though in different figures, formerly, the same mood. For in this scheme, each of the thirty-six moods has its peculiar diagram ; whereas, in all the other geometrical schemes hitherto proposed, (whether by lines, angles, triangles, squares, parallelograms, or circles,) the same (complex) diagram is necessarily employed to represent an indefinite plurality of moods. These schemes thus tend rather to complicate, than to explicate,— rather to darken, than to clear up.—The principle of this notation may be realised in various forms.[a]

The problem, in general, is to manifest by the differences and relations of geometrical quantities (lines or figures), the differences and relations of logical forms. The comparative excellence of any scheme in solution of this problem will be in proportion as it is, 1°, Easy ; 2°, Simple ; 3°, Compendious ; 4°, All-sufficient ; 5°, Consistent ; 6°, Manifest ; 7°, Precise ; 8°, Complete.

In the scheme proposed by me,

1°, I denote *terms* or *notions* by straight lines ; and, as a *syllo-*

[a] Reprinted from *Discussions,* p. 657. For a farther explanation of the rela- | tions denoted by the diagrams, see above, vol. i. p. 180.—Ed.

gism is constituted by *three* related notions, it will, of course, be represented by three related lines.

2°, I indicate the *correlation* of notions by the order and parallel coextension of lines. (The perpendicular order and horizontal extension, here adopted, is arbitrary.)

3°, Lines, like notions, are only *immediately* related to those with which they stand in proximity. Hence the intermediate line in our diagram, representing *the middle term* of a syllogism, is in direct relation with the lines, representing the *extremes,* whereas the latter are only in mutual correlation through it.

4°, The relative *quantity* of notions is expressed by the comparative length of the related lines. In so far as a line commences, (here on the left,) before another, it is *out of relation* with it,—is indefinite and unknown. Where a line terminates under relation, (here towards the right,) it *ceases absolutely to be.* A line beginning and ending in relation, indicates a *whole* notion. A line, beginning before or ending after its correlative, indicates the *part* of a notion.

5°, The kinds of correlation, *Affirmation* and *Negation,* are shown by the connection, or non-connection, of the lines, (here from the left). The connection, (here a perpendicular line,) indicates the identity, or coinclusion, of the connected terms; the absence of this denotes the opposite. The lines in positive or affirmative relation are supposed capable of being slid into each other.

This geometric scheme seems to recommend itself by all the virtues of such a representation, and thus stands favourably contrasted with any other. For it is easy,—simple,—compendious, —all-sufficient,—consistent,—manifest,—precise,—complete.

1°, *Easy.*—Linear diagrams are more easily and rapidly drawn than those of figure; and the lines in this scheme require, in fact, no symbols at all to mark the terminal differences, far less the double letterings found necessary by Lambert.

2°, *Simple.*—Lines denote the quantity and correlation of notions far more simply than do any geometric figures. In those there is nothing redundant; all is significant.

3°, *Compendious.*—In this respect lines, as is evident, are far preferable to figures; but Lambert's linear scheme requires more than double the space sufficient for that here proposed.

4°, *All-sufficient.* — Any scheme by figures, and Lambert's scheme by lines, is, in itself, unintelligible; and depends on the annexation of accidental symbols, to enable it to mark out the differences and relations of terms. Lambert, likewise, endeavours to supply this exigency by another means,—by the fortuitous quality (his dottings) of certain lines. In our scheme lines, simple lines, and lines alone, are sufficient.

5°, *Consistent.*—Lambert's linear scheme is a mere jumble of inconsistencies. Compared with his, those by figures are, in this respect, far preferable.—But the present linear scheme is at once thoroughgoing, unambiguous, and consistent.

6°, *Manifest.*—In this essential condition, all other geometrical illustrations are lamentably defective. In those by figure, each threefold diagram, typifying an indefinite plurality of moods, requires a painful consideration to extract out of it any pertinent elucidation; this is, in fact, only brought to bear by the foreign aid of contingent symbols. Nor can these schemes properly represent to the eye the relation of the toto-total identity of a plurality of terms; the intention requires to be intimated by the external accident of signs. Lambert's lines sink, in general, even below the figures, in this respect.—But as lines are here applied, the sole pertinent inference leaps at once to sense and understanding.

7°, *Precise.* — Ambiguity, vagueness, vacillation, redundancy, and withal inadequacy, prevail in the other schemes. In those by figure, one diagram is sometimes illustrative of as many as a dozen moods, positive and negative; and a single mood may fall to be represented by four diagrams, and perhaps in six several ways. Lambert's lines are even worse.—In our scheme, on the contrary, every mood has a diagram applicable to itself, and to itself exclusively, whilst every possible variety of its import has a corresponding possible variety of linear differénce.

8°, *Complete.*—In this last and all important condition, every scheme, hitherto proposed, is found to fail. A thoroughgoing, adequate, and pliant geometric method ought equally and at once to represent the logical moods in the Unfigured and Figured Syllogism, in the Syllogism Synthetic and Analytic, in Extension and Intension,—this, too, in all their mutual convertibilities, and in all their individual varieties. This our scheme performs; but exclu-

sively. So much, in general. Again, in particular :—Of the
figures, circles and triangles are necessarily inept to represent the
ultra-total coinclusion or coexclusion of terms,—in a word, all the
relations of proportion, except totality and indefinite partiality;
whilst quadrilateral figures are, if not wholly incompetent to this,
operose and clumsy. Lambert's linear method is incompetent to
it in negatives; and such inability ought to have opened his eyes
upon the defects of his whole plan, for this was a problem which
he expressly proposed to accomplish.—The present scheme, on
the other hand, simply and easily performs this, in affirmation
and negation, and with any minuteness of detail.

AUTHOR'S SCHEME OF NOTATION—UNFIGURED AND FIGURED SYLLOGISM—No. II.

(1853.) The following Diagram affords a condensed view of
my other scheme of Syllogistic Notation, fragments of which,
in detail, will be found in Mr Thomson's *Outline of the Laws
of Thought*, and in Mr Baynes' *Essay on the New Analytic of
Logical Forms*. The paragraphs appended will supply the necessary explanations.

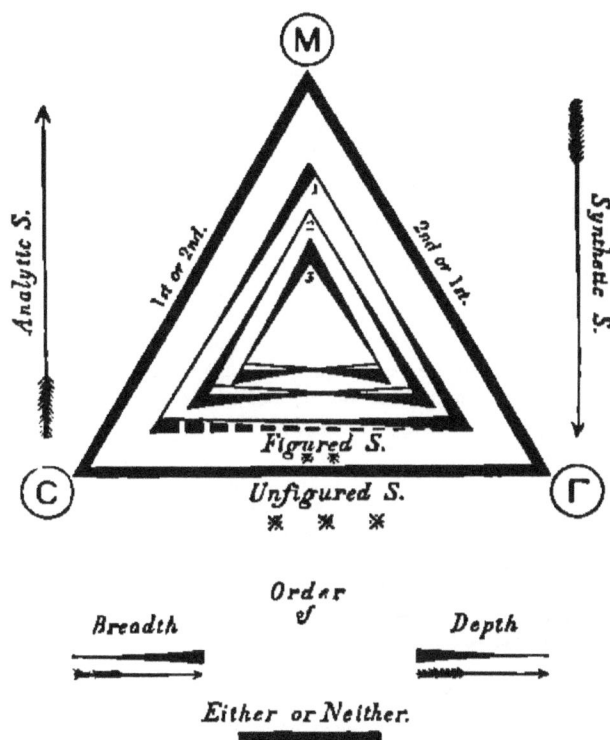

1.) A Proposition, (διάστημα, intercallum, πρότασις, literally *protensio*, the stretching out of a line from point to point), is a mutual relation of two terms (ὅροι) or extremes (ἄκρα). This is therefore well represented,—The two terms, by two letters, and their relation, by a line extended between them.

2.) A Syllogism is a complexus of Three Terms in Three Propositions.—It is, therefore, adequately typified by a Triangle,—by a Figure of three lines or sides.

3.) As upwards and downwards is a procedure arbitrary in the diagram, the diagram indicates that we can, indifferently, either proceed from the Premises (*rationes*) to the Conclusion (*rationatum*), or from the Conclusion to the Premises; the process being only in different points of view, either Synthetic or Analytic. (An exclusive and one-sided view, be it remembered, has given an inadequate name to what are called Premises and Conclusion.)

4.) Rationally and historically, there is no ground for constituting that Premise into Major which is enounced *first*, or that Premise into Minor which is enounced *last*. (See after, p. 697, &c.) The moods of what is called the Fourth Figure, and the Indirect moods of the First Figure, are thus identified.—In the diagram, accordingly, it is shown, that as right or left in the order of position is only accidental, so is first or last in the order of expression.

5.) The diagram truly represents, by its various concentric triangles, the Unfigured Syllogism, as involving the Figured, and, of the latter, the First Figure as involving the two others. (In fact, the whole differences of Figure and Figures are accidental; Moods alone are essential, and in any Figure and in none, these are always the same and the same in number.)

6.) Depth and Breadth, Subject and Predicate, are denoted by the thick and thin ends of the same propositional line.

7.) Depth and Breadth are quantities always coexistent, always correlative, each being always in the inverse ratio of the other.—This is well shown in the connection and contrast of a line gradually diminishing or increasing in thickness from and to end.

8.) But though always coexistent, and consequently, always, to some amount, potentially inferring each other, still we cannot, without the intervention of an actual inference, at once jump from the one quantity to the other,—change, *per saltum*, Predicate into Subject and Subject into Predicate. We must proceed *gra-*

datim. We cannot arbitrarily commute the quantities, in passing
from the Quæsitum to the Premises, or in our transition from the
Premises to the Conclusion. When this is apparently done (as
in the Indirect moods of the First Figure and in all the moods
of the Fourth,) the procedure is not only unnatural, but virtually
complex and mediate ; *the mediacy being concealed by the conceal-
ment of the mental inference which really precedes.*—Indicated by
the line and broken line for the First Figure.

9.) In Syllogism, Figure and the varieties of Figure are determined by the counter relations of Subject and Predicate subsisting between the syllogistic terms,—between the Middle and
Extremes.—All adequately represented.

10.) Figure and the differences of Figures all depending upon
the difference of the mutual contrast of Subject and Predicate
between the syllogistic terms ; consequently, if this relation be
abolished,—if these terms be made all Subjects, (or it may be all
Predicates), the distinction of Figure will be abolished also. (We
do not abolish, be it noted, the Syllogism, but we recall it to one
simple form.)—And this is represented in the diagram. For as
the opposition of Subject and Predicate, of Depth and Breadth,
is shown in the opposition of the thick and thin ends of the
same tapering line ; so where (as in the outmost triangle) the
propositional lines are of uniform breadth, it is hereby shown,
that all such opposition is sublated.

11.) It is manifest, that, as we consider the Predicate or the
Subject, the Breadth or the Depth, as principal, will the one
premise of the Syllogism or the other be Major or Minor; the
Major Premise in the one quantity being Minor Premise in the
other.—Shown out in the diagram.

12.) But as the First Figure is that alone in which there is such
a difference of relation between the Syllogistic Terms,—between
the Middle and Extremes; so in it alone is such a distinction
between the Syllogistic Propositions realised.—By the diagram
this is made apparent to the eye.

13.) In the Unfigured Syllogism, and in the Second and Third Figures, there is no difference between the Major and Minor Terms, and, consequently, no distinction (more than one arbitrary and accidental) of Major and Minor Propositions.—All conspicuously typified.

14.) All Figured Syllogisms have a Double Conclusion; but in the different figures in a different way.—This is well represented.

15.) The Double Conclusions, both equally direct, in the Second and Third Figures, are shown in the crossing of two counter and corresponding lines. The logicians are at fault in allowing Indirect Conclusions in these two figures, nor is Aristotle an exception. (See *Pr. An.*, I., vii. § 4.)

16.) The Direct and Indirect Conclusions in the First Figure are distinctly typified by a common and by a broken line; the broken line is placed immediately under the other, and may thus indicate, that it represents only a reflex of,—a consequence through the other, (κατ᾽ ἀνάκλασιν, *reflexim, per reflexionem.*) —The diagram, therefore, can show, that the Indirect moods of the First Figure, as well as all the moods of the Fourth, ought to be reduced to merely *mediate* inferences;—that is, to conclusions from conclusions of the conjugations or premises of the First Figure.[a]

―――――――

[The following Table affords a view in detail of the Author's Scheme of Syllogistic Notation, and of the valid Syllogistic Moods, (in Figure), on his doctrine of a quantified Predicate. In each Figure, (three only being allowed), there are 12 Affirmative and 24 Negative moods; in all 36 moods. The Table exhibits in detail the 12 Affirmative Moods of each Figure, and the 24 Negative Moods of the First Figure, with the appropriate notation.

[a] Reprinted from *Discussions*, p. 657-661.—ED.

The letters C, Γ, each the third letter in its respective alphabet, denote the extremes; the letter M denotes the middle term
of the syllogism. Definite quantity, (*all, any*), is indicated by
the sign (:); indefinite quantity, (*some*), by the sign (, or). The
horizontal tapering line (▬▬▬) indicates an affirmative relation between the subject and predicate of the proposition.
Negation is marked by a perpendicular line crossing the horizontal (—+—). The negative syllogisms, in all the Figures,
are exactly double the number of the affirmative; for every
affirmative affords a double negative, as each of its premises
may be marked by a negative. In Extension, the broad end of
the line denotes the subject, the pointed end the predicate. In
Comprehension this is reversed; the pointed end indicating the
subject, the broad end the predicate. By the present scheme of
notation, we are thus able to read a Syllogism both in Extension
and in Comprehension. The line beneath the three terms denotes the relation of the extremes of the conclusion. Predesignation of the conclusion is marked only when its terms obtain
a different quantity from what they hold in the premises. Accordingly, when not marked, the quantification of the premises
is held repeated in the conclusion. In the Second and Third
Figures,—a line is inserted above as well as below the terms of
the syllogism, to express the double conclusion in those figures.
The symbol ⌣⌣ shows that when the premises are converted,
the syllogism remains in the same mood; ✕ shows that
the two moods between which it stands are convertible into each
other by conversion of their premises. The middle term is said
to be *Balanced*, when it is taken definitely in both premises.
The extremes are balanced, when both are taken definitely;
unbalanced, when the one is definite, and the other is not.

The Table here given exhibits the Author's final arrangement
of the Syllogistic Moods. The Moods are either A), *Balanced*,
or B), *Unbalanced*. In the former class both Terms and Propositions are Balanced, and it contains two moods,—i.; ii. In the
latter class there are two subdivisions. For either a), the Terms
are Unbalanced,—iii. iv.; or b), both the Terms and Propositions are Unbalanced,—v. vi.; vii. viii.; ix. x.; xi. xii.

It should be observed that the arrangement of the order of
Moods given in the present Table, differs from that of the earlier

scheme printed above, p. 293 *et seq.* The following is the correspondence in the order of moods :—

Present and Final Table.			Earlier Table.
I.	corresponds to		I.
II.	...	...	II.
III.	...	...	XI.
IV.	...	...	XII.
V.	...	...	VII.
VI.	...	...	VIII.
VII.	...	...	III.
VIII.	...	...	IV.
IX.	...	...	V.
X.	...	...	VI.
XI.	...	...	IX.
XII.	...	...	X.

The order of the earlier table is that given by Mr Baynes, in the scheme of notation printed at p. 76 of his *Essay on the New Analytic.* The order of the present table corresponds with that given by Dr Thomson in his *Laws of Thought,* p. 244, 3d edition, 1853.—ED.]

SCHEME OF NOTATION—
TABLE OF SYLLO

A. AFFIRMATIVE MOODS.

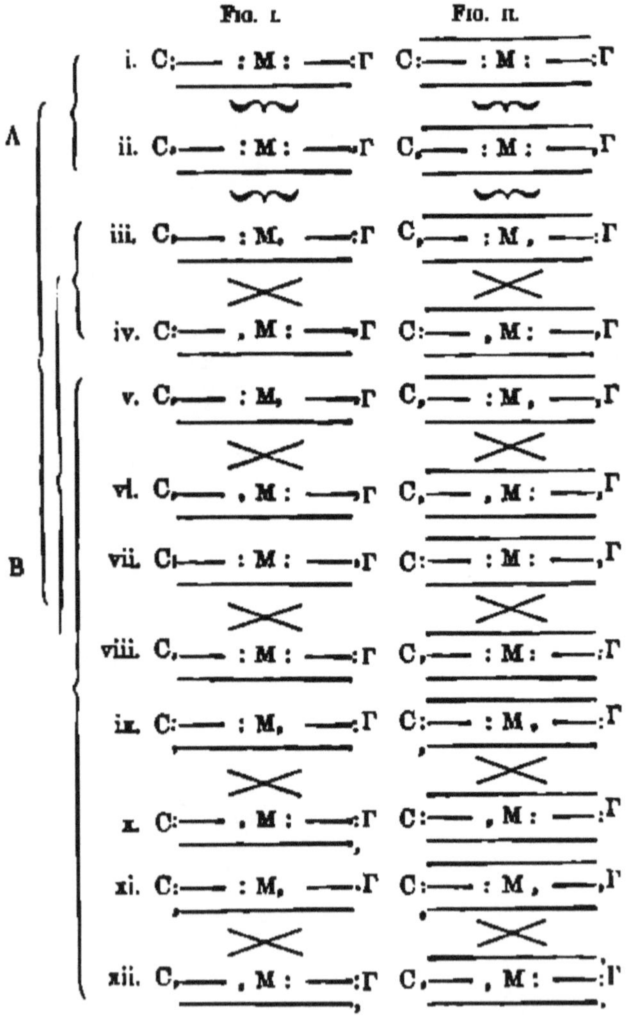

Fig. I. Fig. II.

A. I. and II. are *Balanced*. B. The other moods are *Unbalanced*. Of these, III

–FIGURED SYLLOGISM.

ЗISTIC MOODS.

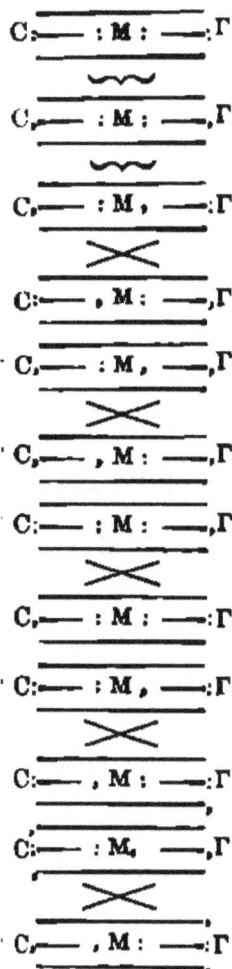

B. NEGATIVE MOODS.

FIG. III.   FIG. 1.

C ———— : M : ————:Γ   i.   a   C:+— : M : ——:Γ
                        b   C:—— : M : +— Γ

C ———— : M : ————,Γ   ii.   a   C,+— : M : ——,Γ
                        b   C,—— : M : +—,Γ

C, ———— : M , ————:Γ   iii.   a   C,+— : M , ——:Γ
                        b   C,—— : M , +—:Γ

C: ———— , M : ————,Γ   iv.   a   C:+— , M : ——,Γ
                        b   C:—— , M : +—,Γ

C, ———— : M , ————,Γ   v.   a   C,+— : M , ——,Γ
                        b   C,—— : M , +—,Γ

C, ———— , M : ————,Γ   vi.   a   C,+— , M : ——,Γ
                        b   C,—— , M : +—,Γ

C: ———— : M : ————,Γ   vii.   a   C:+— : M : ——,Γ
                        b   C:—— : M : +—,Γ

C, ———— : M : ————:Γ   viii.   a   C,+— : M : ——:Γ
                        b   C,—— : M : +—:Γ

C: ———— : M , ————:Γ   ix.   a   C:+— : M , ——:Γ
                        b   C:—— : M , +—:Γ

C: ———— , M : ————:Γ   x.   a   C:+— , M : ——:Γ
                        b   C:—— , M : +—:Γ

C: ———— : M, ————,Γ   xi.   a   C:+— : M , ——,Γ
                        b   C:—— , M , +—,Γ

C, ———— , M : ————:Γ   xii.   a   C,+— , M : ——:Γ
                        b   C,—— , M : +—:Γ

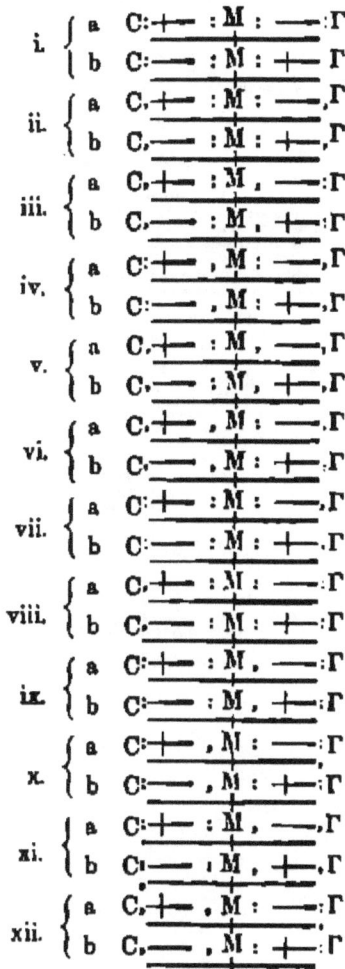

and iv. are unbalanced in terms only, not in propositions ; the rest in both.

INDEX.